£6-00

PHOTOCHEMISTRY
and
SPECTROSCOPY

PHOTOCHEMISTRY
and
SPECTROSCOPY

J. P. SIMONS

Lecturer in Chemistry,
University of Birmingham

WILEY–INTERSCIENCE

a division of John Wiley & Sons Ltd

LONDON · NEW YORK · SYDNEY · TORONTO

Library of Congress Catalog Card No. 71–149581

ISBN 0 471 79202 0

Printed in Great Britain by J. W. Arrowsmith Ltd.,
Winterstoke Road, Bristol BS3 2NT

'The man who makes no mistakes usually does not make anything'.
Edward John Phelps

PREFACE

This book was begun a long time ago. It is already in its fifth edition, but mercifully the first four were either unpublished or uncompleted. The science of photochemistry has expanded so explosively that by now the earlier editions would have been hopelessly out of date. The explosion has been due in large measure, to the rather belated impact of concepts taken from molecular electronic spectroscopy. These are concepts which many chemists find difficult to understand, and for this reason I have dwelt on them in some detail. The book is aimed at teaching the fundamental ideas involved in detailed descriptions of photochemical change. It is not intended as a reference book, and should be regarded as an 'hors d'oeuvre' rather than a 'grand oeuvre'. Its digestion will demand concentrated effort, but it is hoped that it will not cause any pain. It could be read selectively, by final year undergraduate students, and one hopes enthusiastically by graduate students of all ages.

Many friends, both teachers and students, have helped me along the way. Professor R. G. W. Norrish gave me generous encouragement and advice during the book's early stages, as well as transforming me from a callow B.A. to an overconfident Ph.D. Dr. J. K. Brown conscientiously read through the whole of Chapter 1 and some of his modifications are included in it. Drs. J. R. Majer and J. A. Kerr have both read through parts of Chapter 3, and Professor G. Porter and Dr. G. Herzberg were kind enough to let me use Figures 3.12 and 2.38(a) and (b). Mr. B. Emson and Mr. C. R. Boxall produced many of the photographic plates; Mr. Boxall also spotted an error in every section he looked at which was so depressing that his services were dispensed with. Mrs. Marilyn Hill was cheerful, patient and accurate in producing the final typescript. Mrs. A. M. Simons looks forward to being reunited with her husband shortly. They all deserve much more than gratitude.

J. P. SIMONS

vii

CONTENTS

INTRODUCTION

The gradual understanding of photochemical phenomena, and indeed all chemical and physical phenomena is a step-wise process, in which there are three stages of enquiry: namely, 'What has occurred, what has caused it to occur, and how has it occurred?', the questions being asked in that order. Like all other branches of scientific enquiry, photochemistry developed through each of these stages, although being something of a hybrid between physics and chemistry, its early development was slower than most.

The first record of any enquiry into the effect of varying conditions on a photochemical reaction, rather than the mere recording of 'natural' phenomena, was made by the Swedish chemist Scheele in 1777, who observed that violet light was the most effective in darkening silver chloride. Up to that time, several isolated examples of chemical changes induced through exposure to sunlight were known, such as the fading of dyes, the necessity of sunlight for the growth of plants and the photo-sensitivity of certain silver salts, but in all cases, the phenomenon only was recorded.

There was no attempt to vary the conditions with a view to gaining an insight into the nature of the phenomenon. For example, it was not until after several years had elapsed that it was realized that the darkening of silver carbonate was due to exposure to light rather than to exposure to the atmosphere. In the revolutionary fervour of the turn of the 18th century the spirit of enquiry, rather than the mere recording of photo-chemical effects began to take hold, and with the realization in 1817 by Grotthüs, that only the light absorbed was effective in producing photo-chemical change, the first general principle of photochemistry had been discovered.

To a scientist of the mid-twentieth century, such a principle appears, to say the least, obvious, but to realize that it is not the mere passage of a ray of light through the substance, but the actual absorption of a particular colour in the light, that causes a chemical change in it, would certainly not have been obvious to a person with the scientific background of the early 19th century. In fact this principle passed unnoticed until 1841, when it was restated by Draper, and in consequence is now termed the Grotthüs–Draper Law. It is a very worthwhile exercise to tackle the problem of a historian, and try to adopt the mental attitude one might have possessed in past years; in a word, to 'dis-educate' oneself and forget interpretations one now accepts, since it avoids a prejudiced approach to present problems which seem insoluble on the basis of accepted principles. This approach is particularly valuable in photochemistry, since a complete experimental record of all the stages of a photochemical change is rarely, if ever, available and inferences have to be drawn at some stage.

The Grotthüs–Draper Law may be regarded as the first, albeit elementary principle of photochemistry. It was not until the advent of quantum theory, some sixty or so years later, that any understanding developed, of the mechanism by which light absorption could promote chemical change in the absorbing medium. Only the questions, 'What has occurred?', and 'What has caused it to occur?' could be asked; it was the advent of modern physics that allowed photochemistry to emerge from its empirical stage to ask the question 'How has it occurred?'.

The first fundamental concept introduced into photochemistry by modern physics was that light was radiated in discrete quanta, called photons, with an energy proportional to the frequency of the light. Absorption corresponds to the 'capture' of a photon by the molecule. With this concept in mind, Stark and Einstein proposed that there should be a 1:1 equivalence between the number of molecules decomposed and the number of quanta absorbed. On this hypothesis, the **quantum yield** of a photochemical reaction obeying the 'one molecule per photon' law should be unity,

$$\phi = \frac{\text{molecules decomposed}}{\text{photons absorbed}} = 1$$

With the aim of testing this hypothesis, Warburg and Bodenstein undertook the experimental determination of the quantum yields of many photo-initiated reactions; the results spread from 10^{-2} to $\sim 10^{+6}$, and it was obvious that the Einstein–Stark Law was only a part of the story. Bodenstein realized that the absorption of a photon could be followed by secondary chemical reactions, and proposed that the photo-processes be termed **primary**; any subsequent chemical reactions initiated by the primary photo-products are **secondary processes.** The Einstein–Stark Law is only applicable to the primary steps. (In the past few years, the development of high intensity light sources such as lasers, has led to the discovery of biphotonic primary processes. In these the primary quantum yield cannot exceed 0·5.)

Many photochemical reactions have quantum yields less than unity not because of efficient recombination of primary photoproducts, but because of the existence of decay paths which lead the excited molecule back to its ground state without promoting any chemical change. It is convenient to divide the alternative primary steps into **photophysical** (e.g. luminescence), and **photochemical** (e.g. fragmentation). The study of light absorption and the photophysical decay paths takes us into the realms of spectroscopy and to fully understand the behaviour of a photo-excited molecule one has to be both a spectroscopist and a chemist, if not a genius.

Photochemical change is very much a game of chance. The sequence of steps which follow excitation is determined by the relative probabilities of each alternative at each move in the game. The excited system wants to relax to a more stable state and will do it by the quickest possible route. The most probable moves will be the quickest and will 'nip the alternative in the bud'. The competition between alternative primary processes is the key to understanding the physics and chemistry of photochemical change. Their relative rates can be sensitive to changes in the absorbed wavelength and intensity, pressure and concentration, added substrates, temperature, phase and so on, producing a richness of detail which fascinates the experimentalist but clouds the vision of the theorist. The challenge is to break down the nett change into its component parts.

CHAPTER ONE

LIGHT, WAVES, ATOMS AND MOLECULES

1.1 ELECTROMAGNETIC RADIATION

1.1a A classical description

An electric charge has associated with it an electric field which radiates outward uniformly from its centre. If the charge moves with constant velocity or is at rest, then the spherical symmetry of the field is preserved. If the charge accelerates the field tends to lag behind, since it will take time to accommodate itself to the changing velocity of the charge. The change causes a disturbance in the direction of the field, which radiates out as a spherical 'front' into the surrounding space. This disturbance, which is propagated with the velocity of light, constitutes electromagnetic radiation.

If an electric dipole is set into simple harmonic motion, the two charges experience a continuous acceleration (either positive or negative), proportional to their displacement from the mean separation. The associated electric field strength vector varies sinusoidally in phase with the oscillating dipole, and the disturbance is propagated outwards in a direction perpendicular to the axis of the dipole. Energy spreads outward as a continuous train of electromagnetic radiation, polarized in the direction of oscillation. The electric field strength vector, $\mathbf{E}$, varies as

$$\mathbf{E} = \mathbf{E}_0 \sin 2\pi(kx - vt)$$

where $\mathbf{E}_0$ is a vector, constant in time and space. Propagation is in the x-direction of frequency v and wavelength $1/k$, where k is termed the wave number. The velocity of propagation c, the frequency, and the wave number are related by $c = v(1/k)$. Associated with the oscillating electric field there is a magnetic field, lying in the plane perpendicular to the electric vector, and oscillating in phase with it. The amplitude of the two fields are related as

$$E_0/H_0 = (\mu/\varepsilon)^{\frac{1}{2}}$$

where μ is the magnetic permeability and ε the dielectric constant of the medium in which the radiation is propagating.[1]

This simple view of the situation is of value in deciding whether electric dipole absorption or emission can occur in certain instances. For example, if we consider the rotation of a polar molecule, the change in the dipole moment will be solely a change in orientation. If there is no molecular electric dipole moment there can be no interaction with the radiation field. Likewise, the distortion associated with a molecular vibration must give rise to a change in the magnitude of the dipole moment, if radiation is to be absorbed or emitted. In quantum mechanics, as we shall see later, the relevant quantity is the so-called 'transition moment'. The decision as to whether or not this has a finite value when a system undergoes electric dipole transitions between two states forms the basis of all 'selection rules', some of which are far from obvious.

Changes in magnetic dipole moment can give rise to absorption or emission through an interaction with the magnetic vector in the radiation, but the intensity is many orders of magnitude lower than for the electric dipole case. The interaction is of central importance in nuclear magnetic resonance (n.m.r.) and electron spin resonance (e.s.r.) spectroscopy, but it will not concern us here.

1.1b Photons

In the last week of the last century, Planck[2] advanced the revolutionary hypothesis that an oscillating dipole could radiate its energy only in discrete units or 'quanta'. The energy carried by a quantum of radiation was proportional to the frequency of the oscillation, and the two were related by the equation

$$E = h\nu$$

where h is a universal constant, Planck's Constant. With this hypothesis it was possible to generate, for the first time, the correct spectral energy distribution of the radiation emitted by a black body. Note that the continuous nature of this distribution arises from the very large number of oscillating charges (electrons and nuclei) in the macroscopic source, which provides a virtually continuous spread of frequencies.

An extension of this hypothesis was proposed by Einstein in 1905,[3] who suggested that the photoelectric effect (the ejection of electrons from a metal surface by visible or ultraviolet light) could be understood if it were assumed that radiation of frequency ν could be absorbed only in units, or photons, of energy $h\nu$. Thus energy could be transferred from one system to another only in discrete quanta, and not continuously as had been assumed previously in the classical mechanical description.

The 'rate' at which a system absorbs light is governed by the Beer–Lambert Law, which states that the fractional decrease in light intensity (of a given frequency) is proportional to the number of absorbing

molecules in the light path. In quantum theory, the changes occurring in a
system which absorbs or emits electromagnetic radiation are envisaged as
involving transitions between stationary states. For an energy difference
ΔE between such states, the frequency absorbed or emitted is related to
ΔE by $\Delta E = hv$. This relationship sets one condition for the exchange of
quanta between the radiation field and the system; the second is that there
must be a finite transition probability between the states involved.

The intensity of the radiation is measured as the energy falling on unit
area of the system in unit time, and this energy is related directly through
Planck's Constant to the number of quanta and their associated frequency.
The fractional decrease in this intensity is directly proportional to the rate
of absorption of quanta by the sample, and this in its turn depends on the
number of atoms or molecules in the light path and the probabilities of
transition between the states. The direct proportionality between the
fractional absorption and the number of absorbing molecules is expressed
in the widely used Beer–Lambert Law.

Suppose the intensity falls from I to $(I - \delta I)$ as it crosses a section of the
absorbing medium of length δl; then the fractional change in light intensity
is $-\delta I/I$. If the concentration of absorbing molecules is c, then, for a beam
of unit cross-sectional area, the number of molecules it encounters in
crossing the section is $c\delta l$. Thus

$$-\delta I/I \propto c\delta l$$

or

$$-\delta I/I = \alpha_v c\delta l$$

where the proportionality constant α_v is known as the *extinction coefficient*
of the molecule for light of frequency v. For a path of length l, and an
incident and transmitted light intensity of I_0 and I_{trans}, respectively,

$$-\int_{I_{\text{trans}}}^{I_0} \mathrm{d}(\ln I) = \int_{l}^{0} \alpha_v c \,\mathrm{d}l$$

or

$$I_{\text{trans}}/I_0 = \exp(-\alpha_v cl)$$

α_v has the dimensions (area mole^{-1}), and is, in effect, a 'cross-section'
measuring the probability of absorption of a photon of energy hv. In
practice the percent light absorption is often expressed by the equation

$$I_{\text{trans}}/I_0 = 10^{-\varepsilon_v cl}$$

where $\varepsilon_v = \alpha_v/2\cdot303$ is termed the absorptivity. When the concentration is
measured in moles litre^{-1} and the path length in centimetres, the value of
$\varepsilon_v c$ is known as the *optical density* or absorbance. For more than one
absorbing component the optical density is $\sum_i \varepsilon_{v,i} c_i$.

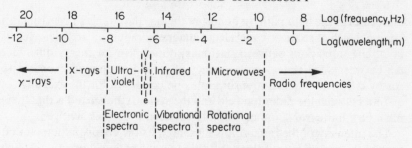

Figure 1.1 The electromagnetic spectrum.

A diagrammatic representation of the electromagnetic spectrum is shown in Figure 1.1. The radiation may be characterized by its wavelength λ, wave number $k = 1/\lambda$, or frequency ν. Frequency is expressed in cycles per second, for which the unit is the hertz (Hz) or multiples thereof. Wavelengths in the ultraviolet and visible have been expressed in nanometres (1 nm $= 10^{-9}$ metre), millimicrons* (1 m$\mu = 10^{-9}$ metre) and Ångstroms* (1 Å $= 10^{-10}$ metre); wavenumber is usually expressed in cm^{-1}* or more recently in μm^{-1}. If a molecule absorbs light in the ultraviolet region, say at 300 nm ($\nu = 10^{15}$ Hz), its energy is increased by $\Delta E = h\nu \simeq 6 \times 10^{-27} \times 10^{15}$ erg per photon absorbed. If one grammolecule absorbs light at this frequency, then $N \simeq 6 \times 10^{23}$ photons will be absorbed (N is the Avogadro Number), corresponding to an energy of $\Delta E \simeq 36 \times 10^{11}$ erg* or ~ 360 kJ mole^{-1}. This energy greatly exceeds the activation energy of the vast majority of chemical reactions and many bond-dissociation energies, and it is for this reason that absorption of ultraviolet or visible light frequently promotes chemical change. The distinguishing feature of a photochemical reaction is that the activation energy required to produce chemically reactive molecules is supplied by absorption of radiation in the visible or ultraviolet regions. When the wavelength of the absorbed light is short enough to promote ionization of isolated molecules, i.e. ejection of an electron, photochemistry merges into the broader field of radiation chemistry.

1.2   ENERGY LEVELS IN ATOMS

1.2a The Bohr–Sommerfeld theory

The first successful model which permitted the calculation of the potential energies of electrons bound to the nuclei of atoms was introduced by Bohr in 1913.[4] It was based on Rutherford's planetary model of the atom, in which the orbiting planets were replaced by electrons, and the sun by the nucleus. Bohr proposed that the frequencies of the spectral lines emitted

* These names or units are being discouraged under the new SI system.

by an atom, from the near-infrared to the far-ultraviolet, could be accounted for, if it were assumed that the electrons could circulate only in certain fixed orbits. If the potential energy of the electron 'jumped' from that associated with one orbit to that associated with another, radiation would be emitted or absorbed depending on the relative energies of the initial and final states. The frequency of the radiation associated with the transition was given by the expression $\Delta E = h\nu$, where ΔE was the difference in the potential energy associated with the two orbits, or *stationary states*. When the spectral lines converged to a continuum, the electron in one of the two states had escaped the influence of the nucleus and no longer circulated in a closed orbit. The frequency of the convergence limit thus corresponds to the ionization energy or ionization potential of the atom. The series of spectral lines converging to this limit are known as Rydberg Series.

Bohr's model required that, although electrons moving in circular orbits would experience a constant acceleration, they did not continuously emit radiation and so spiral into the nucleus; there was no theoretical justification for this, only the practical one that without it atoms and molecules could not exist for longer than a very small fraction of a second! In order to arrive at an expression which accurately predicted the frequencies of the spectral lines of the simplest atom, that of hydrogen, Bohr assumed that the orbital angular momentum of the circulating electron was quantized into units of $nh/2\pi$, where n is a whole number; here too, the justification was one of expediency. The radii of the orbits and the energies of the electrons circulating in them were both proportional to $1/n^2$, and n was termed the principal quantum number. On the basis of these assumptions Bohr was able to calculate, with very great accuracy, the frequencies of 'hydrogen-like' spectra, i.e. those associated with one-electron atoms or ions such as H, He^+, or Li^{2+}.

The model was later extended by Sommerfeld[5] to include the possibility of motion in elliptical orbits; this introduced a second quantum number governing the energy levels of bound electrons, since an electron circulating in an elliptical orbit would have two degrees of freedom. Both the fine structure found to be associated with the spectrum of the hydrogen atom and the gross structure of the alkali metal atom spectra could be interpreted on this basis. Yet, despite this and many other successes, the Bohr theory was based on several very arbitrary assumptions for which there were no theoretical grounds, and so from a logical point of view the theory must be regarded as the first step in the evolution of a more fundamental treatment. This treatment has been provided by the development of wave mechanics and quantum mechanics, which have superseded the older Bohr–Sommerfeld model.

1.2b Wave mechanics

In 1922, Compton[6] established that if radiation carrying a sufficiently high energy interacted with a bound electron it would not be absorbed, but would be deflected or scattered, whilst the electron would be 'shot out' of the atom. However, the energy carried by the scattered radiation (measured by the increase in its wavelength) was slightly reduced, and the deficit increased with the scattering angle. This behaviour could be understood if it were assumed that the radiation behaved as if it were a stream of particles, and, by applying the principle of conservation of momentum to the scattering, Compton demonstrated that the momentum p of the radiation was related to its wavelength λ by the expression $\lambda = h/p$. The following year, de Broglie,[7] with great prescience, proposed that not only photons but *any* moving particles had an associated wavelength, given by Compton's equation.

Subsequent experiments by Davisson and Germer[8] in 1927, and G. P. Thomson[9] in 1928, confirmed that beams of fast and slow electrons, respectively, could be diffracted, in the one case through reflection from a nickel crystal, and in the other on passing through a very thin film of gold foil. A year or two later, Stern[10] showed that atomic and molecular beams could also be diffracted, and in each case the wavelength of the 'material waves' satisfied the expression $\lambda = h/p$. These results, coupled with those of Compton, demonstrated that radiation has momentum and moving particles have an associated wavelength; depending on the experiment, one or other property will be displayed.

A moving electron, a photon, or indeed any moving particle can be considered as a travelling wave group compounded of many wave trains of slightly differing velocity and wavelength. If it were possible to measure the changing position of the particle accurately, the spread of the wave-group would need to be very small; in order to achieve this, the spread in velocities and wavelengths would have to be very large, leading to an extreme uncertainty in the momentum of the particle. Any attempt to measure accurately and simultaneously the position and momentum of the moving particle is bound to fail; this is Heisenberg's Uncertainty Principle[11] and is inherent in the wave character of matter. The accuracy of any measurement is given by the uncertainty relation

$$\Delta x \Delta p \simeq h/2\pi$$

where Δx and Δp are the uncertainties in position and momentum. An alternative formulation can be approached from the other extreme. If the wavelength or frequency of a wave could be specified with complete certainty, the wave would have to be of infinite duration. If its amplitude fell to zero in a finite interval, it would have to be compounded by the superposition of other waves of differing frequency. The shorter the wave train, or the interval Δt during which it was propagated, the greater will be the uncertainty Δv in its frequency, and the two quantities are in inverse proportion $\Delta v \simeq 1/2\pi\Delta t$. This expression enables the natural frequency spread of spectral lines associated with a transition between two states to be related to the lifetimes of the states. If one of the states is very short-lived, the Δv, the *natural line width*, will be very large. The uncertainty in the energy of the state is given by

$$\Delta E \simeq h\Delta v \simeq (h/2\pi)\Delta t$$

In section 2.2c it will be seen how diffuse absorption spectra have been understood and interpreted in terms of the Uncertainty Relationships.

The weakness of the Bohr model is now apparent: it is impossible to specify simultaneously both the momentum and the exact position of the circulating electron. It must not be treated as a particle circulating in definite fixed orbits, but described in terms of a wave whose amplitude ψ is a continuous function of the spatial coordinates. When this is done one finds that the stationary states of the bound electron, introduced as a necessary expedient in the Bohr–Sommerfeld model, are a necessary and natural outcome of the description in terms of wave mechanics.

By analogy with light waves, Born[12] suggested that the 'density' of the particle in a unit of volume about any point where it could be found, be proportional to the square of the amplitude $|\psi|^2 = \psi^*\psi$ of the wave at that point.† If the total integrated density over all space is set to unity, then $\psi^*\psi$ measures the *probability* that the particle will be found in a unit of volume, at that point. For a small element of volume, $d\tau = dx\,dy\,dz$ about (x,y,z), the probability of finding the particle is $\psi^*\psi\,d\tau$ and $\int_{-\infty}^{\infty} \psi^*\psi\,d\tau = 1$. For the analogy to be physically acceptable, this constrains the amplitude ψ to be a single-valued, finite and continuous function of the coordinates.

Returning to a classical oscillating system, for example a vibrating string of infinite length, if any two points on the string are fixed the string will execute a standing wave motion between the two points with amplitude

$$\psi = \psi_f + \psi_b = \psi_0 \sin 2\pi(kx - vt) + \psi_0 \sin 2\pi(-kx - vt)$$

$$= 2\psi_0 \sin 2\pi vt \cos 2\pi kx$$

---

† ψ is complex, i.e. of the form $(a + ib)$, so that its square is of the form $\psi^*\psi = (a - ib) \times (a + ib) = a^2 + b^2$. This preserves the analogy with other vibrating systems where the intensity or energy density is given by a sum of two terms; for example in the case of light, $(\mathbf{E}^2 + \mathbf{H}^2)$ is proportional to the energy density, or the number of photons in unit volume.

Possible wavelengths are restricted to $2d/n$, where d is the distance between the two fixed points and n is an integer reflecting the boundary conditions imposed on the oscillation (see Figure 1.2). If the constraints were removed

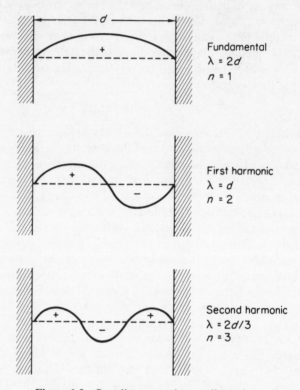

Figure 1.2   Standing waves in one dimension.

so that there were no boundary conditions, the string would oscillate freely over a continuous range of frequencies and wavelengths and the oscillations would propagate along the string. In the one case we have standing or stationary waves with amplitudes that depend on the spatial coordinate only, and in the other we have progressive, travelling waves with amplitudes dependent both on position and time. The one is associated with bound states and the other with unbound states. There is a clear analogy with the discrete, spectral lines emitted and absorbed by atoms, which are associated with changes of the potential energy of the bound electrons, and the continua into which the lines merge at energies beyond the ionization limit, when the electron has sufficient energy to escape the electrostatic field binding it to the nucleus. In this situation it is a free electron that can accept a continuous spectrum of kinetic energies, whereas the bound electron occupies stationary states, associated with standing wave patterns.

By combining the de Broglie equation, $\lambda = h/p$, with the classical differential equation describing the profile of a simple harmonic three-dimensional standing wave, the time-independent Schrödinger equation is obtained (but not 'derived'; the equation represents a postulate to be measured against observation):

$$-\frac{\hbar^2}{2m}\nabla^2\psi + V(x,y,z)\psi = E\psi$$

where m is the mass of the particle moving in a field of potential $V(x,y,z)$, E is its total energy, $\hbar$ is $h/2\pi$, and ∇^2 ('del squared') is the Laplacian operator $(\partial^2/\partial x^2 + \partial^2/\partial y^2 + \partial^2/\partial z^2)$. It may be rewritten in the form

$$\hat{H}\psi = E\psi$$

where $\hat{H}$, the energy or Hamiltonian operator, is given by

$$\hat{H} = -\frac{\hbar^2}{2m}\nabla^2 + V(x,y,z)$$

and is the sum of the kinetic and potential energies of the particle. Since the equation represents a standing wave, with a finite amplitude that is a single-valued continuous function of the coordinates, only certain solutions will be possible, namely those associated with stationary states of the system. Each solution ψ_i will describe a stationary state of constant total energy ε_i. The ψ_i are the *eigenfunctions* of the operator $\hat{H}$, which generates the *eigenvalues* ε_i; compare the equation $\mathrm{d}/\mathrm{d}x\, f(x) = c \times x$, for which $\exp(cx)$ is the eigenfunction and c the eigenvalue.

More generally in quantum mechanics, we assign to any closed physical system a wavefunction from which all the observable dynamical properties can in principle be deduced. For each of the observables (for example energy, angular momentum about a fixed axis), there is an associated operator say $\hat{G}$.[13] If the system is in a particular stationary state (or *eigenstate*), represented by ψ_i, then a measurement of the observable will invariably yield the eigenvalue a_i (a number), where a_i is defined by $\hat{G}\psi_i = a_i\psi_i$. If two or more eigenstates belong to the same eigenvalue, then any state formed by a superposition of them is also an eigenstate with this eigenvalue, and the system is said to be *degenerate*. In these circumstances any number of such states can be constructed, but in a doubly degenerate system, for example, two linearly independent functions suffice for the complete specification. All eigenstates belonging to different eigenvalues are distinct; in the mathematical description they are said to be orthogonal, a condition which will be expressed formally later.

Eigenvalue problems are of central importance in quantum mechanics, as the eigenvalues of an operator associated with an observable specify the possible results of the measurement of the observable. Any state whatsoever of the system can be formulated in terms of a superposition of eigenstates and the result of the measurement is then to give one of the possible values. In a sequence of measurements the probability of the occurrence of any particular value is governed by the 'weighting' of the associated state in the superposition. In this general case the measurements are summarized by an average value which is expressed as

$$\langle a \rangle_{av} = \frac{\displaystyle\int_{-\infty}^{\infty} \psi^* \hat{G} \psi \, d\tau}{\displaystyle\int_{-\infty}^{\infty} \psi^* \psi \, d\tau}$$

If the wave function is normalized, i.e. scaled to unity, the average value is simply given by the numerator. If ψ is an eigenstate ψ_i, then the above average is equal to the appropriate eigenvalue, a_i, as it should be.

In classical vibrating systems, any general state of the motion can be compounded from a superposition of the normal modes of the oscillation and there might once again seem to be a similarity between the classical and quantal 'wave' concepts, but the implication is quite erroneous. For the classical system, the superposition of a state on itself leads to a different state, with a different amplitude of oscillation. In quantum theory however, the superposition of a state on itself leads to the same state. Again the classical state with zero amplitude of oscillation everywhere corresponds to a state of rest; the 'zero' state in quantum mechanics is no state at all!

Solution of the Schrödinger Equation for the electron bound in a hydrogen atom gives the wave functions of its stationary states

$$\psi_{n,l,m_l}(r,\theta,\varphi) = \frac{1}{r} R_{n,l}(r) Y_{l,m_l}(\theta,\varphi)$$

where r,θ,φ are the spherical, polar coordinates of the electron relative to the nucleus at the origin, and n,l and m_l are three integral quantum numbers (reflecting the three spatial dimensions). (The equation is cast into spherical coordinates to mirror the spherical symmetry of the system.) $R(r)$ is the radial wave function, which governs the mean distance of the electron from the nucleus, and $Y(\theta,\varphi)$ are spherical harmonics which govern the angular orientation of the wave functions. The principal quantum number n can take values $n > 0$. The quantum number l determines the orbital angular momentum of the electron about the nucleus $\sqrt{l(l + 1)}\hbar$, and can take the values $0 \leqslant l \leqslant (n - 1)$. The angular momentum of the electron generates a magnetic moment $\mu = \sqrt{l(l + 1)} \, eh/2mc$, where e is the electronic charge; its possible orientations with respect to a magnetic field are governed by the magnetic quantum number m_l, which takes values $-l \leqslant m_l \leqslant l$; so that the components of the magnetic moment in the direction of the field are restricted to $m_l eh/2mc \equiv m_l \mu_B$.*

The set of wave functions $\psi_{n,l,m_l}(r,\theta,\varphi)$ which are solutions of the Schrödinger equation are normalized, and orthogonal to each other, i.e. for any two solutions ψ_i, ψ_j,

$$\int_{-\infty}^{\infty} \psi_i^* \psi_j \, \mathrm{d}\tau = 1 \qquad (i = j)$$
$$= 0 \qquad (i \neq j)$$

and the set is said to be *orthonormal*.† The physical significance of this is that there is an overall unit probability of an electron being in any particular state, but zero probability of its being simultaneously in some other state.

The energy of each stationary state of the atom is

$$E_n \propto -1/n^2$$

independent of the values of l and m_l; states having a given principal quantum number have almost the same energies.‡ In heavier atoms, with more than one circulating electron, the spherical symmetry of the electrostatic field is lost, and the states are split, as shown by the greater complexity of their electronic line spectra. However, the complexity goes beyond that which can be accounted for by the three quantum numbers n, l, and m_i, and Uhlenbeck and Goudsmit in 1925[14] proposed a fourth quantum number, $s = \pm\frac{1}{2}$, to account for the doublet structure of spectral lines emitted by the alkali metal atoms. This quantum number determines the spin angular momentum $\sqrt{s(s + 1)}\hbar$, of the electron about its own axis;

*The quantity $eh/2mc$ is the unit magnetic moment, and is termed the Bohr Magneton, μ_B.

† The integral is, in fact, the scalar product of two functions; cf. the scalar product of two unit cartesian vectors, $\mathbf{e}_i \cdot \mathbf{e}_j = e^2 \cos\theta = 1 \, (i = j, \theta = 0°)$, or $0 \, (i \neq j, \theta = 90°)$.

‡ The relativistic correction splits the degeneracies of levels with the same value of n but different values of l.

coupling of the magnetic moment due to the spin with that due to the orbital angular momentum leads to states of slightly differing energy. The effect is known as *spin–orbit coupling*, and later discussion will illustrate its major role in influencing the course which may be followed in photochemical reactions. When the Schrödinger Equation was modified for relativistic changes in the mass of the electron by Dirac,[15] the spin quantum number emerged naturally in its solution. The complete wave function for the electron in the hydrogen atom must be expressed as a product of a space part, say φ, which is a function of the spatial coordinates only, and a spin part. For the single electron this can only have two values, either α or β (spin coordinate either 'up' or 'down'). The complete wave function for an electron, electron 1, with spin α is then $\psi(1) = \varphi(1)\alpha(1)$.*
The spin wave functions are also orthonormal, so that

$$\int \alpha\alpha \, ds = \int \beta\beta \, ds = 1$$

$$\int \alpha\beta \, . \, ds = \int \beta\alpha \, . \, ds = 0$$

where ds is an element of 'volume' in the space spanned by the spin coordinates.

1.2c Atomic orbitals and electronic states

The region in which the bound electron may be found is the orbital region and the wave functions for stationary states of the atom are termed atomic orbitals, or a.o.'s. Levels with $l = 0, 1, 2, 3, \ldots$, are described as $s, p, d, f, \ldots$, orbitals. For s orbitals the spherical harmonic part of the wave function $Y_{0,0}(\theta,\varphi)$ is unity, and the orbitals are spherically symmetrical. The others have specific angular orientations and are directed in space (see Figure 1.3).

When $n = 1, l = m_l = 0$, only one orbital results, termed 1s. For $n = 2$, either $l = m_l = 0$, or $l = 1$ and $m_l = 0, \pm 1$; the first possibility gives rise to a single orbital termed 2s, and the second produces three degenerate orbitals† termed $2p_{+1}, 2p_0$ and $2p_{-1}$. By taking a linear combination of the wave functions for $2p_{+1}$ and $2p_{-1}$, they can be transformed into two orthogonally directed orbitals $2p_x$ and $2p_y$ which are each orthogonal to the

---

*In heavier atoms, where there is appreciable electron spin–orbit coupling, the two components of the complete wave function are not separable.
† I.e. they have the same energy in the absence of an applied electric or magnetic field.

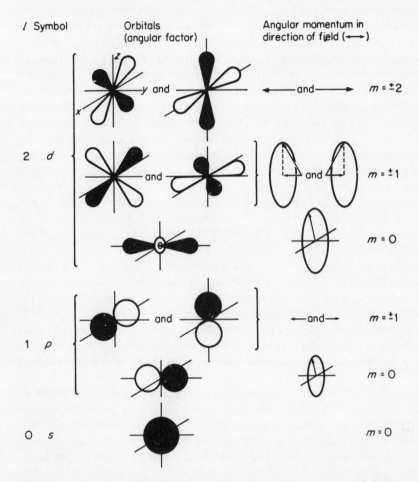

Figure 1.3 Atomic orbitals of type s, p and d, referred to cartesian coordinates. Open and filled lobes represent opposite phases of the wave function.

third $2p_z \equiv 2p_0$ (see Figure 1.3) and the orbital description can be recast in terms of a set of cartesian coordinates with the nucleus at the origin. For $n = 3$, there are five degenerate $3d$ orbitals, three degenerate $3p$ orbitals and one $3s$ orbital, and so on. The sets of degenerate orbitals, or the single orbitals for $l = 0$, are termed 'shells'. When the orbital wave functions are plotted graphically as a function of the coordinates, pictorial representations of the associated standing wave patterns are obtained.

These are determined by the $s, p, d, \ldots,$ character of the orbitals, and the cartesian coordinate representations are shown in Figure 1.3. For a given value of l, the eigenfunction changes sign l times at the origin; thus a d orbital has two nodal planes passing through the origin, a p orbital has one and an s orbital has none. If the wave function is such that its amplitude is antisymmetric (i.e. changes sign) with respect to inversion through the centre of symmetry at the origin, it is said to be 'ungerade' (u); if not, it is 'gerade' (g). Odd orbitals $(l = 1, 3, \ldots)$ have u symmetry and even orbitals $(l = 0, 2, \ldots)$ have g symmetry. In section 1.5 we shall find that the symmetries of the orbitals play a vital role in determining spectral transition probabilities, and the relative lifetimes of atoms (and molecules) in excited electronic states.

In passing to heavier, polyelectronic elements of the Periodic Table, only approximate solutions of the Schrödinger Equation are possible; electron repulsion energy terms enter the Hamiltonian operator and the radial symmetry of the electrostatic field is lost. Nonetheless qualitative similarities remain and the quantum numbers are preserved. As an approximation to an exact solution, a product of 'one-electron' wave functions, can be tried; these would be functions of the position of the chosen electron only and would be independent of the positions of all the others (see ref. 16). For example, in writing the electronic configuration of a lithium atom as $1s(1)1s(2)2s(3)$, one assumes that the nett wave function can be factored into a product of one-electron wave functions (a.o.'s). These are eigenfunctions of assumed one-electron Hamiltonian operators $H(1), H(2), H(3)$, which include interelectronic repulsion in an 'effective' potential energy term, i.e. the 'effective' potential field due to the nucleus and all the other electrons. The complete wave function will also include spin, and to satisfy the Pauli Principle it must be antisymmetric with respect to exchange of any pair of electrons among the occupied orbitals. This is achieved by writing the wave function in the determinantal form which, by permuting the electrons among all the available orbitals, recognizes their indistinguishability.

$$\psi(1, 2, 3) = (1/3!)^{\frac{1}{2}} \begin{vmatrix} 1s(1) & 1\bar{s}(1) & 2s(1) \\ 1s(2) & 1\bar{s}(2) & 2s(2) \\ 1s(3) & 1\bar{s}(3) & 2s(3) \end{vmatrix}$$

$[1s(1) \equiv 1s(1)\alpha(1), 1\bar{s}(1) \equiv 1s(1)\beta(1),$ and $(1/3!)^{\frac{1}{2}}$ is the normalizing factor]. Interchanging any two columns changes the sign of the determinant.

If the first two columns were the same, i.e. the two $1s$ electrons had the same spin, the determinant would vanish. The configuration could not exist, and hence the Pauli Principle is satisfied.

In polyelectronic elements the orbitals are filled in order of increasing energy (Aufbau Principle), subject to the constraint that no two electrons can have an identical set of quantum numbers (a consequence of the Pauli Principle); when a given orbital contains two electrons of opposite spin it can accommodate no more. Where two or more orbitals are degenerate, or nearly so, the orbitals are first filled singly, by electrons with parallel spins. This tendency can be rationalized in simple terms by saying that in different spatial orbitals the mutual repulsion of two electrons is less than if each occupied the same spatial orbital. It provides the basis for Hund's Rule of Maximum Multiplicity (i.e. in the state of least energy the nett spin momentum will be the maximum possible).

The wave functions and energies of electrons bound in atoms are dependent on their quantum numbers n, l, m_l, and s. In a one-electronic atom they define the electronic state of the atom as well, but in poly-electronic atoms the nett angular momenta must be found by vector addition; fortunately, the resultant angular momentum of electrons in closed shells is zero, and only those in partially filled shells need be considered. When these are only singly occupied, the nett values of l and s are those of the odd electron. The magnetic moments due to the spin and orbital angular momenta interact to produce two levels of slightly different energies and total momenta $j = l \pm \frac{1}{2}$. Where the odd electron occupies an s orbital, as in the ground states of the alkali metal atoms, the two levels are degenerate since the electron has no orbital angular momentum. They would be split only in the presence of an externally applied magnetic field.

The electronic states of atoms are denoted by their 'term symbols' $n^{2S+1}L_J$, where S, L and J are the nett spin, orbital and total angular momenta of the electrons, found by vector addition. When there is only one odd electron s, l and j are equivalent to S, L and J, so the term symbol for the ground state of lithium is $2^2S_{\frac{1}{2}}$ ('two doublet S one half'). In its first excited state the $2s$ electron is promoted into the $2p$ orbital, where $j = 1 \pm \frac{1}{2} = \frac{3}{2}$ or $\frac{1}{2}$. Spin–orbit interaction gives the two states very slightly different energies, with the $2^2P_{\frac{3}{2}}$ the lower of the two. The splitting greatly increases in the heavier alkali metals, since the magnitude of the spin–orbit interaction increases rapidly with the atomic number.

When atoms having two electrons in unfilled orbitals are considered, the spin and orbital angular momenta may couple in two alternative ways. In light atoms, where the spin–orbit coupling is small, the two momenta add individually to give resultants L and S, with $(l_1 + l_2) \geqslant L \geqslant (l_1 - l_2)$ and $S = s_1 \pm s_2$. These combine to produce a total resultant angular momentum $(L + S) \geqslant J \geqslant (L - S)$ for $L > S$ (and vice versa for $L < S$); this type of vector addition is termed (l, s) or Russell–Saunders coupling. Since the electron spins may be parallel or opposed, the resultant spin may be 1 or 0. For $S = 0$, $J = L$ only and the state has singlet multiplicity, but a triplet state results for $S = 1$, since $J = (L + 1)$, L or $(L - 1)$. If the electron also has orbital angular momentum in the triplet state, the three levels will be split by the spin–orbit interaction. Otherwise they will be degenerate, i.e. in a ^3S state.

In heavy atoms, and in very highly excited states of some light atoms, where the outer electrons are remote from the nucleus, only the total angular momentum can be measured, since the spin and orbital angular momenta are strongly coupled. The nett angular momentum is J, where $(j_1 + j_2) \geqslant J \geqslant (j_1 - j_2)$, and $j_1 = l_1 \pm s_1$, and $j_2 = l_2 \pm s_2$. This type of vector addition is termed (j, j) coupling, and in those cases where it operates, the atomic state may be represented by the J value alone, since it is now the only 'good' quantum number.

When an atom has two or more *equivalent* electrons, i.e. ones with the same value of n and l, some of the several alternative electronic configurations which may be possible will be excluded, to satisfy the Pauli Principle, (see ref. 17). For example, in its ground electronic state, the carbon atom has the electronic configuration $2s^2(2p_{+1}, 2p_0, 2p_{-1})^2$. Russell–Saunders coupling is appropriate, and would predict six possible terms ^1S, ^1P, ^1D, ^3S, ^3P and ^3D; however application of the Pauli Principle immediately excludes a ^3D state, since if the electrons have the same values of n and l, they must differ either in m_l or s. In the D state, M_L can take the values $+2, +1, \ldots, -2$; if both electrons have their orbital angular momenta directed along the same axis, they both have the same values of m_l, namely $m_l = +1$ or $m_l = -1$, and both will occupy the same $2p_{\pm 1}$ orbital. In this situation they cannot also have parallel spins and only the ^1D term is possible. Similar arguments lead to the exclusion of the ^3S and ^1P terms. Of those which remain, Hund's Rule of Maximum Multiplicity indicates that one of the $2\,^3P_{2, 1\,\text{or}\,0}$ manifold of states will be the ground state. Just which one may be found by reference to the rule that when the valence shell is less than or only half-filled, the lowest level has $J = L - S$, i.e. $J = 0$ in the case of carbon. For oxygen, with four $2p$ electrons, the shell is more than half-filled and the reverse is true; the ground state is now $2\,^3P_2$. In a sodium atom the $3\,^2P_{\frac{1}{2}}$ level lies below $3\,^2P_{\frac{3}{2}}$: in a chlorine atom the order is reversed. The energies of some low lying states in atoms which are of photochemical interest are collected in Table 1.1.

TABLE 1.1 Energy levels of low-lying electronic states of
atoms of photochemical interest
$(1000 \text{ cm}^{-1} \equiv 11\cdot946 \text{ kJ mole}^{-1})$

Atoms	State	Energy above ground state (cm^{-1})
C	2^3P_0	0
	2^1D_2	10,194
	2^1S_0	21,648
N	$2^4S_{\frac{3}{2}}$	0
	$2^2D_{\frac{5}{2}}$	19,223
O	2^3P_2	0
	2^1D_2	15,868
	2^1S_0	33,793
S	3^3P_2	0
	3^1D_2	9,273
	3^1S_0	22,179
Br	$4^2P_{\frac{3}{2}}$	0
	$4^2P_{\frac{1}{2}}$	3,685
I	$5^2P_{\frac{3}{2}}$	0
	$5^2P_{\frac{1}{2}}$	7,603

1.2d The effect of magnetic and electric fields

The influence of magnetic fields on the line spectra emitted by excited
atoms was discovered by Zeeman in 1896.[18] If the source of radiation was
placed in a magnetic field he observed that lines which were orginally
single were split into small groups of multiplets. The cause is closely
related to the splitting promoted by spin–orbit interaction. If a magnetic
field is applied the magnetic moment associated with the resultant angular
momentum of the electrons in the atom, will be oriented in specific directions about the applied field. The energy of the atom, with magnetic
moment μ in the field $\mathbf{H}$ will be $-\mathbf{\mu} \cdot \mathbf{H} = -\mu H \cos \theta$, where θ gives their
relative orientation. If the electronic angular momentum in the absence of
spin is J, then $\mu \cos \theta$ can only take the values

$$M_J \cdot \frac{eh}{2mc} \equiv M_J\mu_B$$

where $J \geqslant M_J \geqslant -J$ (see Figure 1.4). Thus the degeneracy of electronic
states with $J > 0$ will be split, producing $2J + 1$ sub-levels, of energy

$$E_{M_J} = M_J\mu_B H$$

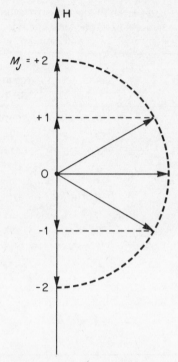

Figure 1.4 Possible orientations of the total angular momentum vector J, along an applied magnetic field **H**. $J = 2$.

If the total angular momentum J includes spin, as well as orbital angular momentum, the nett magnetic moment will be compounded from the two. The unit of magnetic moment due to spin is twice that due to orbital angular momentum i.e. $e\hbar/mc$. Depending on the relative contributions of spin and orbital motion to the nett angular momentum in the particular state concerned, the nett magnetic moment $\mu = gM_J e\hbar/2mc$, where g, the 'Landé g-factor' may lie somewhere between one and two.[19] The possible energies of the atom in the magnetic field are now

$$E_{M_J} = gM_J \mu_B H$$

The first equation accounts for the splittings observed in the original Zeeman experiments, and the latter accounts for the 'anomalous' Zeeman effect (anomalous since the existence of spin angular momentum had not been recognized).

If the applied magnetic field strength is very large, and in excess of the internal spin–orbit magnetic interaction, then the magnetic moments due to spin and orbital angular momentum interact separately with the applied field, and the two quantum numbers M_L and M_S give the separate components of the angular momenta along the field. This phenomenon is the Paschen–Back effect.

External electric fields also split atomic spectral lines into subcomponents, giving rise to the Stark effect.[18] In fields of sufficient strength (typically $> 10^5$ volts cm^{-1}), spectral terms of quantum number J are resolved into levels with angular momentum components along the field of $|M_J| = J, (J - 1), \ldots \frac{1}{2}$ or 0. (M_J is positive since the term splitting by electric fields is proportional to the square of the electric field strength.)

1.3 MOLECULAR ORBITALS AND ELECTRONIC STATES

1.3a Diatomic molecules

Rydberg series have been recognized in the vacuum ultraviolet absorption spectra of molecules, as well as atoms, the convergence limits corresponding to ionization potentials of the molecules. It is reasonable to interpret these series in terms of excitation of an electron into molecular orbitals (m.o.'s) of increasing energy, analogous to the electron orbitals of increasing principal quantum number, in atoms.[16] In principle, the stationary states of the electrons could be found from the exact solution of the Schrödinger Equation for the molecule; the Hamiltonian energy operator would have to include both the interelectronic *and* internuclear repulsion terms, together with the electron–nuclear attraction terms in the potential energy. Unfortunately, the solution requires the initial separation of the equation into equations which are functions of one variable at a time, and this is possible only for one electron atoms. In the case of the simplest H_2^+, endowed with but a single electron, a solution can be found within the limitations of the Born–Oppenheimer approximation; this requires that the electronic wave function does not vary rapidly with a small change in the nuclear coordinates, so that the complete wave function for the molecule can be separated into a product of an electronic part and a nuclear part. It is not an unreasonable approximation to make, since the electronic mass is so much less than that of the nuclei and they can rapidly adjust to any small change in the position of the nuclei. In this situation, exact solutions of the Schrödinger equation for H_2^+ can be found for a series of fixed positions of the two protons. The equilibrium position can be obtained, since it is associated with the minimum total energy, and the stationary states for the H_2^+ molecule located. For polyelectronic molecules, direct integration of the Schrödinger equation is not possible. The *molecular orbital description* provides an approximate method with which to obtain an estimate of the energies and stationary states of polyelectronic molecules (as well as H_2^+).[16]

In the simplest molecular orbital description, it is assumed that the amplitude of the nett molecular wave function is found from the algebraic sum of the amplitudes of the overlapping atomic wave functions which produce it. Molecular orbitals are formed by linear combination of atomic orbitals (LCAO–MO method). Overlapping the two $1s$ a.o.'s of the H_2^+ molecule, will produce two new m.o.'s of amplitudes

$$\psi_+ = \frac{1}{\sqrt{N_+}}(1s_A + 1s_B)$$

or

$$\psi_- = \frac{1}{\sqrt{N_-}}(1s_A - 1s_B)$$

depending whether the overlap is in phase $(+)$ or out of phase $(-)$. The factors N_+, $N_-(= 2(1 \pm S_{AB})$ where $S_{AB} = \int_{-\infty}^{\infty} 1s_A 1s_B \, d\tau$ is the *overlap integral*) ensure that the new wave functions are normalized. Overlap in phase concentrates electron density in the space between the two nuclei (see Figure 1.5), where it can do most good in binding the nuclei together: ψ_+ is a *bonding* orbital. The other orbital ψ_- is *antibonding*, since its probability distribution has a nodal plane between the nuclei:

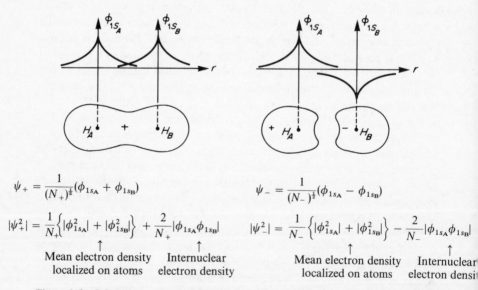

$$\psi_+ = \frac{1}{(N_+)^{\frac{1}{2}}}(\phi_{1s_A} + \phi_{1s_B}) \qquad\qquad \psi_- = \frac{1}{(N_-)^{\frac{1}{2}}}(\phi_{1s_A} - \phi_{1s_B})$$

$$|\psi_+^2| = \frac{1}{N_+}\left\{|\phi_{1s_A}^2| + |\phi_{1s_B}^2|\right\} + \frac{2}{N_+}|\phi_{1s_A}\phi_{1s_B}| \qquad |\psi_-^2| = \frac{1}{N_-}\left\{|\phi_{1s_A}^2| + |\phi_{1s_B}^2|\right\} - \frac{2}{N_-}|\phi_{1s_A}\phi_{1s_B}|$$

     ↑   ↑   ↑   ↑

Mean electron density   Internuclear   Mean electron density   Internuclear
localized on atoms   electron density   localized on atoms   electron densi⟨

Figure 1.5   Schematic representation of molecular orbitals of hydrogen formed from overlap of $1s$ atomic orbitals.

an electron in this orbital will be unable to counter the internuclear repulsion.

These statements are confirmed by estimating the relative energies of the orbitals. Since the m.o.'s ψ_+ and ψ_- do not represent the true eigenstates of the molecule, but approximations thereto, only their average energies can be found; these are

$$\langle E_+ \rangle = \int_{-\infty}^{\infty} \psi_+^* \hat{H} \psi_+ \, d\tau = \frac{1}{N_+} \int_{-\infty}^{\infty} (1s_A + 1s_B)\hat{H}(1s_A + 1s_B) \, d\tau$$

and

$$\langle E_- \rangle = \int_{-\infty}^{\infty} \psi_-^* \hat{H} \psi_- \, d\tau = \frac{1}{N_-} \int_{-\infty}^{\infty} (1s_A - 1s_B)\hat{H}(1s_A - 1s_B) \, d\tau$$

where $\hat{H}$ is the Hamiltonian for the electron circulating in the electrostatic field about the two nuclei. The equations can be rewritten in the form

$$\left. \begin{array}{l} \langle E_+ \rangle = \dfrac{\alpha + \beta}{1 + S} \equiv \alpha + \dfrac{\beta - \alpha S}{1 + S} \simeq \alpha + \beta \\[2ex] \langle E_- \rangle = \dfrac{\alpha - \beta}{1 - S} \equiv \alpha - \dfrac{\beta - \alpha S}{1 - S} \simeq \alpha - \beta \end{array} \right\} \text{ when } S \ll 1$$

where α, the *coulomb integral* $= \int_{-\infty}^{\infty} 1s_A \hat{H} 1s_A \, d\tau = \int_{-\infty}^{\infty} 1s_B \hat{H} 1s_B \, d\tau$ and β, the *resonance integral* $= \int_{-\infty}^{\infty} 1s_A \hat{H} 1s_B \, d\tau = \int_{-\infty}^{\infty} 1s_B \hat{H} 1s_A \, d\tau$. $\langle E_+ \rangle$, $\langle E_- \rangle$ and α approximate to the energies required to remove an electron from the m.o.'s ψ_+ and ψ_- of the H_2^+ molecule, and the $1s$ a.o. of the isolated hydrogen atoms. Since β is negative, they lie in the order $\langle E_+ \rangle < \alpha < \langle E_- \rangle$, and electrons in ψ_+ and ψ_- are respectively bonding and antibonding (see Figure 1.6). In the ground electronic state of H_2^+, the odd

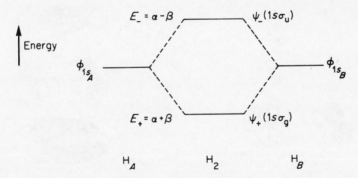

Figure 1.6 σ-orbital energies in H_2.

electron will reside in the most tightly bound m.o., producing a 'one-electron bond'. The dissociation energy of H_2^+ is known to be 255 kJ mole^{-1}.[20] In its first electronically excited state the odd electron will be promoted into ψ_- and the nuclei should now repel each other: the electronic state will be repulsive.

Higher m.o.'s of H_2^+ can be constructed by overlapping higher atomic orbitals of comparable size and energy. Overlapping the $2s$ a.o.'s produces cylindrically symmetrical orbitals similar to the lowest pair, but with an average position rather further removed from the two nuclei. The $2p$ orbitals may overlap axially, or perpendicular to the axis; axial overlap produces two orbitals with the same symmetry as before, but in the other situation two pairs of degenerate orbitals are produced, with zero amplitude at the internuclear axis, (see Figure 1.7). These will be less

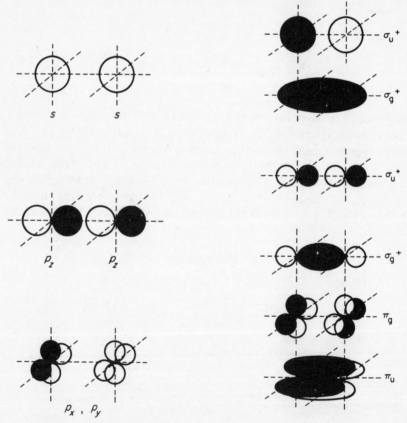

Figure 1.7   Schematic representation of σ and π orbital overlap.

bonding (or antibonding) than the other, since the overlap is much less, and the electrons are concentrated off the central axis.

Suppose the internuclear distance could be reduced, until the nuclei virtually coalesced; the orbitals of the 'united atom' would be those of an electron circulating in the field of a ^{2}He nucleus. If the 'united atom' were slightly deformed, the intense internuclear electrical field would partially split the degeneracy of the orbitals with $l > 0$ (internal Stark effect). Where the orbitals had an angular momentum component of $|m_l|$ (or λ) about the internuclear axis, equal to $0, 1, 2, \ldots$, they would become σ, π, $\delta, \ldots$ orbitals, of which all but the σ orbitals, would be doubly degenerate. Thus ns a.o.'s become $ns\sigma$ orbitals; np a.o.'s become $np\sigma$ and $np\pi$ orbitals; and nd a.o.'s become $nd\sigma$, $nd\pi$ and $nd\delta$ orbitals. The change of symbolism from English to Greek, reflects the change of symmetry from the sphere of the atom to the cylinder of the molecule. The letters indicate not only the orbital angular momenta of the electrons, but also the behaviour of the wave function under the symmetry operations of the point group to which the atom or molecule belongs. Reducing the symmetry by deforming the 'united atom' also reduces the degeneracies in the wave functions. If the internuclear distance is increased further, to the point where the interaction is small, the orbital energies must approach those of the slightly perturbed atomic orbitals, but the symmetry or angular momentum of any given orbital must be preserved between the two extremes. Using this as a 'label', the changing energies can be charted on an orbital correlation diagram.

In describing the electronic configurations of polyelectronic diatomic molecules, the nett wave function is assumed to be a product of 'one-electron' H_2^+-like wave functions which are eigenfunctions of assumed one-electron Hamiltonian operators. The argument is the same as that developed for atomic wave functions on p. 15. The one-electron m.o.'s are filled in order of increasing energy, subject to the constraints of the Pauli Principle and Hund's rule, just as for atoms, and the complete electronic configuration obtained. When the internuclear separation is large, the orbital energies lie in the order

$$\sigma_g(1s) < \sigma_u(1s) < \sigma_g(2s) < \sigma_u(2s) < \sigma_g(2p) < \pi_u(2p) < \pi_g(2p) < \sigma_u(2p)\ldots$$

When the separation is less, a rough estimate of the relative energies of the polyelectronic m.o.'s can be obtained from the appropriate orbital correlation diagram; these have been developed for homo- and hetero-nuclear diatomic molecules by Herzberg, and are presented in Figures 1.8 and 1.9.[20] Each line terminates at the orbital with the lowest energy and the correct symmetry, taken in strict order. It is not possible to 'jump' an orbital satisfying these conditions, since orbitals of the same symmetry do

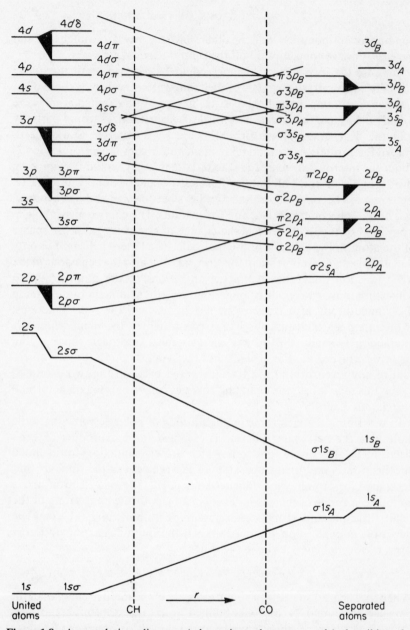

Figure 1.8 A correlation diagram (schematic and not to scale) describing the energies of the electron orbitals of heteronuclear diatomic molecules. (From Herzberg, *Spectra of Diatomic Molecules*, Vol. I, Copyright 1950, by Litton Educational Publishing, Inc., with permission). Note the reversals in the ordering of the orbitals resulting from the symmetry correlation. The values of r at which the intersections occur can only be estimated from spectroscopic data. The dotted lines indicate the observed ordering in CH and CO.

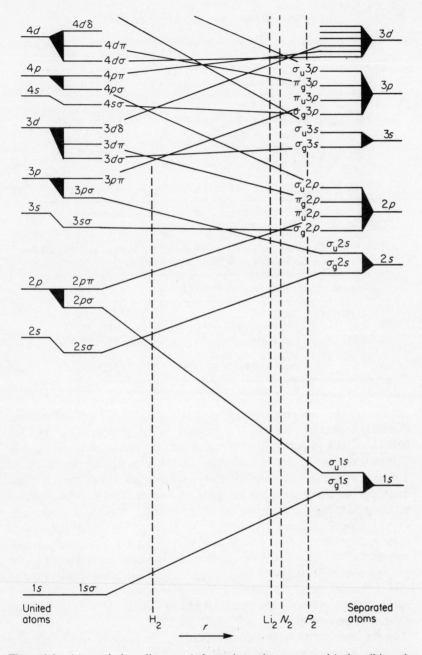

Figure 1.9 A correlation diagram (schematic and not to scale) describing the energies of the electron orbitals of homonuclear diatomic molecules. (From Herzberg, *Spectra of Diatomic Molecules*, Vol. I, Copyright 1950, by Litton Educational Publishing, Inc., with permission).

TABLE 1.2 Electronic states of homonuclear and near homonuclear diatomic molecules (the inner K-shell orbitals are not included)

Molecule	Ground electronic configuration	State(s)[a]
Li_2	$(z\sigma_g)^2$	$^1\Sigma_g^+$
B_2	$(z\sigma_g)^2(y\sigma_u)^2(x\sigma_g)^0(\omega\pi_u)^2$	$^3\Sigma_g^-$, $^1\Delta_g$, $^1\Sigma_g^+$
C_2	$(z\sigma_g)^2(y\sigma_u)^2(x\sigma_g)^0(\omega\pi_u)^4$	$^1\Sigma_g^+$
O_2	$(z\sigma_g)^2(y\sigma_u)^2(x\sigma_g)^2(\omega\pi_u)^4(v\pi_g)^2$	$^3\Sigma_g^-$, $^1\Delta_g$, $^1\Sigma_g^+$
F_2	$(z\sigma_g)^2(y\sigma_u)^2(x\sigma_g)^2(\omega\pi_u)^4(v\pi_g)^4$	$^1\Sigma_g^+$
CN	$(z\sigma)^2(y\sigma)^2(x\sigma)^1(\omega\pi)^4$	$^2\Sigma^+$
NO	$(z\sigma)^2(y\sigma)^2(x\sigma)^2(\omega\pi)^4(v\pi)^1$	$^2\Pi_{\frac{1}{2}}$, $^2\Pi_{\frac{3}{2}}$

[a] Electronic configurations which include equivalent electrons give rise to more than one electronic state (see pp. 29–30).

TABLE 1.3 Electronic states of heteronuclear diatomic molecules (the inner K-shell orbitals are not included)

Molecule	Ground electronic configuration	State(s)
LiH	$(z\sigma)^2$	$^1\Sigma^+$
CH	$(z\sigma)^2(y\sigma)^2(x\sigma)^0(\omega\pi)^1$	$^2\Pi_{\frac{1}{2}}$, $^2\Pi_{\frac{3}{2}}$
NH	$(z\sigma)^2(y\sigma)^2(x\sigma)^0(\omega\pi)^2$	$^3\Sigma^-$, $^1\Delta$, $^1\Sigma^+$
CO	$(z\sigma)^2(y\sigma^2(x\sigma)^2(\omega\pi)^4$	$^1\Sigma^+$
CF	$(z\sigma)^2(y\sigma)^2(x\sigma)^2(\omega\pi)^4(v\pi)^1$	$^2\Pi_{\frac{3}{2}}$, $^2\Pi_{\frac{1}{2}}$

not cross: the non-crossing rule (see ref. 20, p. 295). In homonuclear diatomic molecules, the m.o.'s are either 'u' or 'g' depending on their behaviour with respect to inversion through the centre of symmetry.

Electronic configurations of some diatomic molecules are listed in Tables 1.2 and 1.3. (Except in the 'united atom' limit, the principal quantum numbers have no meaning. A sequence of orbitals such as $\sigma_g(2s)\sigma_u(2s)\sigma_g(2p)\pi_u(2p)\pi_g(2p)\sigma_u(2p)$ is commonly written $z\sigma_g y\sigma_u x\sigma_g w\pi_u v\pi_g u\sigma_u$ (after Mulliken) or $1\sigma_g 1\sigma_u 2\sigma_g 1\pi_u 1\pi_g 2\sigma_u$. The latter notation is used exclusively, in the description of polyatomic molecular orbital sequences. M.o.'s formed by overlap of a.o.'s which are occupied in the ground electronic states of the isolated atoms are termed intravalency shell orbitals; those constructed from higher energy a.o.'s are extravalency shell orbitals. Molecular Rydberg series are associated with excitation into the latter.

In the simplest LCAO–MO approximation, the one-electron m.o.'s of a diatomic molecule are of the form

$$\psi = \sum_v c_v \phi_v$$

where c_v are the coefficients of the a.o.'s ϕ_v centred on atom v. In hetero-nuclear molecules, the more electronegative element will collect more of the electronic charge, and the coefficients will be unequal. The greater the difference the more ionic character in the molecule.

When the spin of the electrons is included, the nett antisymmetrized wave function for the electronic configuration

$$\psi_1(1)\bar\psi_1(2)\psi_2(3)\bar\psi_2(4)\ldots\bar\psi_k(n) \equiv \psi_1^2\psi_2^2\ldots\psi_k^2$$

is

$$\frac{1}{\sqrt{n!}}\begin{vmatrix} \psi_1(1)\bar\psi_1(1)\psi_2(1)\bar\psi_2(1) & \ldots & \bar\psi_k(1) \\ \psi_1(2)\bar\psi_1(2)\psi_2(2)\bar\psi_2(2) & \ldots & \bar\psi_k(2) \\ \cdots & & \\ \cdots & & \\ \cdots & \ldots & \bar\psi_k(n) \end{vmatrix} \equiv |\psi_1\bar\psi_1\psi_2\bar\psi_2\ldots\bar\psi_k|$$

This represents a closed shell configuration in which all the spatial orbitals are doubly occupied. It would be appropriate for the ground electronic state of H_2, which has the electronic configuration $\sigma_g(1)\bar\sigma_g(2)$. In this form the wave function is not antisymmetric with respect to electron exchange, but in the expanded determinantal form

$$\frac{1}{\sqrt{2}}\{\sigma_g(1)\bar\sigma_g(2) - \sigma_g(2)\bar\sigma_g(1)\} \equiv |\sigma_g\bar\sigma_g|$$

it is. There is only one wave function since the electrons must have opposite spin, and it represents a singlet state. In the first excited state of H_2, this constraint no longer applies and four open shell configurations are possible $\sigma_g(1)\sigma_u(2)$, $\bar\sigma_g(1)\sigma_u(2)$, $\sigma_g(1)\bar\sigma_u(2)$ or $\bar\sigma_g(1)\bar\sigma_u(2)$. Separating the space and spin parts of the wave functions, leads to the possibilities

space
$$\left.\begin{array}{l}\sigma_g(1)\sigma_u(2)\\ \sigma_g(2)\sigma_u(1)\end{array}\right\} \to \begin{array}{ll} 1/\sqrt{2}(\sigma_g(1)\sigma_u(2) + \sigma_g(2)\sigma_u(1)) & \text{sym.}\\ 1/\sqrt{2}(\sigma_g(1)\sigma_u(2) - \sigma_g(2)\sigma_u(1)) & \text{antisym.}\end{array}$$

spin
$$\begin{array}{ll}\alpha(1)\alpha(2) \to \alpha(1)\alpha(2) & \text{sym.}\\ \beta(1)\beta(2) \to \beta(1)\beta(2) & \text{sym.}\end{array}$$
$$\left.\begin{array}{l}\alpha(1)\beta(2)\\ \alpha(2)\beta(1)\end{array}\right\} \to \begin{array}{ll} 1/\sqrt{2}(\alpha(1)\beta(2) + \alpha(2)\beta(1)) & \text{sym.}\\ 1/\sqrt{2}(\alpha(1)\beta(2) - \alpha(2)\beta(1)) & \text{antisym.}\end{array}$$

The spin wave functions separate into a triplet of symmetric functions, associated with parallel spins, and a single antisymmetric function, associated with opposed spins. Recombining the space and spin parts, to give antisymmetric nett wave functions only, gives the four states

$$\frac{1}{\sqrt{2}}(|\sigma_g\bar{\sigma}_u| - |\bar{\sigma}_g\sigma_u|) \quad \text{singlet}$$

$$\frac{1}{\sqrt{2}}(|\sigma_g\bar{\sigma}_u| + |\bar{\sigma}_g\sigma_u|)$$

$$|\sigma_g\sigma_u| \qquad\qquad \text{triplet}$$

$$|\bar{\sigma}_g\bar{\sigma}_u|$$

In the absence of a magnetic field (internal via spin–orbit coupling, or external), the triplet states will be degenerate; their energies will lie below that of the singlet state.

The electronic states of diatomic (or linear polyatomic) molecules are described by term symbols which are analogous to those of atoms. They give the spin angular momentum, and the component of the nett orbital angular momentum vector along the internuclear axis. These are denoted Λ and S respectively; their sum is $\Omega = |\Lambda + \Sigma|$, where Σ is the component of S along the internuclear axis, i.e. in the direction of Λ. The term symbol for a particular electronic state is written $^{2S+1}\Lambda_\Omega$.* Instead of the principal quantum number the term is preceded by the letters $\tilde{X}$ (ground state), $\tilde{A}, \tilde{B}, \tilde{C} \ldots$ (successive excited states with the same spin multiplicity as the ground state) or $\tilde{a}, \tilde{b}, \tilde{c} \ldots$ (corresponding states of different multiplicity, e.g. triplets rather than singlets).† States for which $\Lambda = 0, \pm 1, \pm 2, \ldots$ are termed $\Sigma, \Pi, \Delta, \ldots$: the values of Λ may be found by vector addition of the momentum components of the individual electrons, or alternatively, from the symmetries of the occupied molecular orbitals. The nett orbital angular momentum component of H_2 in its ground electronic state, where both electrons occupy the $\sigma_g(1s)$ orbital with $\lambda = 0$, is $\Lambda = 0$; the electrons are paired and $S = 0$. The ground state is therefore $^1\Sigma$. The nett symmetry of the space part of the wave function is obtained from the direct product of symmetry species of the occupied orbitals (see section 1.3b, and Table 1.4). In this case these are both σ_g^+, where the $+$ relates to the symmetric behaviour of the wave function toward reflexion in a plane containing the internuclear (symmetry) axis. The wave functions are symmetric under all the operations of the point group of the molecule, and their product is also totally symmetric, Σ_g^+. The ground electronic state is $\tilde{X}^1\Sigma_g^+$. In general, the product of any doubly filled orbital will be totally symmetric (see section 1.3b), so that

---

* If heavy atoms are present, the spin and orbital angular momenta of each electron couple to give components ω_i, and the total angular momentum is then given by their vector sum. This is termed (ω,ω) coupling, and is analogous to (j,j) coupling in atoms.

† The symbol $\sim$ is used to avoid confusion with group theoretical symbols used to describe polyatomic electronic states.

any linear molecule in which all the orbitals are doubly occupied, will be in a $^1\Sigma_g^+$ electronic state. As a corollary to this, only the half-filled orbitals need to be considered, in determining the symmetries of open shell configurations.

TABLE 1.4 Direct products of symmetry species appropriate to diatomic or linear molecules

Orbital symmetry	Molecular symmetry
$\sigma\sigma$	Σ^+
$\sigma\pi$	Π
$\sigma\delta$	Δ
$\pi\pi$	$\Sigma^+, \Sigma^-, \Delta$
$\pi\delta$	Π, Φ
$\delta\delta$	$\Sigma^+, \Sigma^-, \Gamma$

Note:
(i) For equivalent electrons the multiplicities of possible molecular electronic terms are restricted by the Pauli Principle
(ii) A triple product may be factorized thus

$$\sigma\pi\pi \equiv (\sigma \times \pi) \times \pi \equiv \pi \times \pi \to \Sigma^+, \Sigma^-, \Delta$$

(iii) Where the molecule is centro-symmetric, the u or g character of the product is given by $u \times u = g \times g = g$ or $u \times g = g \times u = u$

When an electron in H_2 is promoted from the $\sigma_g(1s)$ into the $\sigma_u(1s)$ orbital, the symmetry product is now Σ_u^+; the electronic states resulting are $\tilde{A}^1\Sigma_u^+$ and $\tilde{a}^3\Sigma_u^+$, with the triplet state the lower of the two. A more instructive example is provided by the oxygen molecule. In its electronic ground state, the configuration is

$$\underbrace{KK(2s\sigma_g)^2(2s\sigma_u)^2}_{\text{non-bonding}} \underbrace{(2p\sigma_g)^2(2p\pi_u)^4}_{\text{bonding}} \underbrace{(2p\pi_g)^2}_{\text{antibonding}}$$

or

$$KK(z\sigma_g)^2(y\sigma_u)^2(x\sigma_g)^2(w\pi_u)^4(v\pi_g)^2$$

Since the $2p\pi_g$ orbitals are doubly degenerate, the component orbitals are only half filled. The molecule has an open shell configuration, and a triplet ground state in which the two electrons occupy equivalent orbitals. Because of the orbital degeneracy however, several electronic states may be formed with this configuration. The orbital angular momentum of each electron may have components $\lambda = \pm 1$, and the nett angular momentum components are ± 2 or 0, giving Δ or Σ states. When $\Lambda = \pm 2$, both electrons have the same component of orbital angular momentum,

$\lambda = 1$ or $\lambda = -1$; as they cannot also have the same spin, the $^3\Delta$ state does not exist. Both $^1\Sigma$ and $^3\Sigma$ states are possible however, and the ground electronic state must be $^3\Sigma$. Possible symmetry species resulting from the direct product $\pi_g \times \pi_g$ are Σ_g^+, Σ_g^- and Δ_g; the ground electronic state of O_2 is $\tilde{X}^3\Sigma_g^-$. The other two states lie respectively 96 kJ mole^{-1} ($\tilde{a}^1\Delta_g$) and 163 kJ mole^{-1} ($\tilde{b}^1\Sigma_g^+$) above it.[20]

If an electron is promoted from the $2p\pi_u$ to the $2p\pi_g$ orbital, each orbital will possess an odd electron, and the possible electronic states associated with this configuration now have the symmetries $\pi_u \times \pi_g \equiv \Sigma_u^+$, Σ_u^- or Δ_u. Since the two odd electrons occupy separate orbitals they are now non-equivalent and all of the states $^1\Sigma_u^+$, $^1\Sigma_u^-$, $^1\Delta_u$, $^3\Sigma_u^+$, $^3\Sigma_u^-$, and $^3\Delta_u$ may exist. The strong ultraviolet absorption of oxygen, which marks the onset of the 'vacuum ultraviolet region' of the spectrum, is associated with the electronic transition

$$\tilde{B}^3\Sigma_u^- \leftarrow \tilde{X}^3\Sigma_g^-$$

in which the equilibrium bond length is much increased by the promoting of an electron from a bonding into an antibonding orbital. Like O_2, the heteronuclear molecule NH also has two singly occupied π m.o.'s in its ground electronic configuration

$$K(2s\sigma)^2(2p\sigma)^2(2p\pi)^2 \quad \text{or} \quad K(z\sigma)^2(y\sigma)^2(w\pi)^2$$

giving $^3\Sigma^-$, $^1\Delta$ and $^1\Sigma^+$ terms, in ascending order of energy. However, the bonding in the molecule is very different from that in O_2.

In its ground electronic state, the nitrogen atom has the electronic configuration

$$K(2s)^2(2p_x)^1(2p_y)^1(2p_z)^1$$

As the hydrogen atom (in its ground electronic state) approaches the N atom the only appreciable overlap initially, involves the $(1s)_H$ and $(2p_z)_N$ a.o.'s since the other $2p_N$ orbitals are directed perpendicular to the inter-nuclear axis and the $2s_N$ orbital is rather more closely and tightly bound. In this situation, the only bonding orbital is $2p\sigma = c_N(2p_z)_N + c_H(1s)_H$ and the others are non-bonding (see Figure 1.10). [The antibonding combination $2p\sigma^* = c_N(2p_z)_N - c_H(1s)_H$ would be occupied in the highly excited configuration $K(2s\sigma)^2(2p\sigma)^1(2p\pi)^2(2p\sigma^*)^1$, $^5\Sigma^-$ which also correlates with the two ground state atoms.]

If the two atoms approached more closely, appreciable overlap of the $1s_H$ and the $2s_N$ orbitals might be possible to give a combination with the same symmetry, under the molecular point group, as the $2p\sigma$ orbital. In consequence the two combinations may 'mix' to produce a 'hybrid' m.o. Alternatively the situation may be described as an overlap of the $1s_H$ a.o., with a hybrid a.o., centred on the nitrogen atom and given by the linear

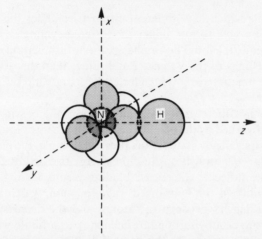

Figure 1.10 Schematic representation of atomic orbital overlap in NH; large internuclear distance.

combination

$$\phi_z = 2s + \lambda 2p_z$$

The other hybrid a.o. is

$$\phi_{-z} = \lambda 2s - 2p_z$$

directed in the opposite sense since the coefficient of $2p_z$ is now negative (see Figure 1.11). λ measures the s or p character of the hybrid orbital, i.e. the degree of hybridization. With this description, the electronic configuration of NH with a small internuclear separation will be

$$K(z\sigma)^2(y\sigma)^2(\omega\pi)^2, \quad \tilde{X}^3\Sigma^-, \quad \tilde{a}^1\Delta, \quad \tilde{b}^1\Sigma^+$$

where

$$z\sigma = c'_N\phi_z + c'_H 1s_H \quad \text{bonding}$$

$$y\sigma = \phi_{-z} \qquad\qquad \text{non-bonding lone pair}$$

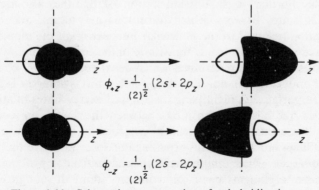

$$\phi_{+z} = \frac{1}{(2)^{\frac{1}{2}}}(2s + 2p_z)$$

$$\phi_{-z} = \frac{1}{(2)^{\frac{1}{2}}}(2s - 2p_z)$$

Figure 1.11 Schematic representation of sp hybridization.

Atomic orbitals which have different symmetries under the point group of the atom, are able to mix under the lowered symmetry of the newly forming molecule. It should be emphasized that this method of constructing orbitals belongs to the Valence Bond rather than the MO description, in which wave functions are written for each separate interatomic bond in the molecule.[21]

Whether or not there actually is any mixing of a.o.'s in this, or any other case, is governed by the balance between the energy required to excite the atom into its valency state (in the present example, the energy needed to excite the $2s \rightarrow 2p$ transition), and the energy released by the increased binding in the molecule. Increased binding results from the increased overlap capability of the hybridized wave function, which concentrates more electron density between the atoms, and the decreased repulsion between electrons occupying bonding and lone pair orbitals, which become widely separated in space. In the case of NH, where only one atom is bonded to the nitrogen atom, and where the $2s$ a.o. is tightly bound relative to $2p$, the molecule derives no 'benefit' from hybridization. In MO terms, there is no mixing of a.o.'s centred on the same atom, in the one electron wave functions. It is the case however, in diatomic molecules such as CN, CO and CF, where the carbon atom has a smaller $2s - 2p$ energy difference, and when we come to discuss polyatomic molecules where one atom may be bonded to several others, we shall see that this is a common situation.

1.3b Molecular symmetry

Frequent reference has been (and will be) made to molecular and orbital symmetries. We have already seen that they determine the number and identity of the electronic states which arise from any given electronic configuration in a diatomic molecule. The same is also true in polyatomic molecules. Further, we shall see in section 1.4, that they also identify the electronic states between which transitions may occur, and even the polarization (referred to the molecular framework) of the dipole change associated with the electronic transition. In the construction of the m.o.'s themselves, symmetry determines which linear combinations of a.o.'s are possible and which are not. For the photochemist, symmetry is of fundamental importance in identifying the allowed energy states of atoms and molecules, the allowed transitions between them, and the correlation between the orbitals of a molecule as its geometry is changed; in general terms, it helps him to relate his experimental observations to the structure of the molecule whose photochemical behaviour he is probing. On the other hand, arguments based on symmetry alone do not provide any information about the detailed forms of the wave functions, or about the way in which the energy levels are ordered. Nonetheless they have far reaching consequences.

The methods by which they are developed involve the mathematics of Group Theory. There are many excellent monographs at all levels of sophistication describing the applications of Group Theory to chemical problems,[22–27] and this is not the place to attempt the production of yet another. However, in order to avoid the impression of 'pulling things out of a hat' in subsequent discussion, we shall deal very briefly with a few general concepts. The application of the results of group theory rarely involves more than the ability to add or multiply small numbers.

A symmetry operation is one which leaves the framework of a molecule unchanged. Five types of symmetry operation may be identified : rotations about an axis of symmetry, reflexions in a plane of symmetry, inversion through a centre of symmetry, improper rotations (i.e. rotation about an axis followed by reflexion in a plane perpendicular to it), and the identity operation (i.e. leaving the molecule alone). The complete set or group of symmetry operations, generated by the symmetry elements (i.e. axes, planes, etc.) of the molecule, constitute its point group. For illustration (see Figure 1.12), the water molecule has a plane of symmetry, $\sigma(yz)$,

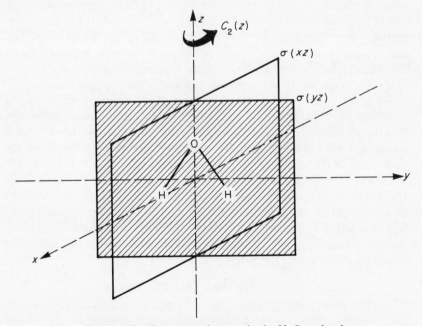

Figure 1.12 Symmetry elements in the H_2O molecule.

another bisecting the HÔH, $\sigma(xz)$, and a twofold rotation axis along the z-axis, $C_2(z)$, all of which leave the molecule unchanged. It belongs to the point group C_{2v} which includes the four operations $C_2(z)$, $\sigma(yz)$, $\sigma(xz)$ and E, the identity operation. (It is necessary that all point groups include the identity operation, in order that they fill the mathematical requirements of a Group.) The $2p_z$ orbital of the oxygen atom is a wave function that is totally symmetric with respect to each of these: the sign of the wave function is unchanged by each operation since it lies along the z-axis where all the symmetry elements intersect.* It is said to belong to the *symmetry species* a_1. If the $2p_z$ orbital were to be used as a basis for the representation of the symmetry operations of the group, the representation would be given by the set of numbers $+1, +1, +1, +1$.

C_{2v}	E	$C_2(z)$	$\sigma(xz)$	$\sigma(yz)$
A_1	$+1$	$+1$	$+1$	$+1$
A_2	$+1$	$+1$	-1	-1
B_1	$+1$	-1	$+1$	-1
B_2	$+1$	-1	-1	$+1$

The $2p_y$ orbital is only symmetric with respect to the operations E and $\sigma(yz)$; it changes sign under the other two and is said to belong to the symmetry species b_2. If the $2p_y$ orbital is used as a basis, the representation would be given by the set of numbers $+1, -1, -1, +1$. It can be shown, that only two other symmetry species exist in this point group, termed a_2 and b_1, and their representations are shown in the table on p. 33. (Note that the symbols A and B always denote symmetric and antisymmetric behaviour respectively, towards rotation about the principle axis.)

Since any of the symmetry operations of the point group carry the molecule into a physically equivalent configuration, any physically observable property of the molecule must remain unchanged by the symmetry operation. It must be invariant with respect to all of the operations of the group, i.e. totally symmetric (put crudely, the molecule has the same physical properties 'upside down' as 'right way up'). Since the energy of the molecule is a physical observable, the Hamiltonian must be unchanged by any symmetry operation of the point group, and in this way it is linked to the symmetry of the molecule. It follows that if there is a wave function ψ_i for which $\hat{H}\psi_i = E_i\psi_i$ and O_R is some symmetry operation of the molecular point group, then

$$O_R\hat{H}\psi_i = O_R E_i\psi_i$$

* We have slid from the z-axis vector, to the $2p_z$ wave function. Since the $2p_z$ function behaves like the z vector, we can transfer our arguments from one to the other with no misgivings.

or

$$\hat{H}(O_R\psi_i) = E_i(O_R\psi_i)$$

since E_i is a constant and $\hat{H}$ is invariant under the operation O_R. Because of the latter $O_R\hat{H} = \hat{H}O_R$; the Hamiltonian operator commutes with all operations of the molecular point group. If both ψ_i and $O_R\psi_i$ are to have the same eigenvalue and ψ_i is nondegenerate, $O_R\psi_i$ can only equal $\pm 1 . \psi_i$. Hence the only possible wave functions of the molecule are those which are either symmetric or antisymmetric toward the symmetry operations of the group (provided the wave functions are non-degenerate); *each one belongs to one of the symmetry species of the group and serves as a basis for its representation.*

Let us now attempt to build up the possible group molecular orbitals of H_2O, utilizing these concepts. The symmetries of the oxygen atom orbitals are

$$2s(a_1), \quad 2p_x(b_1), \quad 2p_y(b_2), \quad 2p_z(a_1)$$

(by convention small letters are used to denote orbital symmetries: capital letters denote molecular symmetries). If the hydrogen atom $1s$ orbitals are taken separately they do not possess the symmetry of the molecule, but taken together, as an H_2 group orbital, there are two possible linear combinations

$$(1s + 1s)(a_1), \quad (1s - 1s)(b_2)$$

Combining these together will produce six m.o.'s, any one of which can only include a.o.'s of the same symmetry species, otherwise the new m.o. will not possess the symmetry of the molecule. Thus $(2p_x)_O$ forms a single orbital of b_1 symmetry,

$$1b_1 = (2p_x)_O \text{ non-bonding}$$

$(2p_y)_O$ and $(1s - 1s)_H$ combine to give two b_2 m.o.'s of the form

$$1b_2 = (2p_y)_O + (1s - 1s)_H \quad \text{bonding}$$

$$2b_2 = (2p_y)_O - (1s - 1s)_H \quad \text{antibonding}$$

and $(2s)_O$, $(2p_z)_O$ and $(1s + 1s)_H$ combine to give three a_1 orbitals, which may plausibly be written

$$2a_1 = (2s)_O - (2p_z)_O + (1s + 1s)_H \quad \text{bonding}$$

$$3a_1 = (2s)_O + (2p_z)_O \qquad\qquad\qquad \text{non-bonding}$$

$$4a_1 = (2s)_O - (2p_z)_O - (1s + 1s)_H \quad \text{antibonding}$$

(The $1a_1$ orbital corresponds to the inner shell $(1s)_O$, contributing little or nothing to the bonding.) They are represented schematically in Figure 1.13.

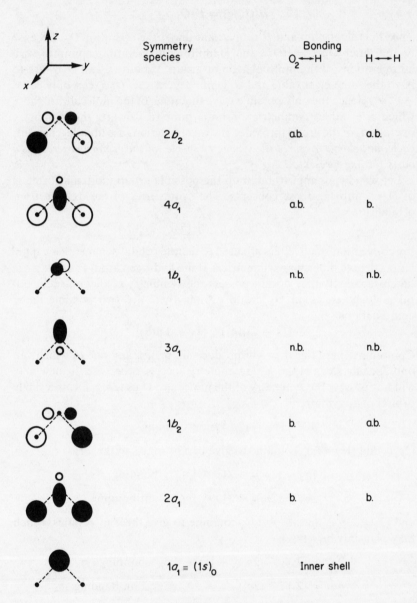

Figure 1.13 Symmetry orbitals of H_2O.

(Note that nothing can be said about the relative contributions made by each a.o.; in the absence of any numerical calculation we cannot know the orbital coefficients.)

The molecule contains eight electrons outside the $(1s)_O$ shell, and the ground electronic configuration may be written

$$\underbrace{(1a_1)^2}_{\substack{\text{inner} \\ \text{shell}}} \underbrace{(2a_1)^2(1b_2)^2}_{\substack{H_2\text{–}O \\ \text{bonding}}} \underbrace{(3a_1)^2(1b_1)^2}_{\substack{\text{non-} \\ \text{bonding}}}$$

The nett symmetry is obtained from the direct product of the symmetry species of the occupied orbitals. It can be shown that the direct product representation is found by simply multiplying the corresponding elements of each representation. Thus

$$a_1 \times a_1 = (+1)^2(+1)^2(+1)^2(+1)^2 \equiv a_1$$
$$b_1 \times b_1 = (+1)^2(-1)^2(+1)^2(-1)^2 \equiv a_1$$

Obviously all doubly occupied orbitals have a_1 symmetry and the overall symmetry of H_2O is A_1. All the electrons are paired, so the electron configuration gives the state $\tilde{X}^1A_1$. In general, any molecule which has a closed-shell electronic configuration (which includes the vast majority of stable molecules) will have a totally symmetric singlet ground state (cf. $^1\Sigma_g^+$ in the H_2 molecule).

If an electron is promoted from the $1b_1 \rightarrow 4a_1$ orbital, the overall symmetry will be $b_1 \times a_1 = b_1$; two states are possible, $\tilde{A}^1B_1$ and $\tilde{a}^3B_1$. In both of these an electron has been promoted from a non-bonding to an antibonding orbital, and the stability of the molecule should be reduced. In fact, when the $\tilde{A}^1B_1$ state is populated by light absorption, dissociation into H and OH is the immediate result.[28]

In point groups that contain greater than twofold axes of symmetry, some of the symmetry species are degenerate. This means that certain symmetry operations have the effect of 'mixing' some of the basis functions. For example, NH_3 has a 3-fold rotation axis lying along the intersection of three vertical planes, each of which contains an N–H bond; it belongs to the point group C_{3v}. With C_3 lying along the z axis, the wave function $(2p_z)_N$ provides a basis for the totally symmetric representation of the group (a_1), since it is symmetric under all the operations. Now suppose we were to try $(2p_x)_N$ as a basis. Operating $C_3(z)$ on $2p_x$, or $2p_y$ gives

$$C_3(z)(2p_x)_N = -\tfrac{1}{2}(2p_x)_N - \sqrt{\tfrac{3}{2}}(2p_y)_N$$
$$C_3(z)(2p_y)_N = \sqrt{\tfrac{3}{2}}(2p_x)_N - \tfrac{1}{2}(2p_y)_N$$

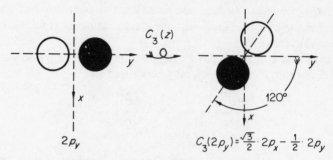

$$2p_y \qquad C_3(2p_y) = \frac{\sqrt{3}}{2} \cdot 2p_x - \frac{1}{2} \cdot 2p_y$$

Figure 1.14   Effect of performing the operation $C_3(z)$ on the $2p_y$ wave function.

i.e. new functions which are linear combinations of the pair of old functions (see Figure 1.14). This also happens with all the other operations of the group, and the *set* of functions $\{(2p_x)_N, (2p_y)_N\}$, is said to form a basis for a two-dimensional representation of the group, usually denoted by the letter E. In this situation, the effect of the operation $C_3(z)$ on the degenerate set of functions must be expressed by the two dimensional matrix

$$C_3 \equiv \begin{vmatrix} -\frac{1}{2} & \sqrt{\frac{3}{2}} \\ -\sqrt{\frac{3}{2}} & -\frac{1}{2} \end{vmatrix}$$

rather than by the numbers (one dimensional matrices) ± 1 used with non-degenerate functions. Each operation of the group must convert any member of a degenerate set of wave functions, into another member, or a linear combination of members of the set. The *character* χ, of the matrix is the sum of its diagonal elements, and $\chi(C_3) = -1$; that of the three reflexions $\chi(\sigma_v) = 0$, and $\chi(E) = 2$. It is these numbers which are listed in collections of group character tables; that for the point group C_{3v} is given below.

C_{3v}	E	$2C_3$	$3\sigma_v$
A_1	1	1	1
A_2	1	1	-1
E	2	-1	0

Group m.o.'s of NH_3 can be constructed in the same way as for H_2O. The nitrogen a.o.'s belong to the symmetry species

$$(2s)_N - a_1, \quad (2p_z)_N - a_1, \quad \{(2p_x)_N, (2p_y)_N\} - e$$

and the hydrogen a.o.'s generate group orbitals of the form*

$$\phi_{H_3} \equiv (1s + 1s' + 1s'') - a_1, \qquad \phi_{H_3} \equiv \{(1s' - 1s''),(1s - 1s' - 1s'')\} - e$$

These can be combined to give three totally symmetric m.o.'s for which combinations of the form

$$2a_1 = (2s)_N - (2p_z)_N + \phi_{H_3} \quad \text{bonding}$$

$$3a_1 = (2s)_N + (2p_z)_N \qquad\qquad \text{non-bonding}$$

$$4a_1 = (2s)_N - (2p_z)_N - \phi_{H_3} \quad \text{antibonding}$$

are plausible, and two doubly degenerate m.o.'s of the form

$$1e = \{(2p_x)_N,(2p_y)_N\} + \phi'_{H_3} \quad \text{bonding}$$

$$2e = \{(2p_x)_N,(2p_y)_N\} - \phi'_{H_3} \quad \text{antibonding}$$

(As with H_2O, we have not discussed the orbital coefficients.) The group m.o.'s are represented in Figure 1.15.

The ground electronic configuration of NH_3 will be

$$\underbrace{(1a_1)^2}_{\substack{\text{inner}\\\text{shell}}} \underbrace{(2a_1)^2(1e)^4}_{\substack{\text{N-H}\\\text{bonding}}} \underbrace{(3a_1)^2}_{\substack{\text{non-}\\\text{bonding}}} , \tilde{X}^1A_1$$

To conclude this section, we consider the direct product of two degenerate representations. Suppose, for example, a molecule with C_{3v} symmetry had its two outermost electrons half filling an m.o. of symmetry species e, (for example, the probable ground state of the nitrene $:NCH_3$). Possible associated electronic states will be those generated by the direct product $e \times e$. The characters of its matrix representatives are $(2)^2 (-1)^2 (0)^2 \equiv 4\ 1\ 0$. The same set of numbers is obtained if the characters of A_1, A_2 and E are added together. By simply adding the characters of the appropriate irreducible representations listed in the group character table, the direct product representation can be reduced into the sum†

$$E \times E = A_1 + A_2 + E$$

This simple rule is sufficient to determine the representations generated by any direct product; the reduction of the product

$$\Pi \times \Pi = \Sigma^+ + \Sigma^- + \Delta$$

---

* Note that the orbitals have not been normalized—we are concerned only with their symmetry properties in the present discussion.

† The ground electronic state will be a triplet; since the triplet spin wave functions are symmetric (see p. 27), and the nett wave function must be antisymmetric (Pauli) the ground state must be $\tilde{X}^3A_2$. $^1A_2$, $^3A_1$ and ^3E are excluded by the Pauli Principle.

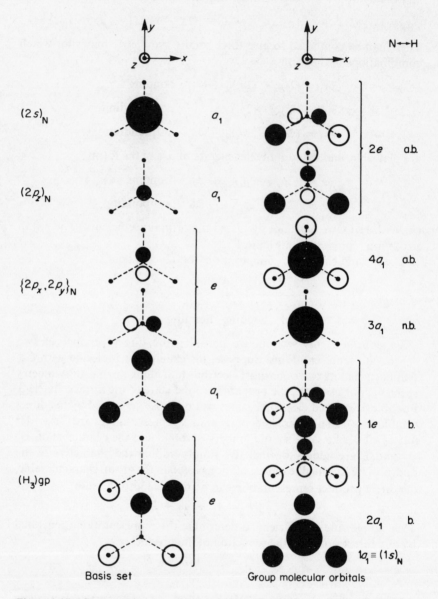

Figure 1.15 Group molecular orbitals of NH_3 (or non-planar AH_3). Viewed from above with H_3-plane below N-atom.

in linear molecules which belong to the point group $C_{\infty v}$, was effected in just this way. (Note that the product of a pair of two-dimensional representations gives a sum of one two-dimensional and two one-dimensional representations. The nett dimension of four is preserved.) The rule is invaluable in the discussion of selection rules.

1.3c Polyatomic molecules

Construction of orbitals in polyatomic molecules can be effected by an extension of the methods employed for diatomic molecules. M.o.'s are formed by linear combination of a basis set of a.o.'s, using only those a.o.'s which are occupied in the ground states of the free atoms. Where the molecule has some symmetry elements (or perhaps local symmetry in a large molecule), the possible combinations must conform to the molecular point group. These 'rules' have already been illustrated in the previous section. In an alternative description, orbitals of a given atom may be mixed by linear combination, to produce hybrid orbitals of the required spatial orientation.

Orbitals which are formed by overlap in the molecular plane, or along internuclear axes and which are symmetric with respect to the plane are termed σ-orbitals; those formed by sideways overlap and which are antisymmetric to reflexion in the molecular plane are termed π-orbitals. It is commonly assumed that the π-electrons can be treated quite separately from the remainder, i.e. the σ and π-wave functions are separable (Hückel approximation). Correlation diagrams can be set up, showing how the energies of the m.o.'s vary with the geometry of the molecule, as was done for diatomic molecules; the bond angle now provides an additional parameter. It is found that the shapes, spectra and chemistry of iso-electronic molecules are often comparable; (iso-electronic includes only the electrons in the outermost shells of the combining atoms). Differences are due principally to the presence of nuclei of differing electronegativity in the molecular skeleton which weight the orbital coefficients in the LCAO–MO, and to interelectronic repulsion which the simple LCAO–MO treatment ignores. There will be no discussion of the methods which have been developed for the calculation of the eigenstates of molecules; the reader must find them elsewhere (see refs. 16, 21, 26, 29). The following discussion presents only a qualitative description of the orbitals, in molecules which are representative of a particular class. The aim is to provide a conceptual model of the electronic configurations of molecules, through which their photochemical behaviour may be interpreted.

A rough classification distinguishes four groups of molecules:

(i) Saturated
- σ-m.o.'s only (e.g. paraffin hydrocarbons)
- σ- and non-bonding m.o.'s (e.g. H_2O, NH_3, CH_3I)

(ii) Unsaturated
- σ- and π-m.o.'s (e.g. CO_2, C_2H_4, aromatic hydrocarbons)
- σ-, π- and non-bonding m.o.'s (e.g. aldehydes and ketones, pyridine)

CH_4

The methane molecule has tetrahedral symmetry, and belongs to the point group T_d. Under the operations of the group, the $(2s)_C$ a.o. is totally symmetric (a_1), while the three $(2p)_C$ orbitals form a triply degenerate set (t_2). The four hydrogen atoms lie at alternate corners of a cube with a carbon atom at its centre (see Figure 1.16). Their $1s$ orbitals can be combined into four group orbitals. Combination in phase produces a totally symmetric orbital, which has the correct symmetry to overlap the $(2s)_C$ a.o. (see Figure 1.16)

$$\phi_1(a_1) = (2s)_C + \lambda[(1s)_1 + (1s)_2 + (1s)_3 + (1s)_4]_{H_4}$$

The other H_4 group orbitals are triply degenerate; the three combinations which have the correct symmetry to overlap the $2p_x$, $2p_y$ and $2p_z$ a.o.'s of the carbon atom are represented in Figure 1.16. They give the m.o.'s

$$\phi_2(t_2) = (2p_x)_C + \lambda'[(1s)_1 + (1s)_2 - (1s)_3 - (1s)_4]_{H_4}$$

$$\phi_3(t_2) = (2p_y)_C + \lambda'[(1s)_1 - (1s)_2 + (1s)_3 - (1s)_4]_{H_4}$$

$$\phi_4(t_2) = (2p_z)_C + \lambda'[(1s)_1 - (1s)_2 - (1s)_3 + (1s)_4]_{H_4}$$

Each of these is σ_{C-H} bonding and doubly occupied in the ground electronic state.

$$\underbrace{(1a_1)^2}_{(1s)_C} \quad \underbrace{(2a_1)^2(1t_2)^6}_{\text{C–H bonding}} \quad \tilde{X}^1A_1$$

Out of phase overlap generates four σ^*_{C-H} antibonding orbitals, into which electrons will be promoted by ultraviolet excitation.†

* Antibonding orbitals are conventionally denoted by an asterisk.

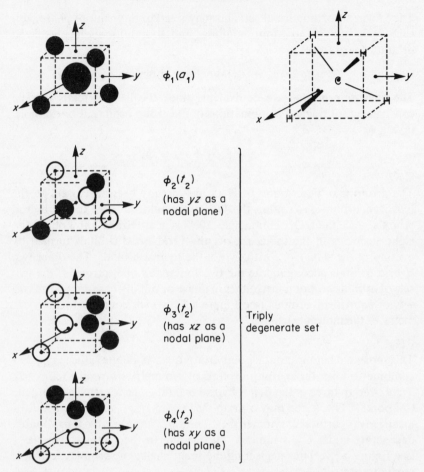

$\phi_1(a_1)$

$\phi_2(t_2)$
(has yz as a nodal plane)

$\phi_3(t_2)$
(has xz as a nodal plane)

Triply degenerate set

$\phi_4(t_2)$
(has xy as a nodal plane)

Figure 1.16 Representation of group orbitals in CH_4 (only bonding combinations are shown).

As an alternative to the group orbital description, we may mix the $2s$ and $2p$ a.o.'s of the carbon atom to give four equivalent tetrahedrally directed hybrid orbitals, sp^3 hybrids,

$$t_1 = \tfrac{1}{2}(2s + 2p_x + 2p_y + 2p_z) \quad \text{directed towards } H_1$$

$$t_2 = \tfrac{1}{2}(2s + 2p_x - 2p_y - 2p_z) \quad \text{directed towards } H_2$$

$$t_3 = \tfrac{1}{2}(2s - 2p_x + 2p_y - 2p_z) \quad \text{directed towards } H_3$$

$$t_4 = \tfrac{1}{2}(2s - 2p_x - 2p_y + 2p_z) \quad \text{directed towards } H_4$$

They form an orthonormal set, and may overlap in or out of phase with each $(1s)_H$ a.o. to give four bonding, and four antibonding localized orbitals, for example

$$\sigma'_1 = t_1 + \lambda''(1s)_H - \sigma_{C-H} \text{ bonding}$$

The two descriptions are equivalent, since localized orbitals can be constructed from linear combinations of the group orbitals. For example, taking $4\lambda' = \lambda''$

$$\phi_1 + \phi_2 + \phi_3 + \phi_4 \equiv \phi'_1$$

H_2O

The group m.o. description of H_2O has already been discussed. In the localized orbital description the O a.o.'s would be close to sp^3 hybrid orbitals, since the $H\hat{O}H$ bond angle (105°) is near tetrahedral. Four of the eight electrons in the molecule, occupy O–H bond orbitals formed by overlap of the $(1s)_H$ a.o.'s with two of the hybrid orbitals. The other two hybrid orbitals accommodate the two lone pairs of electrons. If the two sets of orbitals are combined either in phase or out of phase, they form two sets of equivalent orbitals (see Figure 1.17) which conform to the symmetry of the molecule (cf. p. 37).

CH_3I

This molecule belongs to the point group C_{3v}. As in NH_3, the $(1s)_H$ a.o.'s combine to form three group orbitals of symmetries a_1 and e (cf. p. 38). They overlap three of the $C(sp^3)$ hybrid orbitals. The fourth lies along the C–I bond ($\parallel {}^cC_3(z)$), and may overlap the $(5s)_I$ and $(5p_z)_I$ a.o.'s, to produce a localized σ_{C-I} orbital of symmetry a_1. The $(5p_x)_I$ and $(5p_y)_I$ a.o.'s are doubly degenerate under C_{3v} symmetry, and belong to the symmetry species e (see Figure 1.18). Thus neglecting the inner shells, the ground electronic configuration is

$$\underbrace{(1a_1)^2(1e)^4}_{\substack{\text{C–H} \\ \text{bonding}}} \underbrace{(2a_1)^2}_{\substack{\text{C–I} \\ \text{bonding}}} \underbrace{(3a_1)^2(2e)^4}_{\substack{\text{I non-} \\ \text{bonding}}}, \quad \tilde{X}^1A_1$$

The first electronically excited state has an electron in the $4a_1(\sigma^*_{C-I})$ orbital

$$\ldots (3a_1)^2(2e)^3(4a_1)^1, \quad \tilde{A}^1E \text{ (or } \tilde{a}^3E)$$

and should lead to dissociation at the C–I bond, as indeed it does.[30]

Walsh has shown that this description can be improved by mixing some $(6s)_I$ a.o. character into the $4a_1$ orbital, since it has the same symmetry and its energy lies relatively close to that of the $(5p)_I$ orbitals (the levels close up as the principal quantum number increases). Thus the $4a_1$ orbital acquires

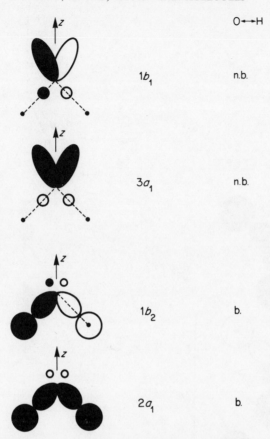

$O \leftrightarrow H$

$1b_1$ n.b.

$3\sigma_1$ n.b.

$1b_2$ b.

$2\sigma_1$ b.

Figure 1.17   Representation of equivalent localized orbitals in H_2O, occupied in ground electronic state.

some $(6s)_I$ character, and vice versa.[31] There is said to be configuration interaction between electronic configurations of the same symmetry.

In polyhalomethanes such as CH_2I_2, interaction between the non-bonding electrons on the I atoms, leads to a slight splitting of their energies.

Molecules of the type AB_2

CO_2 is a linear molecule, and under the molecular symmetry (point group $D_{\infty h}$), the $2s$ and $2p_z$ a.o.'s of the central carbon atom belong to the symmetry species σ_g and σ_u, respectively (see Figure 1.19). The $(2p_x)_C$ and $(2p_y)_C$ a.o.'s form a degenerate set of π_u symmetry.

In constructing the O_2 group orbitals, the $(2s)_O$ a.o.'s can safely be left out of the basis set, since the higher nuclear charge of the oxygen atom

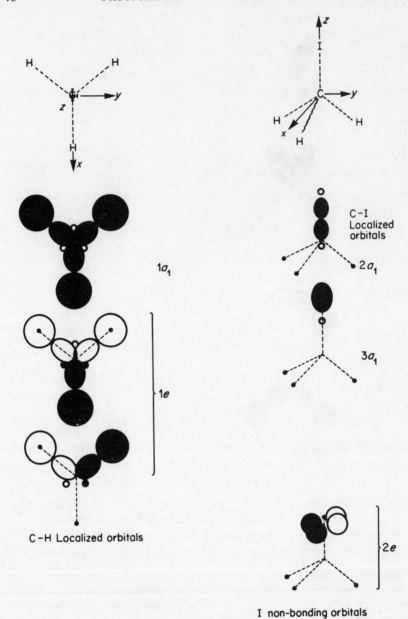

Figure 1.18 Representation of localized orbitals of CH_3I (showing those occupied in the ground electronic state).

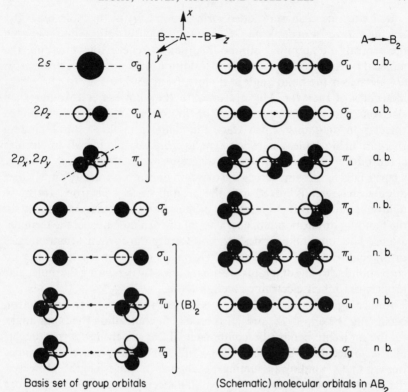

Figure 1.19 Representation of molecular orbitals in a linear AB_2 molecule. (In an unsymmetrical molecule, BAC, the 'u' and 'g' subscripts disappear, and the mixing of the basis orbital functions is less restricted.)

binds them more tightly than in the carbon atom. Combination of the $(2p)_O$ a.o.'s gives four sets of group orbitals with symmetries σ_g, σ_u, π_u and π_g, as shown in Figure 1.19.

The nett molecular orbitals can now be constructed by combination either in or out of phase, to produce two pairs of σ- and four pairs of π-orbitals which are either bonding or antibonding, and one pair of π-orbitals which is C–O non-bonding. The 22 electrons (6 inner + 16 valence shell) fill the lowest of these in the ground electronic state, to give the configuration

$$\underbrace{(1\sigma_g)^2(1\sigma_u)^2(2\sigma_g)^2}_{1s \text{ inner shell}}\ \underbrace{(3\sigma_g)^2(2\sigma_u)^2}_{(2s)_O}\ \underbrace{(4\sigma_g)^2(3\sigma_u)^2}_{\substack{\text{C–O} \\ \text{bonding}}}\ \underbrace{(1\pi_u)^4}_{\substack{\text{C–O} \\ \text{bonding}}}\ \underbrace{(1\pi_g)^4}_{\substack{\text{C–O non-} \\ \text{bonding}}},\ \tilde{X}^1\Sigma_g^+$$

If an extra electron were added, as in the ion CO_2^- or the molecule NO_2, it would have to occupy the $2\pi_u$ orbital, which is antibonding between the central and terminal atoms. In order to relieve this situation, the molecule 'responds' by bending.[32] Addition of a further electron as in O_3, decreases the bond angle still more, from 180° to 134° to 117°.[26] All molecules of the type AB_2 are linear if they possess $\leqslant 16$ outer shell electrons, and bent if they possess 17 or 18 electrons. Walsh has constructed correlation diagrams which trace the changes in the 'orbital binding energies' in molecules as the bond angles decrease; that for AB_2 molecules is reproduced in Figure 1.20. The diagrams are semi-empirical: whether a curve falls in moving across the diagram depends primarily on whether more s character is mixed into the orbital, since s electrons are more tightly bound than the p electrons of the same shell, and secondarily on the bonding or antibonding character of the orbital between the terminal atoms. The 'binding energies' do not correspond to ionization energies or one electron orbital energies.[33] However, whatever it is that they do correspond with, the diagrams are most successful in relating the molecular properties to their electronic configurations.

Returning to the case of NO_2, it is evident that the outermost electron occupying the $6a_1 - 2\pi_u$ orbital is crucial in controlling the bond angle. As the molecule bends, the component of $2\pi_u$ which lies perpendicular to the molecular plane $2b_1 - 2\pi_u$, will fall slightly since it becomes weakly O–O bonding (assuming no change in bond length). However, the other component $6a_1 - 2\pi_u$, will fall steeply, since it acquires more and more $(2s)_N$ character, and in consequence becomes far more tightly bound. If the bond angle were reduced to 90°, there would be no hybridization of the $2s$ and $2p$ orbitals at the central atom, and the orbital would lose all $(2p)_N$ character. The $1\pi_g$ orbitals become less tightly bound as the angle decreases, since the terminal atoms are brought closer together, and their $2p$ orbitals lying perpendicular to the direction of the bonds overlap out of phase.

The first electronic transition in CO_2 should promote an electron from the highest occupied orbital $1\pi_g$ into the lowest vacant one $2\pi_u$. The Walsh diagram predicts that the molecule should now bend to decrease the outermost orbital energy, and indeed it does. The first excited state has the configuration

$$\ldots (1a_2)^2(4b_2)^1(6a_1)^1, \ \tilde{A}^1B_2$$

correlating with the $^1\Delta_u$ component of the linear configuration

$$\ldots (1\pi_g)^3(2\pi_u)^1,$$

and an equilibrium bond angle of 122°.[26] In general, the geometry of the

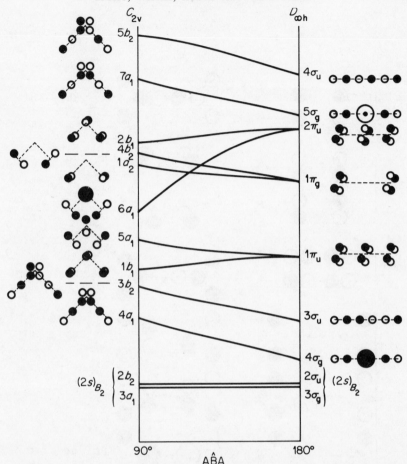

Figure 1.20 Walsh orbital correlation diagram for molecules of the type AB_2. (The orbital diagrams are given only to show the form of the m.o's and are not meant to represent the actual contribution of each a.o. to the basis set. Any orbital of a given symmetry will have contributions from other orbitals of the same symmetry, mixed into it.)

first excited states of many molecules, approximates to that of their anion (for example, CO_2^-) or the molecule that is 'one electron up' in the series.[32]

C_2H_4

The molecule is planar with an HĈH bond angle of $118°$;[26] it belongs to the point group D_{2h}. The group orbitals are shown schematically in Figure 1.21. With the exception of the $1b_{3u}$ and $1b_{2g}$ m.o.'s, which are respectively C–C π-bonding and antibonding, they all lie in the molecular plane and form the σ-framework. The molecule's twelve outer shell

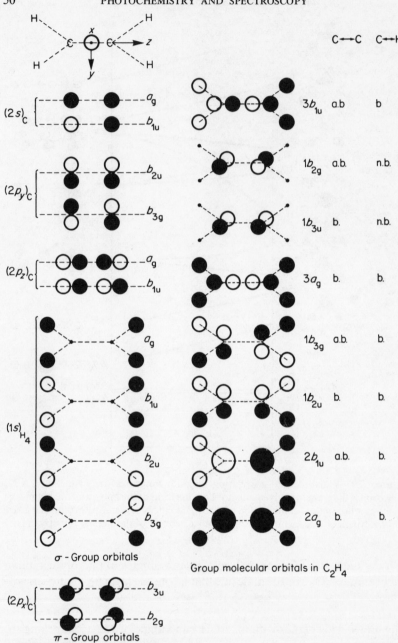

σ - Group orbitals

Group molecular orbitals in C_2H_4

π - Group orbitals

Basis set of group orbitals

Figure 1.21 Representation of low-lying group molecular orbitals in C_2H_4.

electrons fill the orbitals in ascending order to give the ground electronic configuration

$$\ldots (2a_g)^2(2b_{1u})^2(1b_{2u})^2(1b_{3g})^2(3a_g)^2(1b_{3u})^2, \ \tilde{X}^1A_g$$

in which the outermost pair of electrons occupy the bonding π-m.o.

Since the HĈH bond angle is close to 120°, the C atoms are near to sp^2 hybridization, in which the $(2s, 2p_y$ and $2p_z)_C$ a.o.'s mix to produce three equivalent, coplanar trigonal hybrid orbitals. In a localized m.o. description, one pair overlap to give a C–C σ-bond while the remainder overlap the $(1s)_H$ a.o.'s. The unhybridized $(2p_x)_C$ orbitals overlap in a plane perpendicular to that of the molecule, to give a localized π-orbital. This imposes a planar configuration in the ground electronic state, since torsion about the C–C axis reduces the π-orbital overlap and hence the molecular binding energy.

The first electronic transition transfers an electron from the bonding, to the antibonding π-orbital to give the configuration

$$\ldots (3a_g)^2(1b_{3u})^1(1b_{2g})^1, \ \tilde{A}^1B_{1u} \text{ or } \tilde{a}^3B_{1u}$$

The C–C π-bonding is now destroyed and there is no restraint against rotation about the remaining σ-bond. In fact, rotation into a 90° twisted conformation is favoured, since it eases the mutual repulsion of the electrons in the singly occupied orbitals. The further away from each other they can get, the better. This has important photochemical consequences, since $\pi \rightarrow \pi^*$ excitation in substituted olefins can lead to *cis–trans* isomerization following rotation in an excited state. Indeed, the very mechanism by which you see this page is believed to involve *cis–trans* isomerization of a polyolefin in the retina of your eye![34]

Benzene

The planar hexagonal symmetry of benzene permits separation of the $2p\pi$ a.o.'s centred on the carbon atoms, from their trigonal hybrid orbitals whose overlap forms the σ-framework of the molecule. The six equivalent $2p\pi$ orbitals overlap to form six group molecular orbitals; the possible combinations are readily found using group theoretical methods.[26]

Under D_{6h} symmetry, the six equivalent orbitals split to form two non-degenerate group orbitals with symmetries a_{1u} and b_{2g} and two doubly degenerate orbitals with symmetries e_{1g} and e_{2u} (note that the symmetry species are those for the z-axis lying along the six-fold symmetry axis). Their construction can be visualized, by considering the overlap of the π-orbitals of a pair of C_3 fragments (see Figure 1.22). Corresponding orbitals may combine in or out of phase to give the six π-orbitals of the

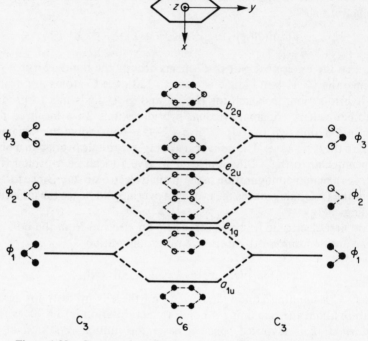

Figure 1.22   Construction of π-group molecular orbitals in benzene.

benzene ring. These are the only possible combinations; any others, say $(\phi_1 + \phi_2)$, do not conform to the molecular symmetry. The orbital energy increases with the number of nodal planes, parallel with their increasing antibonding character. The ground electronic state of benzene has the configuration

$$\ldots (a_{1u})^2 (e_{1g})^4, \; \tilde{X}^1 A_{1g}$$

Promotion of an electron from $e_{1g} \rightarrow e_{2u}$ gives the configuration

$$\ldots (a_{1u})^2 (e_{1g})^3 (e_{2u})^1$$

associated with three states resulting from the direct product

$$E_{1g} \times E_{2u} = B_{1u} + B_{2u} + E_{1u}$$

Each of these may have singlet or triplet multiplicity and none are forbidden by the Pauli Principle, since the two odd electrons are not equivalent. The lowest of the singlet states is $\tilde{A}^1 B_{2u}$; in it the hexagonal equilibrium geometry is preserved but the C–C bond length is slightly increased because of the decreased π-bonding.[26] The relative energies of the states of B_{1u} and B_{2u} symmetry are reversed in the triplet manifold, and the lowest triplet state is $\tilde{a}^3 B_{1u}$. Until recently it was thought that the molecular symmetry of the triplet was altered from hexagonal (D_{6h}) to quinoidal (D_{2h}),[35b] but recent experiments indicate that hexagonal symmetry is retained in an isolated molecule, free of environmental perturbations.[35b]

CH_2O

Formaldehyde is planar in its ground electronic state with an $H\hat{C}H$ bond angle of $121°$,[26] and it belongs to the point group C_{2v}. Its low-lying orbitals are depicted in Figure 1.23. A total of 16 electrons (4 inner and 12 outer shell), occupy the lowest of these in the ground electronic state, to give the configuration

$$\ldots \underbrace{(3a_1)^2 (1b_2)^2}_{\text{C–H}} \quad \underbrace{(4a_1)^2}_{\text{C–O}} \quad \underbrace{(5a_1)^2}_{n_O} \quad \underbrace{(1b_1)^2}_{\text{C–O}} \quad \underbrace{(2b_2)^2}_{n_O}, \quad \tilde{X}^1 A_1$$

$$\sigma\text{-bonding } \sigma\text{-bonding} \qquad \pi\text{-bonding}$$

The notation n_O indicates a non-bonding orbital centred on the oxygen atom.

In a localized orbital description, the σ-bonds are formed from overlapping sp^2 trigonal hybrid orbitals centred on the carbon atom, with the two $(1s)_H$ a.o.'s, and with a partially hybridized sp orbital centred on the oxygen atom (hybridization is more difficult here since the $2s - 2p$ energy difference is increased by the greater nuclear charge). The π-bond is formed by overlapping the $2p_x$ a.o.'s, while the remaining $(2p_y)_O$ and $(sp$ hybrid$)_O$ orbitals accommodate the two lone pairs of electrons.

The lowest unoccupied orbital $2b_1$, is C–O π-antibonding; if an electron is promoted into it, to give the configuration

$$\ldots (1b_1)^2 (2b_2)^1 (2b_1)^1, \quad \tilde{A}^1 A_2 \text{ or } \tilde{a}^3 A_2$$

the molecule loses its planar conformation. Walsh has emphasized that this has the effect of relieving the antibonding character of the newly occupied orbital, and increasing the binding energy of the molecule.[36]

As the $H_2\hat{C}O$ angle falls (to $149°$ in the singlet state and $145°$ in the triplet[26]), the $2b_1$ orbital acquires some $(2s)_C$ character, and the promoted electron becomes more tightly bound.[36] There is a close parallel with the bending of CO_2 in its first excited state, discussed earlier.

Pyridine

Pyridine is iso-electronic with benzene, but the substitution of a nitrogen atom for one of the CH groups requires two of the electrons which used to

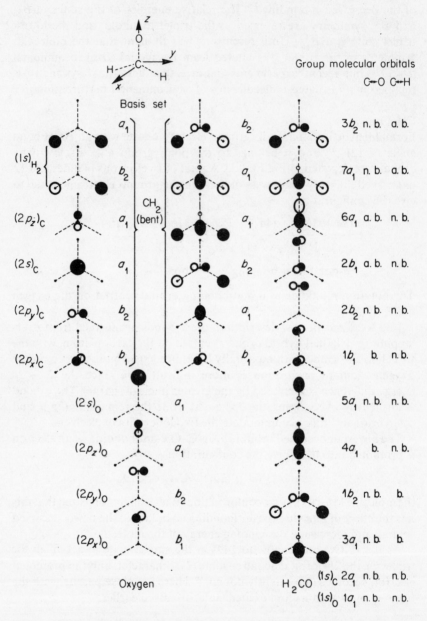

Figure 1.23 Representation of low-lying group molecular orbitals in CH_2 and $H_2C{=}O$

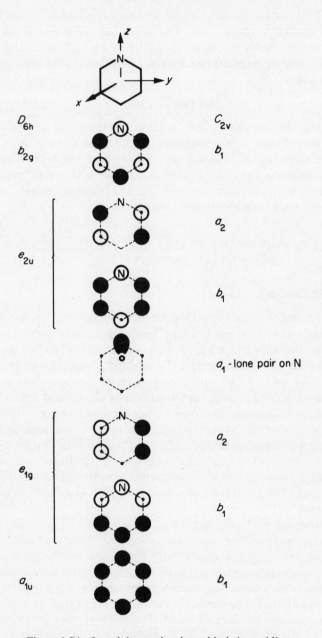

Figure 1.24   Low-lying molecular orbitals in pyridine.

be C–H σ-bonding, to be localized in a non-bonding sp^2 hybrid orbital centred on the nitrogen atom. The π-orbitals remain qualitatively the same, though the lowered symmetry $(D_{6h} \to C_{2v})$ splits the degeneracies of the e_{1g} and e_{2u} orbitals (see Figure 1.24). The ground state electronic configuration is now

$$\ldots (b_1)^2(b_1)^2(a_1)^2(a_2)^2, \tilde{X}^1A_1$$

Those orbitals which have electron density on the nitrogen atom are more tightly bound than in benzene, since the nitrogen atom is more electronegative than carbon. The non-bonding orbital lies a little below the highest occupied π-orbital. Electronic excitation may now promote an electron from the π-system or from the non-bonding orbitals, and both transitions have been identified experimentally.[29]

1.4   VIBRATIONAL AND ROTATIONAL STATES: POTENTIAL ENERGY FUNCTIONS

1.4a Diatomic molecules

Within the terms of the Born–Oppenheimer approximation, we were able to separate the electronic motion from the nuclear motion within a molecule; the complete wave function was separable into the product of an electronic, and a nuclear part. The electronic eigenfunctions could be obtained by solving the Schrödinger equation for any given set of instantaneously fixed nuclear coordinates. The choice of a different set of coordinates will change the energies associated with each electronic state; in a bound electronic state, the energy will vary about a minimum corresponding to the equilibrium nuclear configuration in that state. At that point, the attraction between the nuclei and the electrons is just counterbalanced by the internuclear coulombic repulsion. The variation gives the potential energy $V(q)$ for the nuclear motion as a function of the coordinates q.

Of course, the nuclei will not remain fixed in space but will vibrate about the equilibrium position in a bound electronic state, and also rotate in space about the centre of mass. In a diatomic molecule the positions of the nuclei can be expressed in polar coordinates. r will represent the length of the radius vector connecting the two atoms, and θ and φ will determine its spatial orientation. The nuclear wave function can then be separated in its turn, into the product of a radial part $R(r)/r$, and an angular part $Y(\theta, \varphi)$, in just the same way as the electronic wave function was in the spherically symmetric hydrogen atom. The radial function depends only on the internuclear separation, and is the vibrational wave function ψ_v, and the angular part is the rotational wave function, ψ_r. The complete wave function is

$$\psi = \underbrace{\psi_e \psi_{\text{spin}} \psi_v \psi_r}$$
$$\text{electronic} \quad \text{nuclear}$$

and the total internal energy is

$$E = E_e + E_v + E_r$$

assuming the motions to be completely independent of each other.* (The assumption can break down when two potential energy functions corresponding to different electronic states cross each other. In this situation it is possible for the vibrational motion to induce a transition from one electronic state to the other; in later discussion we shall see that this phenomenon is of major importance in photochemistry.)

With the vibrational and rotational parts of the nuclear wave function separated, the vibrational eigenfunctions and energy levels can be found from the solutions of the Schrödinger equation

$$\frac{\partial^2 \psi_v}{\partial r^2} + \frac{2\mu}{\hbar^2}[E - V(r)]\psi_v = 0$$

for a molecule of reduced mass μ, oscillating in the field of potential energy $V(r)$. For a molecule in a stable electronic state, $V(r)$ decreases with $r > r_e$, the equilibrium distance, but when $r < r_e$ the electrostatic repulsion between the nuclei gradually overcomes the attraction, and as $r \rightarrow 0$, $V(r) \rightarrow \infty$. In an unstable state, the electrostatic attraction of the nuclei to the electronic charge between them is insufficient to overcome the internuclear repulsion, and $V(r)$ is a continuously increasing function of r; the potential energy function is said to be repulsive.

The simplest approximation to $V(r)$ in a bound electronic state is to assume the potential energy is that of a simple harmonic oscillator

$$V(r) = -\tfrac{1}{2}k(r - r_e)^2$$

where k is the force constant. When $r \gg r_e$ this is entirely unrealistic of course, since with $V(r) \rightarrow \infty$ the molecule could never dissociate. It is equally unrealistic when $r \ll r_e$, since under these conditions the internuclear repulsion is dominant and $V(r) \propto 1/r$. However, it represents a tolerable first approximation when $r \simeq r_e$. Possible solutions of the Schrödinger equation are restricted to those which are single-valued, finite, continuous functions of r, and which fall to zero as $r \rightarrow \infty$; under

* Their relative magnitudes are in the order $E_e(\sim 10^4 - 10^5 \text{ cm}^{-1}) \simeq 10^2 E_v \simeq 10^4 E_r$.

the harmonic oscillator approximation they turn out to be the set of oscillating orthogonal functions known as Hermite polynomials. They are symmetrical about the equilibrium position and are represented schematically in Figure 1.25. In the ground level, the function is Gaussian. The

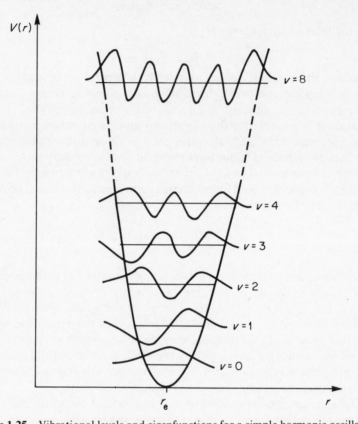

Figure 1.25   Vibrational levels and eigenfunctions for a simple harmonic oscillator.

associated vibrational energy levels are equally spaced, with energies

$$E_v = (v + \tfrac{1}{2})h\omega$$

where $v = 0, 1, 2, \ldots$ is the vibrational quantum number, and $\omega = \sqrt{k/\mu}/2\pi$, is the oscillation frequency. If the molecule occupies a given vibrational level v, the value of $|\psi_v(r)|^2 dr$ gives the probability of its having an internuclear distance lying between r and $r + dr$. The probability of its occupying the given level depends on the distribution of molecules among the available vibrational states; if the system is in thermal equilibrium, this will be a Boltzmann distribution, $n_{v=v} = n_{v=0} \cdot \exp[-(E_v - E_0)/RT]$.

At high vibrational levels, most molecules will be fully extended or compressed. This approaches the classical description which requires that the molecule spend most of its time at the turning points of the vibration, and very little time at the equilibrium position where kinetic energy is a maximum and the potential energy is zero. At the ground level however, the situation is reversed and the most probable separation is the equilibrium one. In addition, the molecule can never rest at the equilibrium position where its position and momentum would be known simultaneously; it *must* have zero point energy.*

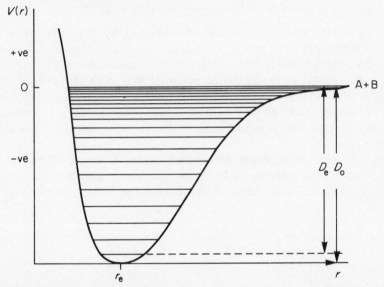

Figure 1.26 Schematic representation of morse potential energy function, and its associated vibrational energy levels.

A more realistic potential energy function will be anharmonic, with the restoring force falling to zero as $r \rightarrow \infty$. The simplest of these is due to Morse, and is represented in Figure 1.26. In the Morse function

$$V(r) = D_e[1 - e^{-a(r-r_e)}]^2$$

where D_e is the depth of the potential well measured from its minimum, and the constant $a = (2\pi^2\mu c/D_e h)\omega_e$.[37] ω_e is the vibration frequency the molecule would have, if it could oscillate with vanishingly small amplitude at the minimum of the curve. Because of the anharmonicity, the frequency varies with the vibrational energy.

* Note also, that all the wave functions extend beyond the classical turning points; this has important consequences when potential energy functions cross.

Apart from the fact that the Morse function contains only three adjustable parameters, its great virtue is that when inserted into the Schrödinger equation, exact solutions can still be found, associated with the energy levels

$$E_v = h\omega_e[(v + \tfrac{1}{2}) - (v + \tfrac{1}{2})^2 x_e]$$

The factor $x_e = \omega_e/4D_e$, is an anharmonicity constant, and the product $x_e\omega_e$ is termed the anharmonicity. The equation represents the first approximation to the general equation

$$E_v = h\omega_e[(v + \tfrac{1}{2}) - (v + \tfrac{1}{2})^2 x_e + (v + \tfrac{1}{2})^3 y_e - \ldots]$$

for the vibrational energy levels, which is obtained by expressing $V(r)$ as an empirical power series in $(r - r_e)$, in order to 'nibble' closer and closer to observation. The levels are no longer equally spaced, but slowly converge to a series limit at the lip of the potential well. At higher energies the molecule dissociates, and the levels merge into a continuum appropriate to an unbound state. In the bound state, the potential energy is quantized, while the continuum is associated with the kinetic energy of the dissociating atoms.

As the vibrational quantum number increases, the vibrational wave functions for the anharmonic oscillator are increasingly weighted toward the extended molecule, reflecting the approach to dissociation (see Figure 1.27).

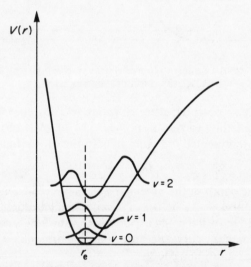

Figure 1.27   Schematic representation of vibrational eigenfunctions of an anharmonic oscillator.

The disadvantages of the Morse function include its approximate character, and its rate of convergence which is too slow for covalently bonded molecules (for example, O_2) and too fast for molecules which have considerable ionic character. Also $V(r) \rightarrow \infty$ as $r \rightarrow 0$. Alternative potential energy functions are discussed critically by Gaydon.[38]

If the potential energy function of the molecule is repulsive, or if its potential energy lies in the continuous part of an attractive state, the 'vibrational' eigenfunctions will have a large amplitude at only one turning point, where the approaching atoms are reflected back by the potential barrier. As an approximation (a remarkably successful one) the eigenfunctions have been represented by δ-functions which have unit amplitudes at points on the curve but zero amplitude at all other points. Ultraviolet absorption continua associated with electronic transitions into repulsive states have been quantitatively interpreted on this basis[39] (see section 2.2b).

Solutions of the Schrödinger equation for the rotational motion of a rigid diatomic molecule (i.e. $r = r_e$), can be found if the factor $2\mu E/\hbar^2$ is set equal to $J(J + 1)/r_e^2$, where $J = 0, 1, 2 \ldots$. Thus the rotational energy takes the values

$$E_r = \frac{\hbar^2}{2\mu r_e^2}J(J + 1) = \frac{\hbar^2}{2I}J(J + 1)$$

and the rotational terms are given by

$$F(J) = 2BJ(J + 1)$$

where B, the rotational constant, is $h/8\pi^2 cI$, I is the moment of inertia of the molecule, and J is the rotational quantum number.* (The rotational angular momentum is $p_r = \hbar\sqrt{J(J + 1)}$, so that $E_r \equiv p_r^2/2I$, the classical result.) The rotational eigenfunctions are the spherical harmonics discussed in section 1.2b.

1.4b Polyatomic molecules

To a good approximation the separation of the vibrational and rotational parts of the nett nuclear wave function can still be made in a polyatomic molecule. However, the vibrational motion is now far more complex, and the rotational motion can be resolved into components about three principle axes. If the vibrations are assumed to be simple harmonic, with the restoring force proportional to the displacement, it can be shown that the nett vibrational motion of an N-atomic molecule can be resolved

---

* If the molecule has electronic (spin or orbital) angular momentum, other contributions must be added; the equation is strictly accurate only for $^1\Sigma$ states.

into $3N - 6$ components termed normal modes.* Each of these is bound by a potential energy of the form $V(Q_i) = 1/2(k_i Q_i^2)$, so that the nett potential energy is

$$V(Q_i) = \tfrac{1}{2} \sum_i^{3N-6} k_i Q_i^2$$

where the normal coordinates Q_i, are linear combinations of the displacements of each nucleus. Each normal mode has a wave function $\psi_i(Q_i)$ which conforms to the symmetry of the molecule, and unless degenerate is either symmetric or antisymmetric to all the operations of the molecular point group. Each one is analogous to that of a diatomic molecule, with the appropriate normal coordinate replacing the single displacement from the equilibrium bond length. For example, NO_2 has $3 \times 3 - 6 = 3$ normal modes; they are represented in Figure 1.28. Two

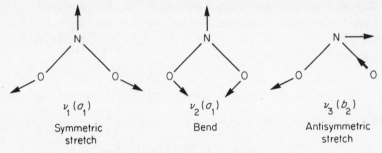

$$\nu_1\,(a_1)$$

Symmetric
stretch

$$\nu_2\,(a_1)$$

Bend

$$\nu_3\,(b_2)$$

Antisymmetric
stretch

Figure 1.28 Normal modes of a bent triatomic molecule (NO_2).

are stretching modes, with one totally symmetric (symmetry species a_1 of point group C_{2v}) and the other antisymmetric (b_2), while the third leaves the bond lengths unchanged and alters the bond angle; the bending mode is also totally symmetric (a_1). The complete wave function is now

$$\prod_i^{3N-6} \psi_i(Q_i)$$

The Schrödinger equation can be separated into a sum of $3N - 6$ independent equations, which can be solved individually to give the nett vibrational energy

$$E_v = \sum_i^{3N-6} E_i = h \sum_i^{3N-6} (v_i + \tfrac{1}{2})\omega_i$$

*Or $3N - 5$ in linear molecules, since its rigid displacements include three translational components, but only two of rotation. In each mode all the atoms move in phase, with the same frequency, about the equilibrium configuration.

In the simple harmonic oscillator approximation, the wave functions form an orthogonal set, and each normal mode is independent of the others; transfer of energy from one excited mode to another is impossible. In reality this is untrue, since the vibrations are anharmonic; they are only weakly so in the ground vibrational level, but as the classical vibrational amplitude increases and higher vibrational levels are populated they become increasingly anharmonic. The separation into normal modes is no longer strictly followed, and the mixing permits redistribution of vibrational energy. The phenomenon of intramolecular vibrational energy transfer, or 'flow', is of major importance in photochemical kinetics (and indeed all kinetics) since the initial act of light absorption frequently produces a molecule endowed with excess vibrational, as well as electronic energy (see section 1.5b).

Since the potential energy is now a function of $3N - 6$ normal coordinates, its variation is described by a $3N - 5$-dimensional potential energy hypersurface (one extra dimension for the potential energy). In any discussion of dissociation or chemical kinetic processes, it is very helpful to have some notion of their form, but unfortunately their representation demands rather more dimensions than we can visualize. Nonetheless, approximate representations in two or three dimensions where the vibrational motion in all but one or two modes is 'frozen' are of great utility. They correspond to sections through the potential energy hypersurface. The approximation is least severe in bent triatomic molecules, where the potential energy surface is plotted as a function of the two bond lengths and the bond angle is held constant. Figure 1.29 shows a contour surface for a molecule of the type AB_2; because of its symmetry the surface is symmetric about the line $r_{A-B} = r'_{A-B}$. The simplest approximation to the potential function is to assume that it is a sum of two diatomic Morse functions[40]

$$V(r, r') = D_e[(1 - e^{-a(r - r_e)})^2 + (1 - e^{-a(r' - r_e)})^2]$$

where D_e is the dissociation energy $D(BA-B)$ measured from the potential minimum, and r_e is the equilibrium A–B bond length. A repulsive potential energy surface is represented in Figure 1.30. Much more can (and will) be said on the subject of potential energy surfaces in subsequent discussion; for the moment, the aim is restricted to introducing basic concepts.

The rotational motion of polyatomic molecules is analysed in terms of the angular momentum about three cartesian principal axes passing through its centre of mass. The moments of inertia about the three axes are $I_A < I_B < I_C$. The methods by which these are located will not be discussed, though we may note in passing that in an axially symmetric molecule one of them will be the symmetry axis; if the molecule possesses

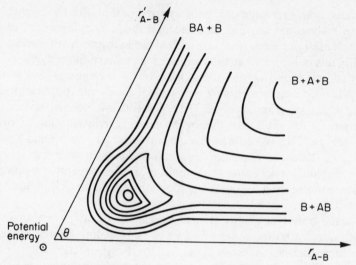

Figure 1.29 Potential energy contours for a triatomic molecule, AB_2. (i) It is assumed that the $B\hat{A}B$ bond angle remains constant; the surface actually represents a section through a 4-dimensional hypersurface; (ii) oblique axes are chosen since if they are set at an angle θ, where

$$\theta = \cos^{-1}\left\{\frac{m_B m_B}{(m_A + m_B)^2}\right\}^{\frac{1}{2}} = \cos^{-1}\left(\frac{m_B}{m_A + m_B}\right)$$

and m_A, m_B are atomic masses, the vibrational motion can be represented by the trajectory of a frictionless mass sliding on the surface.

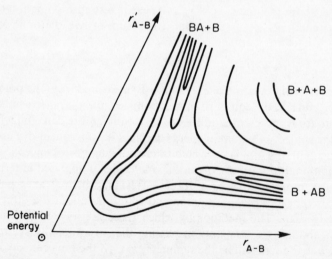

Figure 1.30 Repulsive potential energy surface for a triatomic molecule, AB_2.

a plane of symmetry, two of the axes will lie in the plane. Rigid molecular rotors are classified into four general types,

(i) linear rotors $\quad I_A = 0, I_B = I_C \quad$ e.g. HCl, CO_2

(ii) spherical tops $\quad I_A = I_B = I_C \quad$ e.g. CH_4

(iii) symmetric tops

$\quad$ oblate $\qquad I_A = I_B < I_C \qquad$ e.g. $CHCl_3$

$\quad$ prolate $\qquad I_A < I_B = I_C \qquad$ e.g. CH_3Cl

(iv) asymmetric tops $\;\; I_A < I_B < I_C \qquad$ e.g. H_2O

If the rotational angular momentum about the three axes is p_A, p_B and p_C, the nett rotational energy is

$$E_r = \frac{p_A^2}{2I_A} + \frac{p_B^2}{2I_B} + \frac{p_C^2}{2I_C}$$

The total angular momentum is restricted to $\hbar\sqrt{J(J+1)}$, and with this restriction the rotational energy levels of a prolate symmetric top turn out to be

$$F(J,K) = BJ(J+1) + (A-B)K^2$$

where $A = h/8\pi^2 cI_A$, $B = h/8\pi^2 cI_B$ and $K = 0, \pm1, \pm2, \ldots \pm J$. K relates to the component of the total angular momentum about the unique axis; the positive and negative values reflect clockwise and counter-clockwise rotation. In an oblate symmetric top

$$F(J,K) = BJ(J+1) + (C-B)K^2$$

and in a spherical top the equation reduces to

$$F(J) = BJ(J+1)$$

The equation for the rotational energy of an asymmetric rotor is more complex and is of the form

$$F(J, W_{J_\tau}) = \tfrac{1}{2}(B+C)J(J+1) + \{A - \tfrac{1}{2}(B+C)\}W_{J_\tau}$$

where W_{J_τ} is dimensionless but depends upon the moments of inertia of the molecule.

All the equations represent first approximations to 'real-life' molecules. They have to be modified to take account of centrifugal distortion which increases with the rotational angular momentum; the moments of inertia depend on the vibrational energy since the anharmonicity alters the equilibrium geometry a little; in linear molecules, the bending vibrations are degenerate, and they contribute orbital angular momentum about the internuclear axis (cf. degenerate electronic states which possess electronic orbital angular momentum); electron spin and orbital momentum must also be added to the total. However, these matters are not often of direct importance in photochemistry and will not be pursued here. Herzberg's Volume III is the place to find them.[26]

1.5 ELECTRONIC TRANSITIONS: SELECTION RULES

1.5a Transition moments

The opening section of this chapter touched on the intimate relation between the absorption and emission of radiation, and a changing electric or magnetic dipole in the emitter or receiver. At the molecular level magnetic dipole changes generate electron or nuclear spin magnetic resonances, and need not concern us for the present. Electronic transitions, which promote photochemical reaction, are associated with electric dipole changes.

If an atom or molecule makes a transition from a state where its complete wave function is Ψ_i to another, Ψ_f the transitional dipole moment of the molecule is defined by the equation

$$\mu_{i \to f} = \int_{-\infty}^{\infty} \Psi_f^* \hat{M} \Psi_i \, d\tau$$

The dipole moment operator $\hat{M}$, which has the same form as the classical dipole moment vector, is the resultant of an electronic and a nuclear part,

$$\hat{M} = \hat{M}_e + \hat{M}_n = e \sum_\mu \mathbf{r}_\mu - e \sum_\nu z_\nu \mathbf{r}_\nu$$

$\mathbf{r}_\mu$ and $\mathbf{r}_\nu$ are the position vectors of the μth electron and the νth nucleus, of charges e and $z_\nu e$ respectively. The complete wave function Ψ depends not only on the electronic and nuclear coordinates, but also on time; it is separable into the product[20]

$$\Psi = \psi_{en} \cdot e^{-2\pi i E t/h}$$

Since the dipole moment operator has no time dependence, the transition moment integral can be rewritten

$$\mu_{i \to f} = \int_{-\infty}^{\infty} \Psi_{en,f}^* \hat{M} \psi_{en,i} \, d\tau_{en} \cdot e^{-2\pi i (E_f - E_i) t/h}$$

with the first term giving the amplitude of the dipole change and the second its time dependence. The transition dipole moment oscillates with a frequency,

$$\nu_{if} = (E_f - E_i)/h$$

the Bohr frequency. The probability of the transition depends on the square of the amplitude: if the transition i $\leftrightarrow$ f is allowed, the amplitude must be non-zero. If the initial and final states are one and the same, i = f and the time dependence disappears; there is destructive interference and the amplitude is simply the permanent dipole moment of the molecule in the stationary state. ν_{if} can be regarded as a beat frequency between two different states.

Assuming that the motions of the electrons and nuclei are separable (Born–Oppenheimer approximation), $\psi_{en} = \psi_e \psi_n$, and the amplitude can be written

$$R_{if} = \int_{-\infty}^{\infty} \psi_{e,f}^* \hat{M}_e \psi_{e,i} \, d\tau_e \int_{-\infty}^{\infty} \psi_{n,f}^* \psi_{n,i} \, d\tau_n$$

$$+ \int_{-\infty}^{\infty} \psi_{n,f}^* \hat{M}_n \psi_{n,i} \, d\tau_n \int_{-\infty}^{\infty} \psi_{e,f}^* \psi_{e,i} \, d\tau_e$$

($\hat{M}_e$ operates only on the electronic, and $\hat{M}_n$ only on the nuclear wave functions.) The second term can be neglected, since the set of electronic wave functions are orthogonal and the amplitude of the transition moment reduces to

$$R_{if} = e \int_{-\infty}^{\infty} \psi_{e,f}^* \sum_\mu \mathbf{r}_\mu \psi_{e,i} \, d\tau_e \int_{-\infty}^{\infty} \psi_{n,f}^* \psi_{n,i} \, d\tau_n$$

The electronic transition $i \leftrightarrow f$ *is allowed so long as* $R_{if} \neq 0$; this is the basis of all the selection rules for absorption or emission of electric dipole radiation. All we have to do is work out the integrals in R_{if}! It looks awful, but there is a short cut: so long as we are prepared to ask only, 'Is the transition allowed or forbidden?' and not worry about the quantitative answer, all we need to know about the wave functions is their symmetry. The transition moment is a molecular property, and must be invariant under all the symmetry operations of the molecular point group. For example, rotation of a molecule of water through 180° should have no influence on the likelihood of its absorbing a photon: *the integrand in* R_{if} *must be totally symmetric or* $R_{if} = 0$.

If the electronic wave function is separable into the product of a space and a spin part, (i.e. in the absence of spin–orbit coupling), the electronic part of R_{if} reduces to

$$(R_{if})_{es} = e \int_{-\infty}^{\infty} \psi_{e,f}^* \sum_\mu \mathbf{r}_\mu \psi_{e,i} \, d\tau_e \int_{-\infty}^{\infty} \psi_{s,f}^* \psi_{s,i} \, d\tau_s$$

since $\mathbf{r}_\mu$ cannot operate on the spin coordinates. The second integral is zero if the combining states have different spins, and unity if not, because the set of spin wave functions is orthonormal. Transitions with $\Delta S \neq 0$ are said to be spin-forbidden: this is the spin selection rule. It becomes less and less rigorous as the degree of spin–orbit coupling increases (with increasing atomic number).

If the molecule contains only a single atom, the nuclear overlap integral drops out; the transition probability is determined solely by the square of the integral

$$(R_{if})_e = e \int_{-\infty}^{\infty} \psi_{e,f}^* \sum_\mu \mathbf{r}_\mu \psi_{e,i} \, d\tau_e$$

In general, the transition moment can be resolved into three cartesian components of the type

$$(R_{if})_x = e \int_{-\infty}^{\infty} \psi_{e,f}^* \sum_{\mu} \hat{x}_{\mu} \psi_{e,i} \, dx$$

where

$$(R_{if})_e^2 = (R_{if})_x^2 + (R_{if})_y^2 + (R_{if})_z^2$$

For $(R_{if})_e$ to be non-zero, at least one of these components must be totally symmetric, and this will define the direction in which the transition dipole moment is polarized.*

In an atom, where the symmetry is spherical, the x, y and z axes are equivalent and only one component need be considered. Consider two possible transitions, $1s \rightarrow 2s$ and $1s \rightarrow 2p_x$, of the single electron in a hydrogen atom. The transition probabilities are determined by the integrals

$$(R)_{1s \rightarrow 2s} = e \int_{-\infty}^{\infty} 2s \, . \, \hat{x} \, . \, 1s \, dx$$

$$(R)_{1s \rightarrow 2p_x} = e \int_{-\infty}^{\infty} 2p \, . \, \hat{x} \, . \, 1s \, dx$$

The product functions are represented graphically in Figure 1.31; only the second is totally symmetric, and therefore allowed. Wave functions with $l = 1, 3, \ldots$ change sign at the origin and are odd, the others are even. Since the components of the dipole moment operator $\hat{x}$, $\hat{y}$ and $\hat{z}$ are all odd functions, one can arrive at a symmetric transition moment only by combining even and odd states. Transitions with $\Delta l = 0$ are said to be symmetry-forbidden: this is the origin of the so-called Laporte selection rule in atomic spectra. Since the dipole moment operator is a u function, transitions between states of u and g symmetry are allowed, while u $\leftrightarrow$ u or g $\leftrightarrow$ g transitions are forbidden.

In molecular oxygen, the electronic transition

$$\tilde{B}^3 \Sigma_u^- \leftarrow \tilde{X}^3 \Sigma_g^-$$

gives rise to intense absorption in the far ultraviolet (< 200 nm). Taking the z direction parallel to the internuclear axis, the symmetry species of x, y and z will be $(x,y) \equiv \Pi_u$, $z \equiv \Sigma_u^+$. The symmetries of the three components of the transition moment will be given by the direct products

$$(R)_x, (R)_y \equiv \Sigma_u^- \times \Pi_u \times \Sigma_g^- = (\Pi_g) \times \Sigma_g^- = \Pi_g$$

$$(R)_z \equiv \Sigma_u^- \times \Sigma_u^+ \times \Sigma_g^- = (\Sigma_g^-) \times \Sigma_g^- = \Sigma_g^+$$

* This leads to the optical dichroism of oriented molecules.

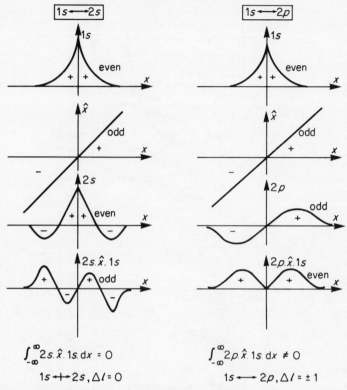

Figure 1.31 Graphical integration of transition moments of a hydrogen atom.

(obtained from the group character table for the point group $D_{\infty h}$). The second of these is totally symmetric, and the transition is symmetry-allowed for radiation polarized with its electric vector oscillating parallel to the z axis. The spin conservation rule is also satisfied and the transition should occur with high probability, as indeed it does. On the other hand an oxygen molecule excited into the $\tilde{b}^1 \Sigma_g^+$ electronic state, for example, will be metastable since its return to the ground state is both spin and symmetry-forbidden. None of the components

$$(R)_x, (R)_y \equiv \Sigma_g^+ \times \Pi_u \times \Sigma_g^- = (\Pi_u) \times \Sigma_g^- = \Pi_u$$

$$(R)_z \equiv \Sigma_g^+ \times \Sigma_u^+ \times \Sigma_g^- = (\Sigma_u^+) \times \Sigma_g^- = \Sigma_u^-$$

is totally symmetric.

The ultraviolet absorption spectrum associated with the allowed transition is shown in Figure 1.32. It is not just a single line, but a long

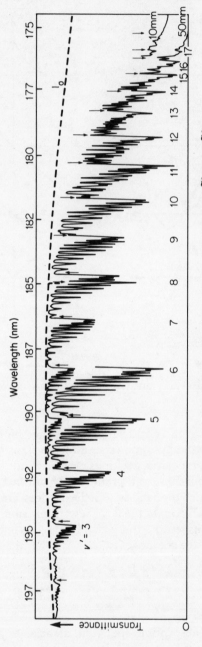

Figure 1.32 Schumann–Runge absorption bands of oxygen ($\tilde{B}^3\Sigma_u^- \leftarrow \tilde{X}^3\Sigma_g^-$), showing extended ($v'$, 0) vibrational progression, and associated rotational structure. (From G. W. Bethke, *J. Chem. Phys.*, **31**, 670 (1959).)

progression of bands, (known as the Schumann–Runge system); each one is associated with a change in vibrational energy and their fine structure is due to changes in rotational energy, both of which have so far been neglected. Line spectra are displayed only by atoms. The vibrational structure is governed by the Franck–Condon Principle.

1.5b The Franck–Condon Principle

When there is a change in the nuclear, as well as the electronic coordinates, the transition moment must include the factor

$$(R_{if})_n = \int_{-\infty}^{\infty} \psi_{n,f}^* \psi_{n,i} \, d\tau_n$$

If the vibrational and rotational motion can be separated, this can be resolved into the product of two components. From the point of view of photochemical changes, we can neglect the rotational part and consider only (electronic + vibrational) excitation, for which

$$(R_{if})_{ev} = (R_{if})_e (R_{if})_v = (R_{if})_e \int_{-\infty}^{\infty} \psi_{v,f}^* \psi_{v,i} \, d\tau_v$$

The intensity of any vibrational band in the electronic transition is determined by the magnitude of the vibrational overlap integral $(R_{if})_v$.

This is a quantum mechanical expression of the Franck–Condon Principle, which states that since the frequency of the radiation promoting an electronic transition ($\sim 10^{15}$–10^{16} Hz) is very much greater than that of the molecular vibrations ($\sim 10^{12}$–10^{13} Hz), there can be very little change in either the molecular geometry or the nuclear kinetic energy in the time of the transition. Effectively, the absorbed photon takes a 'snapshot' of the molecule, in which its vibrational (and rotational) motion is 'frozen'. At room temperature, most molecules will occupy the ground vibrational level, $v'' = 0$, where the equilibrium geometry is the most probable. If the molecule is excited into an electronic state with a much larger equilibrium geometry, it will be endowed with considerable vibrational energy. At the instant of excitation it will be highly compressed relative to the new equilibrium position.

In Figure 1.33 transitions between two attractive electronic states in a diatomic molecule are represented schematically; in one case the equilibrium geometry increases and in the other it remains unchanged. Since the nuclei are 'frozen', the transitions are represented by vertical lines which all begin from the most populated level, $v'' = 0$. When $\Delta r_e \simeq 0$, the vibrational overlap integral is large for transitions where $\Delta v = 0$

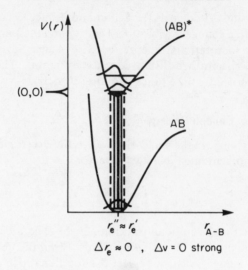

$r_e'' \approx r_e'$

$\Delta r_e \approx 0$, $\Delta v = 0$ strong

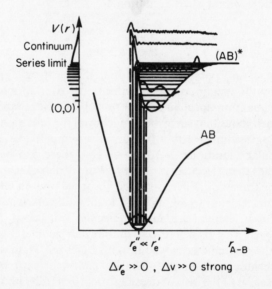

$r_e'' \ll r_e'$

$\Delta r_e \gg 0$, $\Delta v \gg 0$ strong

Figure 1.33 Electronic transitions from the ground vibrational level in the ground electronic state of diatomic molecules : application of the Franck–Condon Principle.

but very small when $\Delta v \neq 0$. The $(0, 0)$ band will be the most intense in absorption. When $\Delta r_e \gg 0$, the most probable transitions will populate high vibrational levels of the upper electronic state. The vertical lines terminate near the classical turning point of the vibration where the vibrationally excited molecules are most likely to be found. Because of the anharmonicity, the levels are closely spaced and many vibrational bands will be possible; the absorption spectrum will consist of a long series of the type $(v', 0)$, which will begin weakly but progress with increasing intensity toward the violet end of the spectrum. The strongest bands will be those for which the square of the overlap integrals, or Franck–Condon factors are largest. This is the situation in the Schumann–Runge band system of O_2, where $\Delta r_e = 0.397$ Å; in contrast the change in bond length following the transition $\tilde{A}^3\Pi \leftarrow \tilde{X}^3\Sigma^-$ in NH, is only 0.001 Å, and the only bands observed in absorption or emission, belong to the $\Delta v = 0$ sequence.[20]

1.5c Polyatomic molecules

In N-atomic molecules, there are $(3N - 6)$ normal modes of vibration. However, if the molecule has elements of symmetry and a totally symmetric ground state allowed electronic transitions from the ground vibrational level (of which there is, of course, only one), can only excite totally symmetric modes in the final state. This follows directly from the fact that the transition moment integral can only be non-zero if the integrand is totally symmetric. This will be true for the product of the vibrational wave functions only if the initial and final vibrational levels belong to the same symmetry species. If one of them is the ground level, which is already totally symmetric, the other level is forced to be totally symmetric also.

For example, the first electronic transition in the planar radical $\cdot N{=}CF_2$, produces the absorption spectrum shown in Figure 1.34 (see plate I, facing page 114). The radical has C_{2v} symmetry in both electronic states, and the absorption has been assigned to the transition $\tilde{A}^2A_1 \leftarrow \tilde{X}^2B_2$.[42] It has B_2 polarization (since the direct product $A_1 \times B_2 \times B_2 = A_1$ is totally symmetric), and is thus allowed for a dipole change lying in the molecular plane, perpendicular to the C_2 axis (see p. 33). The radical has six normal modes, of which three are totally symmetric, and they are represented in Figure 1.35. Although there is only a small change in the equilibrium geometry on excitation, (the biggest change is $\Delta r_{C-N} = 0.04$ Å), each of the three vibrations is excited weakly, in the upper state, with v'_2 the most intense. This is just the result that would be predicted on the basis of the Franck–Condon Principle, and the symmetry requirements for an allowed electronic transition.

In this example, the transition was symmetry allowed. Symmetry-forbidden transitions also occur (though with much reduced intensity) when vibrational modes of the appropriate symmetry are excited in the final state. They are said to be *vibronically* allowed. Consider again the

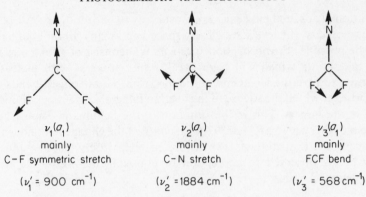

Figure 1.35   Totally symmetric normal modes in $\dot{N}{=}CF_2$.

full expression for the transition moment

$$(R_{if})_{ev} = \int_{-\infty}^{\infty} \psi_{e,f}^* \hat{M} \psi_{e,i} \, d\tau_e \int_{-\infty}^{\infty} \psi_{v,f}^* \psi_{v,i} \, d\tau_v$$

If the transition is electronically allowed one of the three components $\psi_{e,f}^* \hat{M}_x \psi_{e,i}$, $\psi_{e,f}^* \hat{M}_y \psi_{e,i}$ or $\psi_{e,f}^* \hat{M}_z \psi_{e,i}$ must be totally symmetric, and if the nett transition moment $(R_{if})_{ev} \neq 0$, so also must be the product $\psi_{v,f}^* \psi_{v,i}$. If the transition is electronically forbidden none of the components will be totally symmetric. In order that $(R_{if})_{ev} \neq 0$ it is essential that the product $\psi_{v,f}^* \psi_{v,i}$ has the same symmetry as one of the components of the transition dipole. Alternatively, a transition between vibronic states (wave function ψ_{ev}) is allowed if a component of the product $\psi_{ev,f}^* \hat{M} \psi_{ev,i}$ is totally symmetric (ψ_{ev} has the same symmetry as the product $\psi_e \psi_v$).

The classical example of a vibronically allowed transition is provided by benzene, whose near ultraviolet absorption arises from the transition

$$\tilde{A}^1 B_{2u} \leftarrow \tilde{X}^1 A_{1g} \qquad \text{(see p. 51–52)}$$

If this were symmetry allowed, the transition dipole would have to have a component of B_{2u} symmetry, since the totally symmetric representation is generated only by the direct product

$$B_{2u} \times B_{2u} \times A_{1g} = A_{1g}$$

It is a sad fact that neither $\hat{M}_x$ or $\hat{M}_y(E_{1u})$ or $\hat{M}_z(A_{2u})$ satisfy this requirement,

$$x \text{ or } y \text{ polarization (in plane)}: B_{2u} \times E_{1u} \times A_{1g} = E_{2g}$$

$$z \text{ polarization (perp. to plane)}: B_{2u} \times A_{2u} \times A_{1g} = B_{1g}$$

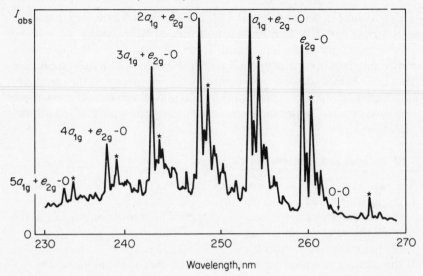

Figure 1.36 Absorption spectrum of benzene vapour ($\tilde{A} \leftarrow \tilde{X}$ transition). The main progression in the upper state 'breathing' mode, a_{1g} also carries one quantum of a degenerate, skeletal deformation mode, e_{2g}. The other strong progression marked *, carries one quantum of the corresponding e_{2g} vibration of the ground electronic state. It is a progression of 'hot' bands.

However, if a vibrational mode of e_{2g} or b_{1g} symmetry were excited in the transition, all would be well.* In fact benzene has four normal modes with e_{2g} symmetry (though none of b_{1g} symmetry). Its near ultraviolet absorption is shown in Figure 1.36.

The (0,0) band, where vibronic interaction is impossible, does not appear. There is however, a prominent progression of bands, whose separation corresponds to the totally symmetric a_{1g} 'breathing' vibration of the upper state. The striking observation is that these do not start at the origin (where there is no absorption band), but are displaced from it by an energy which corresponds to one quantum of an e_{2g} mode in the upper state. This mode distorts the hexagonal shape of the ring. The progression in the 'breathing' frequency reflects the increase in C–C bond length in the $^1B_{2u}$ state ($1.397 \rightarrow 1.434$ Å).[26] The transition populates vibronic levels in the upper state, and is polarized in the molecular plane.

The near ultraviolet absorption of formaldehyde arises from the transition

$$\tilde{A}^1A_2 \leftarrow \tilde{X}^1A_1 \qquad \text{(see p. 53)}$$

* The symmetry species of normal vibrational modes are conventionally denoted by small letters.

This is forbidden since neither $\hat{M}_x$, $\hat{M}_y$ nor $\hat{M}_z$ belong to the A_2 representation (see p. 33). The transition would become vibronically allowed by excitation of any of the non-totally symmetric modes in the upper state. However, the arguments applied to benzene are not strictly applicable here since the molecule is non-planar in the $^1A_2$ state and the symmetry is lowered (see p. 53). This allows excitation of vibrational modes not totally symmetric under the operations of the point group of the ground electronic state.

1.5d Induced and spontaneous transitions

The rate at which light of frequency v_{if} induces the electronic transition $f \leftarrow i$, is determined by

(i) the radiation energy density $\rho(v_{if})$ per unit frequency interval in the neighborhood of v_{if}, and

(ii) the transition moment.

If the number of atoms or molecules which can absorb light at the frequency v_{if} is n_i, the average number of transitions per second when a beam of light passes through them is

$$P_{if} = n_i g_f B_{if} \rho(v_{if})$$

where B_{if} is known as the Einstein coefficient for induced absorption and is related to the transition moment R_{if} through the equation

$$B_{if} = (8\pi^3/3h^2)R_{if}^2$$

g_f is the degeneracy of the final state.[20] The magnitude of B_{if} will depend on the 'allowedness' of the transition, which will be reflected in the observed absorption coefficient.

Electronic transition probabilities are often expressed in terms of the oscillator strength, f. This is defined as the ratio between the experimental transition probability and that of a molecule in which a single electron, bound in a spherically symmetrical field, is able to oscillate with simple harmonic motion under the influence of the electromagnetic radiation. In the latter idealized situation exact calculation of the transition moment is possible and it can be used as a reference standard. Its value is

$$(R_{if})^2_{ideal} = \frac{3e^2h}{8\pi^2mc\tilde{v}_{if}}$$

where $\tilde{v}_{if} = v_{if}/c$ is the frequency expressed in cm^{-1} and m is the mass of the electron. Thus the oscillator strength is given by

$$f_{if} = \frac{8\pi^2mc\tilde{v}_{if}}{3e^2h}(R_{if})^2$$

In a molecular spectrum the absorption spreads over a range of frequencies because of the changes in vibrational and rotational energy accompanying the electronic transition. Its nett probability is measured by the integrated absorption coefficient $\int \varepsilon(\tilde{v}_{if}) \, d\tilde{v}_{if}$ of the complete spectrum, summed over all the vibrational and rotational transitions; it is a kind of integrated photon capture cross-section, measured by the area under the absorption curve. The integrated absorption and the oscillator strength are of course, related, and it can be shown[20] that

$$f_{if} = \frac{10^3 \ln 10 mc^2}{\pi N e^2} \int \varepsilon(\tilde{v}_{if}) \, d\tilde{v}_{if} \equiv 4 \cdot 32 \times 10^{-9} \int \varepsilon(\tilde{v}_{if}) \, dv_{if}$$

where N is the Avogadro number. Thus a transition which has a very high probability has $f_{if} \simeq 1*$ and its integrated absorption is $\sim 2 \cdot 5 \times 10^8 \, \mathrm{l \, mole^{-1} \, cm^{-2}}$. If the absorption profile is symmetrical, with a half-width $\Delta \tilde{v}_{\frac{1}{2}}$ and a maximum extinction coefficient ε_{max}, the peak area approximates to

$$\varepsilon_{max} \Delta \tilde{v}_{\frac{1}{2}} \simeq 2 \cdot 5 \times 10^8 \, \mathrm{l \, mole^{-1} \, cm^{-2}}$$

A typical continuous, polyatomic molecular absorption spectrum has $\Delta \tilde{v}_{\frac{1}{2}} \simeq 5,000 \, \mathrm{cm^{-1}}$, so that for a fully allowed transition $\varepsilon_{max} \simeq 10^5 \, \mathrm{l \, mole^{-1} \, cm^{-1}}$. The widths of atomic line spectra are typically $\sim 0 \cdot 5 \, \mathrm{cm^{-1}}$, and for them $\varepsilon_{max} \simeq 10^8 - 10^9 \, \mathrm{l \, mole^{-1} \, cm^{-1}}$. The greater the spread in the spectrum, the more the intensity is diluted over the broad spectral range.

Forbidden absorption bands have oscillator strengths very much less than unity and very much smaller values of ε_{max}. For example, the forbidden near-ultraviolet absorption in benzene and formaldehyde has $\varepsilon_{max} = 204$ and $10 \, \mathrm{l \, mole^{-1} \, cm^{-1}}$ respectively.

Having induced the atom or molecule to jump to a higher electronic state, unless it suffers some other fate whilst in that elevated condition (see the next chapter, and much of the rest of the book for the alternative options), it will re-emit its energy and return eventually to the ground state. The rate at which the reverse transition f $\rightarrow$ i may occur is

$$P_{fi} = n_f g_i B_{fi} \rho(v_{fi}) + n_f g_i A_{fi}$$

where there is now the possibility of either induced (or stimulated) emission for which the Einstein coefficient is B_{fi}, or spontaneous emission for which the coefficient is A_{fi}. The former is induced by light at the resonant frequency $v_{if} = v_{fi}$, while the latter will occur in isolation. n_f is

---

* If absorption is measured in solvent of refractive index n, a slight correction factor must be introduced, particularly if $n \gg 1$; Rubinowicz divides f_{if} by the factor $9n/(n^2 + 2)^2$.[43]

the population of molecules in the upper state and g_i the degeneracy of the lower state.

The coefficient of spontaneous emission is given by[20]

$$A_{fi} = 8\pi h \tilde{v}_{if}^3 B_{if} = \frac{64\pi^4}{3h} \tilde{v}_{if}^3 (R_{if})^2$$

and for the usual conditions of absorption and emission in an open cell (i.e. one which has transparent walls) $A_{fi} \gg B_{fi}\rho(v_{fi})$. Microscopic reversibility requires that $B_{fi} = B_{if}$, so that when there is a steady state population of excited atoms

$$n_i g_f B_{if}\rho(v_{if}) = n_f g_i A_{fi}$$

or

$$\frac{n_i}{n_f} = \frac{g_i A_{fi}}{g_f B_{fi}\rho(v_{if})} \gg 1$$

The population of the upper state is going to be very small (note that $A_{fi} \propto \tilde{v}_{if}^3$, and in the spectroscopies which occur at much lower frequencies it is not difficult to reach the situation when $n_i g_f \simeq n_f g_i$; under this condition, there is a dynamic balance between absorption and emission; since there is no nett absorption, the system is saturated).

The coefficient for spontaneous emission is a rate constant for the, unimolecular decay of the isolated atom or molecule, in the absence of any competing processes (for example, dissociation). Its reciprocal gives the natural radiative lifetime of the excited state, $\tau_0 = 1/A_{fi}g_i$. Since $A_{fi} \propto B_{if}$ and $B_{if} \propto \int \varepsilon(\tilde{v}_{if}) \, d\tilde{v}_{if}$, the natural radiative lifetime can be estimated from the area under the absorption curve. With $g_i = g_f = 1$

$$\tau_0 = \frac{N}{10^3 \ln 10} \left\{ 8\pi c \tilde{v}_{if}^2 \int \varepsilon(\tilde{v}_{if}) \, d\tilde{v}_{if} \right\}^{-1}$$

$$\simeq 3.3 \times 10^8 \left\{ \tilde{v}_{if}^2 \int \varepsilon(\tilde{v}_{if}) \, d\tilde{v}_{if} \right\}^{-1} \sec^{-1}$$

Our typical polyatomic molecule which has $\Delta \tilde{v}_{\frac{1}{2}} \simeq 5{,}000 \text{ cm}^{-1}$, would have

$$\tau_0 \simeq 3.3 \times 10^8 \{ \tilde{v}_{if}^2 \Delta \tilde{v}_{\frac{1}{2}} \varepsilon_{max} \}^{-1} \simeq 10^5 / \tilde{v}_{if}^2 \varepsilon_{max} \sec$$

If the transitions were fully allowed, with $\varepsilon_{max} \simeq 10^5 \text{ l mole}^{-1} \text{ cm}^{-1}$, and the absorption were centred around $\tilde{v}_{if} \simeq 30{,}000 \text{ cm}^{-1}$ in the ultraviolet, its radiative lifetime would be $\sim 10^{-9}$ sec. In general, $\tau_0 \simeq 10^{-4}/\varepsilon_{max}$ sec, for a wide range of polyatomic molecules absorbing in the near-ultraviolet. This represents an estimate of the calculated lifetime in the absence of any other competing processes and as such it is an upper limit. The transition populating the $\tilde{A}^1 B_{2u}$ state of benzene has $\varepsilon_{max} \simeq 200 \, 1 \, mole^{-1} \, cm^{-1}$: its natural radiative lifetime must be $\leqslant 5 \times 10^{-7}$ sec. In fact a considerable proportion of the excited benzene decays via the lowest triplet state and possibly through the intermediate formation of isomers, so that its actual lifetime must be shorter than this. If the radiative transition back to the ground state (or to any intermediate state) is forbidden, the electronic state is said to be metastable. In this situation, it will be possible to maintain a rather larger population of excited atoms or molecules for a given absorbed light intensity. If a means of increasing the radiation density could be found, so that the direction of the inequality $A_{fi} \gg B_{if}\rho(v_{if})$ could be reversed, the rate of depopulation of the upper state would be $n_f g_i B_{if}\rho(v_{if})$. Under these conditions, the nett rate of excitation at the steady state is

$$(n_i g_f B_{if} - n_f g_i B_{if})\rho(v_{if}) = 0$$

and the system is saturated,

$$n_i g_f = n_f g_i$$

If it were possible to find a system in which there was an intermediate level k, into which the upper state could readily decay, but which was metastable with respect to the ground state, it would be possible to maintain a population inversion between the intermediate and ground levels. With a sufficiently high rate of pumping from the level i $\rightarrow$f, and sufficiently high rate of transition from f $\rightarrow$k, the intermediate level would acquire a larger and larger population. (Note that it is impossible to obtain a population inversion between states i and f, since absorption ceases when the system saturates, and that is as far as one can go. If the level k provides a 'sink' for the excited species, the populations of i and f will fall, while that of k builds up. Its maximum population is independent of the onset of saturation between i and f.) As soon as the point is reached at which $n_k g_i > n_i g_k$, the nett rate of induced absorption from the state i $\rightarrow$k, becomes

$$(n_i g_k B_{ik} - n_k g_i B_{ik})\rho(v_{ik}) < 0$$

The rate is negative and we observe induced or stimulated emission instead. This is the principle on which the three level laser is based.

The oldest, and best example of this is the ruby laser. The levels, and spontaneous transition probabilities are shown in Figure 1.37. A cylindrical crystal of ruby is exposed to an intense flash of visible light, which pumps a large number of the Cr^{3+} ions in the crystal via the short-lived $^4F_2$ levels, into the metastable ^2E state, in a time which is short compared

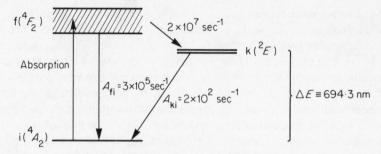

Figure 1.37   Electronic states associated with laser action of Cr^{3+} ions in ruby.

with the natural radiative lifetime. As soon as there is a population inversion between the metastable ^2E state and the ground $^4A_2$ state, the crystal will be in a 'lasing' condition. In order to take advantage of this, the crystal is placed in a Fabry–Perot interferometer which consists of two planar or confocal mirrors whose optical axes are parallel with the cylindrical axis of the crystal. Photons emitted by the metastable ions remain trapped in the multiply reflecting system, and traverse back and forth along the crystal stimulating further emission until the population inversion is removed. If one of the mirrors is partially silvered only, an intense parallel beam of coherent, monochromatic light emerges as a brief pulse in the red. As soon as the pumping flash has restored the inversion, the whole process begins anew and a sequence of bursts of laser light are observed for as long as the flash is sufficiently intense to regenerate the inversion. By this device, the radiation energy density is amplified enormously, hence the word laser—Light Amplification through Stimulated Emission of Radiation. The maintenance of a high energy density ensures that the rate of induced emission exceeds the rate of spontaneous emission. The latter is a random, 'wastage' process, which reduces the population of the upper state and the coherence of the laser light.

Very high intensity pulses of monochromatic light, of nanosecond (10^{-9} sec) duration, can be produced in 'giant-pulse laser systems'. These are obtained by optical pumping of the laser crystal in a 'Q-spoiled cavity', i.e. one which cannot resonate because of the insertion of a shutter in the optical cavity. If the shutter is allowed to open, only when the population inversion has reached a maximum, and it opens fast enough, a very intense and very brief burst of laser light is emitted. The simplest 'shutter' is a saturable optical filter, opaque to low light intensities, but becoming transparent under very high intensity illumination: Solutions of crypto-cyanine dyes operate on this principle. At very high intensities, so many of the molecules are promoted into the lowest triplet level that the normal absorption associated with excitation out of the ground state, disappears. We shall see that 'Q-spoiled laser pulses' have great potential for studying the transient stages of a photochemical reaction.

REFERENCES

1. See R. P. Feynmann, *Lectures on Physics*, Vol. 1, Addison-Wesley, Reading, Mass., 1965, for an entertaining discussion
2. M. Planck, *Verhandl. deut. physik. Ges.*, **2**, 202 (1900)
3. A. Einstein, *Ann. Physik.*, **17**, 132 (1905)
4. N. Bohr, *Phil. Mag.*, **26**, 1, 476, 857 (1913)
5. A. Sommerfeld, *Ann. Physik.*, **51**, 1 (1916)
6. A. H. Compton, *Phys. Revs.*, **22**, 409 (1927)
7. L. de Broglie, *Ann. de phys.*, (10), **3**, 22 (1925)
8. C. Davisson and L. H. Germer, *Phys. Revs.*, **30**, 705 (1927)
9. G. P. Thompson, *Proc. Roy. Soc.*, *A*, **117**, 600 (1928)
10. I. Estermann, R. Frisch and O. Stern, *Z. Physik.*, **73**, 348 (1931)
11. W. Heisenberg, *Z. Physik*, **43**, 172 (1927)
12. M. Born, *Z. Physik.*, **37**, 863 (1926)
13. R. P. Feynmann, *Lectures on Physics*, Vol. 3, Chapter 20, Addison-Wesley, Reading, Mass., 1965
14. G. E. Uhlenbeck and S. Goudsmit, *Naturwiss.*, **13**, 953 (1925)
15. P. A. M. Dirac, *Quantum Mechanics*, 3rd Ed., Oxford University Press, Oxford, 1947
16. R. G. Parr, *Quantum Theory of Molecular Electronic Structure*, Benjamin, New York, 1963
17. G. Herzberg, *Atomic Spectra and Atomic Structure*, Dover Publications, New York, 1944, p. 130
18. E. U. Condon and G. H. Shortley, *The Theory of Atomic Spectra*, Cambridge University Press, Cambridge, 1935
19. A. Landé, *Z. Physik.*, **15**, 189 (1923)
20. G. Herzberg, *Molecular Spectra and Molecular Structure*, Vol. 1, 2nd Ed., Van Nostrand, New York, 1950
21. L. Pauling, *Nature of the Chemical Bond*, 3rd Ed., Cornell University Press, 1961
22. D. S. Schonland, *Molecular Symmetry*, Van Nostrand, London, 1965
23. H. H. Jaffé and M. Orchin, *Symmetry in Chemistry*, Wiley, New York, 1965
24. F. A. Cotton, *Chemical Applications of Group Theory*, Interscience, New York, 1963
25. R. M. Hochstrasser, *Molecular Aspects of Symmetry*, Benjamin, New York, 1966
26. G. Herzberg, *Molecular Spectra and Molecular Structure*, Vol. 3, Van Nostrand, New York, 1966
27. V. Heine, *Group Theory in Quantum Mechanics*, Pergamon, London, 1960
28. D. H. Volman, *Advances in Photochemistry*, Vol. 1, Interscience, New York, 1963, p. 43

29. J. N. Murrell, *The Theory of the Electronic Spectra of Organic Molecules*, Methuen, London, 1963
30. J. R. Majer and J. P. Simons, *Advances in Photochemistry*, Vol. 2, Interscience, New York, 1964, p. 137
31. A. D. Walsh, *J. Chem. Soc.*, 2321 (1953)
32. A. D. Walsh, *J. Chem. Soc.*, 2266 (1953)
33. C. A. Coulson and A. H. Neilson, *Discussions Faraday Soc.*, **35**, 71 (1963)
34. E. W. Abrahamson and S. E. Ostroy, *Prog. Biophys. Mol. Biol.*, **17**, 179 (1967)
35a. M. S. de Groot and J. H. van der Waals, *Mol. Phys.*, **6**, 545 (1963)
35b. M. S. de Groot, I. A. M. Hesselman and J. H. van der Waals, *Mol. Phys.*, **16**, 45 (1969)
36. A. D. Walsh, *J. Chem. Soc.*, 2306 (1953)
37. P. M. Morse, *Phys. Rev.*, **34**, 507 (1929)
38. A. G. Gaydon, *Dissociation Energies and Spectra of Diatomic Molecules*, 2nd Ed., Chapman and Hall, London, 1953
39. C. F. Goodeve and A. W. C. Taylor, *Proc. Roy. Soc. A*, **157**, 181 (1936); see also A. S. Coolidge, H. M. James and R. D. Present, *J. Chem. Phys.*, **4**, 143, (1936) and N. S. Bayliss, *Proc. Roy. Soc. A*, **158**, 551 (1937)
40. E. Thiele and D. J. Wilson, *J. Chem. Phys.*, **35**, 1256 (1961)
41. J. Franck, *Trans. Faraday Soc.*, **21**, 536 (1925); E. U. Condon, *Phys. Revs.*, **32**, 858 (1928)
42. R. N. Dixon, G. Duxbury, R. C. Mitchell and J. P. Simons, *Proc. Roy. Soc., A*, **300**, 405 (1967)
43. A. Rubinowicz, *Repts. Progr. in Phys.*, **12**, 233 (1948–9)

PROBLEMS

1. Oxygen atoms have the electronic configuration $K(2s)^2(2p)^4$; show that this can generate the six electronic terms $2^3P_{2,1,0}$, 2^1D_2 and 2^1S_0.
(Hint: One way of distributing the four $2p$ electrons is

$2p_{+1}$	$2p_0$	$2p_{-1}$

$$\uparrow\downarrow \quad \uparrow\downarrow \qquad \begin{cases} M_L = 2(+1) + 2(0) = +2 \\ M_S = 2(+\tfrac{1}{2}) + 2(-\tfrac{1}{2}) = 0 \end{cases}$$

another is

$$\uparrow\downarrow \quad \uparrow \quad \uparrow \qquad \begin{cases} M_L = 2(+1) + 0 + (-1) = +1 \\ M_S = 3(+\tfrac{1}{2}) + (-\tfrac{1}{2}) = +1 \end{cases}$$

In all there are fifteen possible distributions.)

2. The first two electronic transitions in NO_2 have been assigned as $\tilde{A}^2B_1 \leftarrow \tilde{X}^2A_1$ and $\tilde{B}^2B_2 \leftarrow \tilde{X}^2A_1$, respectively. In which direction will the electronic transition moments be polarized? (Use the C_{2v} group character table given in section 1.3b; under the operations of the group, the Cartesian vectors $\overrightarrow{Ox}$, $\overrightarrow{Oy}$ and $\overrightarrow{Oz}$ provide bases for the B_1, B_2 and A_1 representations respectively.)

3. Using the Walsh diagram (Figure 1.20), suggest the probable equilibrium geometry and electronic configuration of NF_2, and the probable assignment and polarization of its lowest electronic transition.

4. Use the C_{3v} group character table to establish whether the first electronic transition in CH_3Br is symmetry allowed or forbidden, and discuss the low intensity of the associated absorption spectrum.
(Under the symmetry operations C_{3v}, $\overrightarrow{Ox}$ and $\overrightarrow{Oy}$ form a basis for the doubly degenerate representation E, while $\overrightarrow{Oz}$ is totally symmetric, A_1.)

5. The radicals NO_3 and NCF_2, both with 23 valence electrons, are known to be planar both in their ground and first excited electronic states. What would you expect for the equilibrium conformation of OCF_2 (24 valence electrons), in its ground state, and in its first excited state? Do you think the CF_3 radical will be planar in its ground state? How might an $n \rightarrow \pi^*$ transition in OCF_2 differ from one in OCH_2?

6. The photochemical dissociation of chlorine yields a normal atom $(^2P_{\frac{3}{2}})$ and an excited atom $(^2P_{\frac{1}{2}})$. Analysis of the electronic band spectrum indicates that continuous absorption begins at $20,893$ cm^{-1}, and that the transition from the lowest vibrational level in the ground state to the lowest in the excited state has a value of $17,710$ cm^{-1}. From the atomic spectrum of chlorine, the excitation $^2P_{\frac{1}{2}} \leftarrow {}^2P_{\frac{3}{2}}$ has a value of 881 cm^{-1}.
Calculate the dissociation energy in (a) the ground state and (b) the excited state of the chlorine molecule. Define the ground state dissociation energy.

$(h = 6{\cdot}624 \times 10^{-27}$ erg sec, $c = 3 \times 10^{10}$ cm sec^{-1}, 1 joule $= 10^7$ ergs)

7. When CO is produced in a photodissociation it often carries excess vibrational energy. If its bond length reached a maximum value of $1{\cdot}23$ Å during its vibrational motion (assuming a classical description), how many vibrational quanta might the molecule be expected to carry? (Assume the potential function approximates to that of a Morse oscillator, and use the following data: reduced mass $\mu = 6{\cdot}858$ atomic units, $D_e = 90{,}000$ cm^{-1}, $\omega_e = 2{,}170$ cm^{-1}, $r_e = 1{\cdot}13$ Å, $\omega_e x_e = 13{\cdot}46$ cm^{-1}, $N = 6{\cdot}02 \times 10^{23}$, $h = 6{\cdot}624 \times 10^{-27}$ erg sec.)

CHAPTER TWO

LIGHT ABSORPTION AND ITS PHYSICAL CONSEQUENCES

What Goes Up Must Come Down

The 'complete photochemist' should be at ease both in the company of spectroscopists and chemists. He must understand the 'rules' which govern light absorption and emission by atoms and molecules and be familiar with the possible physical and chemical consequences of electronic excitation. Ideally, he wants to correlate the photochemistry with the spectroscopy, although this is usually very difficult to achieve in detail. The structure of the absorption spectrum of a molecule reflects at least a part of what can happen to it in the upper state. If the rotational and/or vibrational structure can be resolved, information can be obtained relating to the relevant equilibrium geometries and chemical bonding in the combining states. The pattern of rotational fine structure is very sensitive to the symmetries of the two states; when the structure can be resolved the symmetries, and hence the nature of the electronic states, can be identified. However, the molecules which interest the photochemist are usually either dissociating, dissolved in solution, or undergoing rapid radiationless processes in their upper states; each of these processes broadens the rotational levels into a continuum. The nature of the excited state can then be inferred from the vibrational structure (or lack of it) combined with the application of the Franck–Condon Principle and/or from a qualitative, and sometimes quantitative, molecular orbital description of the molecular electronic states. If the molecule is dissolved in solution, the effect of the solvent polarity on the absorption wavelengths reflects the electronic distribution in the excited state, and in particular it reveals any separation of charge. If the molecule is oriented in a crystal matrix, the polarization of the transition moment can be found from its optical dichroism under polarized light.

Having identified the excited state populated by light absorption, the next, and more formidable problem, is to trace the path of its subsequent decay. This chapter deals with both problems and gives an account of the kind of background knowledge that is necessary if one is to have a reasonable chance of success. Let us start at the beginning and talk about atoms.

2.1   ATOMIC SPECTRA

2.1a   Selection rules: spin–orbit coupling

Atomic absorption spectra consist of progressions or series of spectral lines, which are associated with changes in the electronic angular momentum and the principal quantum numbers of the populated atomic orbitals; these ultimately converge at the ionization potential. Changes in the angular momentum must satisfy the selection rules for electric dipole radiation. In the ground electronic states of light atoms (atomic number $\lesssim 20$) the orbital and spin angular momenta can be treated separately (Russell–Saunders coupling) and for the allowed transitions the changes in nett angular momentum are restricted to those where

$$\Delta L = 0, \pm 1$$

$$\Delta S = 0$$

$$\Delta J = 0, \pm 1 \text{ but } J = 0 \nrightarrow 0$$

but for the electron excited $\Delta l = \pm 1$. In heavy atoms, the total angular momentum J is the only observable and only the selection rule for ΔJ is rigorous $((j,j)$ coupling). Intercombinations with $\Delta S = \pm 1$, become increasingly probable as the atomic number increases.

The simplest spectrum is that of the H atom, in which all levels with the same principal quantum number n are very nearly degenerate. The first electronic transition in absorption, $2p \leftarrow 1s$, lies in the far ultraviolet at 121·566 nm, and is the first member of the Lyman series (see Figure 2.1): it is known as the Lyman-α spectral line and in emission is a useful source of monochromatic radiation in the far ultraviolet. In absorption, all transitions originate from the $1s$ level, but in emission many series are possible,

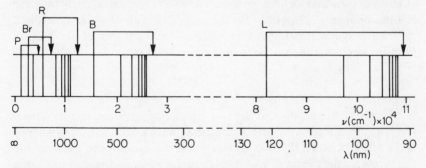

Figure 2.1   Schematic diagram of the Rydberg series that comprise the spectrum of atomic hydrogen. P: Pfund; Br: Brackett; R: Ritz–Paschen; B: Balmer; L: Lyman. Arrows indicate series limits. (After R. N. Dixon, *Spectroscopy and Structure*, Methuen, London, 1965.)

since they terminate at any values of n. In general, the spectral frequencies are

$$\tilde{v} = R_H \left\{ \frac{1}{n_2^2} - \frac{1}{n_1^2} \right\} \text{cm}^{-1}$$

where R_H is the Rydberg constant for the hydrogen atom $(109,677 \cdot 581 \text{ cm}^{-1})$, and n_1 and n_2 refer to the upper and lower states.

In the alkali metal atoms, the outermost electron is much less tightly bound and their absorption spectra lie at much longer wavelengths. The lines at longest wavelength correspond to the transitions $np \leftarrow ns$, where $n = 2$ (Li), 3 (Na) and so on. They are the first members of the so-called principal series, $mp \leftarrow ns$ (where $m \geqslant n$), for which the frequencies are

$$\tilde{v} = T_{ps} - \frac{R}{(m + p)^2} \text{cm}^{-1}$$

T_{ps} is the series limit, p is a correction factor <1 and R is a Rydberg constant. The simple equation for the hydrogen atom is inappropriate since the outermost electrons circulate in a field resulting from the inner electrons as well as the nucleus. Since the atoms have one odd electron, each of the levels with $L > 0$ is split into a doublet, with $J = L \pm \frac{1}{2}$; the $n^2 S_{\frac{1}{2}}$ terms are doubly degenerate. The principal series are composed of pairs of lines associated with the transitions $m^2 P_{\frac{3}{2}} \leftarrow n^2 S_{\frac{1}{2}}$, $m^2 P_{\frac{1}{2}} \leftarrow n^2 S_{\frac{1}{2}}$, both of which are allowed ($\Delta J = 1,0$). For example, the yellow sodium D lines at $599 \cdot 0$ nm and $599 \cdot 6$ nm are associated with the transitions $3^2 P_{\frac{3}{2}, \frac{1}{2}} \leftarrow 3^2 S_{\frac{1}{2}}$. The term splitting depends on the degree of spin–orbit coupling. We shall discuss this coupling in some detail since its photochemical consequences are very important in polyatomic molecules (see section 2.5d).

The weak interaction between the magnetic moments due to the electronic spin and orbital motion introduces an extra term into the Hamiltonian energy operator for the atom. If the electron circulates in a spherically symmetrical field $V(r)$, where r is the distance of the electron from the nucleus, the spin–orbit interaction term is[1]

$$\hat{H}_{so} = \frac{1}{2m^2 c^2} \cdot \frac{1}{r} \frac{\partial V(r)}{\partial r} \mathbf{L} . \mathbf{S} = \xi(r) \mathbf{L} . \mathbf{S}$$

$\mathbf{L} . \mathbf{S}$ is the scalar product of the two angular momentum vectors (effectively the projection of the one on the other, since $\mathbf{L} . \mathbf{S} = LS \cos \theta$ where θ is the angle between the two vectors): m is the mass of the electron and $\xi(r)$ measures the degree of spin–orbit coupling. For hydrogen-like atoms, where $V = -ze^2/r$,

$$\frac{1}{r} \cdot \frac{\partial V(r)}{\partial r} = \frac{ze^2}{r^3}$$

and

$$\xi(r) = \frac{ze^2}{2m^2c^2r^3}$$

The energy of the interaction in a level of given n and l is

$$E_{so} = \int \psi^*_{n,l,m_l,m_s} \hat{H}_{so} \psi_{n,l,m_l,m_s} \, d\tau$$

The radial part can be extracted from the net wave function to give

$$E_{so} = \int R^*_{n,l}(r)\xi(r)R_{n,l}(r) \, dr \int \phi^*_{l,m_l,m_s}(\mathbf{L} \cdot \mathbf{S})\phi_{l,m_l,m_s} \, d\tau$$

For hydrogen-like orbitals, the radial part

$$\int R^*_{n,l}(r)\xi(r)R_{n,l}(r) \, dr \equiv \xi_{n,l}$$

is given by

$$\xi_{nl} = \frac{h^2 e^2 z^4}{2m^2 c^2 a^3 n^3} \cdot \frac{1}{l(l+1)(l+\frac{1}{2})}$$

where 'a' is the mean orbital radius. $\xi_{nl}/\hbar^2 \equiv \zeta_{nl}$ is known as the spin–orbit coupling constant for the levels nl; note that it increases as z^4, but decreases as n^3.[1] For the levels with $j = l \pm \frac{1}{2}$, one can show that the angular part of the integral takes the values $+\frac{1}{2}l$ and $-\frac{1}{2}(l+1)$, so that their energies are

$$E_J = E_{nl} + \tfrac{1}{2}l\zeta_{nl} ; \qquad E_{J-1} = E_{nl} - \tfrac{1}{2}(l+1)\zeta_{nl}$$

(E_{nl} would be the energy in the absence of spin–orbit interaction). The levels are split by

$$\Delta E_{so} = (l + \tfrac{1}{2})\zeta_{nl} \equiv J\zeta_{nl}$$

In polyelectronic atoms, the assumption is often made that the spin–orbit term in the Hamiltonian energy operator is approximately equal to the algebraic sum

$$\hat{H}_{so} = \sum_{i=1}^{N} \xi(r_i)\mathbf{L}_i \cdot \mathbf{S}_i$$

summed over the N electrons. However, this is not strictly valid since the force field is no longer coulombic, and the cross interaction between the spin in one level and the orbital angular momentum in another is ignored.[2,3] In a molecule, the approximation has no theoretical justification, although it is still the case that the spin–orbit interaction is a strong function of the atomic number both in polyelectronic atoms and in molecules (for those orbitals which have a large amplitude on the atom in question). Where there is just one odd electron outside a closed shell, since the nett spin and orbital angular momentum in a closed shell is zero, it is only necessary to consider the outermost electron in the approximation for H_{so}. This brings us back to the alkali metal atoms: if they can be treated as approximations to a hydrogen-like system, we should expect the spacing between the doublet components of their absorption spectra to increase with z; this agrees with observation.[1] In addition it is also observed that the splittings in a given atom fall with increasing n as predicted for the hydrogen-like atoms. Spin–orbit coupling constants calculated on the assumption $n^2P_{\frac{3}{2}} - n^2P_{\frac{1}{2}} = (l + \frac{1}{2})\zeta_{np}$ are $27\,\text{cm}^{-1}$ (Na, $3p$), $87\,\text{cm}^{-1}$ (K, $4p$), $360\,\text{cm}^{-1}$ (Rb, $5p$) and $840\,\text{cm}^{-1}$ (Cs, $6p$) (ref. 1, p. 145).

The spectra of the alkali metal vapours are far richer in emission and include series originating from ns (sharp), nd (diffuse) and nf (fundamental) orbitals, in addition to the principal series observed in absorption. The term diagram for Li is shown in Figure 2.2.

Atoms in Group II of the Periodic Table have two electrons outside a closed shell configuration. In their ground electronic states, these fill the outermost s orbital, and the atoms have $^1S_0$ ground states. Allowed electronic transitions will again be of the type $np \leftarrow ns$, but the excited configurations can give four possible states, n^1P_1, or $n^3P_{2,1\,or\,0}$; the two electrons are in separate orbitals and both singlet or triplet states are possible. Of the four electronic transitions

$$n^1P_1 \leftarrow n^1S_0 \quad \text{(a)}$$

$$n^3P_2 \leftarrow \quad\quad\quad \text{(b)}$$

$$n^3P_1 \leftarrow \quad\quad\quad \text{(c)}$$

$$n^3P_0 \leftarrow \quad\quad\quad \text{(d)}$$

(a) is allowed and (b–d) are spin-forbidden. In addition, (b) and (d) contravene the selection rule for J, since (b) has $\Delta J = 2$, and (d) has $J = 0 \rightarrow 0$.

No photochemical laboratory is complete without a low pressure mercury vapour lamp, and the more adventurous may also possess a cadmium vapour lamp. These emit intense monochromatic light in the ultraviolet regions, in particular intense spectral lines at $253.7\,\text{nm}$ and $184.9\,\text{nm}$ (Hg), and $326.1\,\text{nm}$ and $228.8\,\text{nm}$ (Cd); the same lines also occur in absorption in the vapour, and are termed *resonance* lines (see section

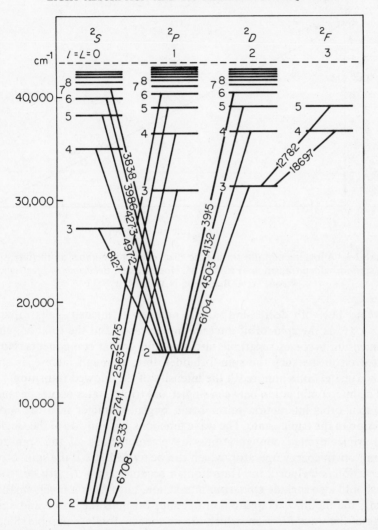

Figure 2.2 Energy level diagram for the Li atom (Grotian diagram). The number beside each level is the principle quantum number n for the outer electron. Wavelengths in nm are indicated on the connecting lines representing transitions. (From R. N. Dixon, *Spectroscopy and Structure*, Methuen, London, 1965.)

2.1b). Those at the shorter wavelength are associated with the transition $n^1P_1 \leftrightarrow n^1S_0$, where $n = 5(\text{Cd})$ or $6(\text{Hg})$. The other two correspond to the transition $n^3P_1 \leftrightarrow n^1S_0$ which is spin-forbidden, but allowed under (j, j) coupling. Figure 2.3 shows the relative intensities of the two transitions, for all the atoms in Group II. The forbidden line is most intense for

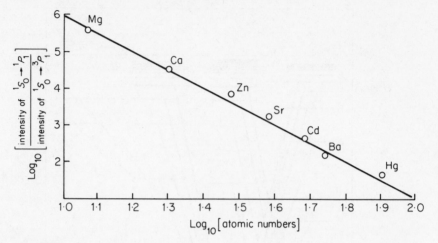

Figure 2.3 A log–log plot illustrating the effect of atomic number on the intensity of a spin-forbidden transition. (From R. M. Hochstrasser, *Behaviour of Electrons in Atoms*, W. A. Benjamin, N.Y., 1964, p. 103.)

mercury but with decreasing atomic number its intensity falls rapidly ($\propto z^5$), i.e. as the spin–orbit interaction decreases, and the total angular momentum becomes separable into spin and orbital components. **Note** that even in mercury the spin-forbidden line, although intense, is still approximately one hundredth the intensity of the allowed transition.

The intercombination between singlet and triplet states occurs because the spin–orbit interaction mixes some singlet character into the wave function of the triplet state. The wave functions $^1\psi_n$ and $^3\psi_n$ of the singlet and triplet excited configurations are eigenfunctions of the spin-free Hamiltonian energy operator, which can be termed $\hat{H}_0$. If the spin–orbit interaction is included, the Hamiltonian becomes $\hat{H}_0 + \hat{H}_{so}$ (to be strict we should also include spin–spin interaction, but this is a much smaller term); the modified Hamiltonian operates on both the space and spin parts of the wave function, which are no longer strictly separable. Since $\hat{H}_{so} \ll \hat{H}_0$, this amounts to a small perturbation of the original wave functions, and its effect can be found by a simple first order perturbation treatment.[4] The important result is that the corrected triplet state wave function is of the form

$$[^3\psi_n]_{corr} = {}^3\psi_n + \left\{ \frac{\displaystyle\int {}^3\psi_n^* \hat{H}_{so} {}^1\psi_n \, d\tau}{{}^1E_n - {}^3E_n} \right\} {}^1\psi_n \equiv {}^3\psi_n + \lambda\, {}^1\psi_n$$

where ^1E_n and ^3E_n are the energies of the unperturbed triplet and singlet states. The factor λ determines the degree of singlet character mixed into the triplet state by the spin–orbit interaction. It rises as the spin–orbit interaction constant ζ increases and as the separation between the two unperturbed energy levels decreases; thus other nearby singlet states may also perturb the triplet state and introduce further singlet character into the corrected wave function. By the same token, low-lying triplet states may also perturb the ground singlet state and introduce some triplet character, but since the ground state usually lies far below the triplet levels of atoms, this can be neglected. The corrected transition moment for the intercombination is of the form

$$e \int [^3\psi_n]_{corr} \sum_i \mathbf{r}_i\, ^1\psi_0 \cdot d\mathbf{r} = e \int\, ^3\psi_n \sum_i \mathbf{r}_i\, ^1\psi_0\, dr + \lambda\, e \int\, ^1\psi_n \sum_i \mathbf{r}_i\, ^1\psi_0 \cdot dr$$

the first term is zero but the second term is not. Depending on the amount of singlet character introduced the singlet–triplet transition will be more or less intense. The transition can be said to have 'borrowed' intensity from the allowed transition(s) into the nearby singlet state(s); in the present case, the $n^3P_1 \leftrightarrow n^1S_0$ transition borrows intensity from $n^1P_1 \leftrightarrow n^1S_0$.

2.1b Line width[5]

Although the electronic transition in atoms give rise to spectral lines, their profiles do not present a sharp contour, but are smoothly curved, and spread over a frequency range about a central maximum. Several factors may contribute to line broadening; in order of increasing effect these include the following:

(i) Natural broadening
Present in all spectral lines under all conditions. It arises from the finite lifetime of the upper state excited by absorption from, or decaying by emission to, the ground state. The lifetime determines the duration of the wave train emitted or absorbed; the shorter the duration, the greater the uncertainty in the wavelength (see section 1.2b). The spread in the spectral line is given by

$$\Delta v \simeq \frac{1}{2\pi\tau}$$

where τ is the mean lifetime of the upper state. For $Hg(6^3P_1)$, $\tau \simeq 10^{-7}$ sec and $\Delta v \simeq 1.4 \times 10^6\ \text{sec}^{-1}$, which is equivalent to $5 \times 10^{-5}\ \text{cm}^{-1}$. The natural bandwidth is very small, unless the lifetime of the upper state is reduced by some competing unimolecular decay process (for example, the process of pre-ionization: if a polyelectronic atom absorbs light of sufficiently short wavelength—say, in the x-ray region—it can be excited into a discrete level which has an energy lying above the first ionization limit. Instead of re-emitting the absorbed light, it may decay via a radiationless transition into the ionization continuum, where it immediately dissociates into a positive ion and an electron; this shortens the natural lifetime of the upper state and broadens the absorption line. Emission is prevented altogether if the rate of pre-ionization is sufficiently fast. The phenomenon is known as the Auger Effect, and is analogous to the predissociation of molecules into atoms (see section 2.2c).

(ii) *Doppler broadening*

The atoms in a vapour are in perpetual thermal motion; any movement of an absorbing or emitting atom along the line of sight of an external observer will lead to a slight alteration in the frequency of the observed spectral line. In a large assembly of atoms, where a Maxwell–Boltzmann distribution of velocities is maintained, there is a symmetrical broadening of the line, and its half-width is given by

$$\Delta v_{D} = \frac{v_0}{c}\sqrt{\frac{8RT \ln 2}{M}} \ \sec^{-1}$$

where v_0 is the central frequency and M is the atomic weight of the vapour. Doppler broadening thus increases with $T^{\frac{1}{2}}$ and decreases with $M^{\frac{1}{2}}$. For $Hg(6^3P_1)$ at 20°C, $\Delta v_D \sim 10^9 \ \sec^{-1} \equiv 3 \times 10^{-2} \ cm^{-1}$, a thousand times greater than the natural width (but still small, nonetheless). The extinction coefficient for a Doppler broadened spectral line varies as

$$\varepsilon_v = \varepsilon_{v_0} \exp \left\{ -\left(\frac{2(v - v_0)}{\Delta v_D} \right)^2 \ln 2 \right\}$$

where ε_{v_0} is the extinction coefficient at the central frequency v_0. This imposes a Gaussian profile on the spectral line.

(iii) *Pressure broadening*

If the interval between collisions with other atoms or foreign molecules in the vapour is less than the mean natural lifetime of the upper state, then under a classical description there will be a variation in the phase of the absorbed or emitted wave train. This reduces the length of the unperturbed wave train and increases the uncertainty in its frequency. The effect has been discussed by Lorentz, Holtsmark and others, and alternative models are clearly explained by Kuhn.[5] If there is a direct impact of short duration, there will be an abrupt change in the phase of the wave train, which is equivalent to a 'chopping' of the unperturbed wave. This leads to a symmetrical broadening of the spectral line about the central frequency. On the other hand, a 'near' collision of longer duration leads to a gradual and temporary change in the phase of the

wave train; the result is an asymmetric broadening of the line and a shift
of the central maximum to the red.

If there are n atoms or foreign molecules per cc, and the mean relative
velocity of the colliding atoms or molecules is $\bar{v}$, the interval between
collisions is

$$\Delta t = \frac{1}{2n\bar{v}\pi\sigma^2}$$

where $\pi\sigma^2$ is the mean collision cross-section. If the natural lifetime were
$\sim 10^{-8}$ sec (for example, for a fully allowed radiative transition), collisional
broadening would set in at ~ 10 torr. The larger the lifetime, the collision
cross-section and the pressure, the more susceptible is the atom to collisional perturbation. The collision cross-section depends on the form of
the interaction between the atom and its collision partner. If the interaction
is attractive and the collision is 'sticky', the broadening may be very
considerable ($\gg 0.1$ nm), since the absorbing or emitting 'atom' is in
effect a 'quasi-molecule'. Its energy levels will vary with the 'bond' length,
which varies with time, and the transition energies will spread about a
mean corresponding to the average bond length at the instant of the transition. If the excited atom encounters an ion, as is most probable if the vapour
is excited in an electrical discharge, the spectral lines are broadened by the
intense electrical field centred on the ion, i.e. the lines exhibit the Stark
Effect.

(iv) Self-absorption

If the spontaneous emission to the ground state has a high probability,
then the induced absorption from the ground state will also have a high
probability; the transition is associated with a resonance line. The absorption coefficients of such lines are very large, since all the absorption is
concentrated into a very narrow frequency range. Unless the pressure is
very low, emitted resonance radiation will be re-absorbed by unexcited
atoms within the confines of the containing vessel. The radiation is said
to be 'imprisoned', since the light cannot readily escape. If there is an
alternative route for the decay of the excited atom, (for example, a collisionally induced transition to an intermediate metastable state, such as
the 6^3P_0 level of mercury), the resonance line may be 'lost' during the
succession of emissions and reabsorptions within the vapour. The emission
line outside the vapour is said to be reversed.

Generally, the width of the emission line from a discharge lamp is
greater than the absorption lines of the ground state atoms, which occupy
the cooler regions of the lamp near the wall and away from the central
discharge. The nett result is that as the temperature and pressure of the

vapour in the discharge increases (because of increasing current density), the resonance emission at the central frequency disappears from view and only the broadened 'wings' of the line are emitted from the lamp. The effect on the emission profile is shown in Figure 2.4 (see plate II of plate section). A particularly striking example of this effect is shown in flash discharges through the rare gases in silica capillary tubes. The very high current density in the discharge results in broadening of all the emission lines, which merge into a very useful white light continuum. However, sharp, reversed atomic silicon absorption lines stand out against the continuum (see Figure 2.5 (plate II of plate section)). The discharge is hot enough to vaporize the silica surface (melting point, 1710°C), and dissociate some of the vapour into silicon atoms, which absorb light emitted from the intensely hot centre of the discharge.

2.2   DIATOMIC MOLECULAR SPECTRA

A broad classification distinguishes banded, diffuse and continuous electronic spectra; the natural width of the spectral lines increases as the lifetime of the upper state is reduced, and diffuseness results from the overlap of adjacent lines. The classification therefore relates to the lifetime or stability of the upper state. In addition, one can distinguish between intra- and extravalency shell transitions. The former involve molecular orbitals constructed from intravalency shell atomic orbitals, while in the latter there has been an increase in the principal quantum number of a component atomic orbital. The resulting absorption bands are associated with Rydberg transitions and generally these lie in the far ultraviolet. We shall only concern ourselves with the intravalency shell transitions lying at longer wavelengths, where until very recently the great bulk of photochemical studies have been made.

2.2a   Band spectra

If the combining electronic states are both attractive, then each one will have discrete vibrational and rotational energy levels. An electronic transition in any given molecule will transfer it from the rotational and vibrational level it occupied in the initial state to some other rotational and vibrational level of the final electronic state. Assuming that the electronic, vibrational and rotational motions are separable, the nett change in the energy of a diatomic molecule effected by absorption of light of frequency $\tilde{v}$ (cm^{-1}) is

$$\Delta E = hc\tilde{v} = \Delta E_e + \Delta E_v + \Delta E_r$$

Since $\Delta E_e \simeq 10^2 \Delta E_v \simeq 10^4 \Delta E_r$, the gross structure of the spectrum of a large assembly of molecules is associated with the vibrational transitions, and the detail is provided by the rotational fine structure.

To a first approximation the vibrational energy levels of a given electronic state are given by

$$E_v = (v + \tfrac{1}{2})h\omega_e - (v + \tfrac{1}{2})^2 x_e h\omega_e$$

(see section 1.4a). Transitions between vibrational levels v'' in the lower state and v' in the upper state will be associated with spectral lines of frequency (in cm^{-1}) given by

$$\tilde{v}_{v'v''} = \tilde{v}_e + [(v' + \tfrac{1}{2})\tilde{\omega}_e' - (v' + \tfrac{1}{2})^2 x_e'\tilde{\omega}_e'] - [(v'' + \tfrac{1}{2})\tilde{\omega}_e'' - (v'' + \tfrac{1}{2})^2 x_e''\tilde{\omega}_e'']$$

where $\tilde{\omega}_e = \omega_e/c$ and $hc\tilde{v}_e$ refers to the energy difference between the minima of the two potential energy curves. In practice, of course, it is not possible to observe this energy difference directly because of the zero point energies; the nearest approach is the (0,0) transition for which

$$\tilde{v}_{0,0} = \tilde{v}_e + \frac{1}{2}\left[\tilde{\omega}_e'\left(1 - \frac{x_e'}{2}\right) - \tilde{\omega}_e''\left(1 - \frac{x_e''}{2}\right) \right]$$

The frequency $\tilde{v}_{v'v''}$ can now be expressed in the form

$$\tilde{v}_{v'v''} = \tilde{v}_{0,0} + a'v' - b'(v')^2 - a''v'' + b''(v'')^2$$

where $a = (1 - x_e)\tilde{\omega}_e$ and $b = x_e\tilde{\omega}_e$. If the molecule is excited from the zero vibrational level of the lower state, the vibrational bands form a simple progression at frequencies*

$$\tilde{v}_{v'0} = \tilde{v}_{0,0} + a'v' - b'(v')^2$$

and the vibrational spacing decreases with v'. There is no restriction on the magnitude of Δv and the relative intensities depend on the relative equilibrium bond lengths in the initial and final states. The discussion of the Franck–Condon Principle in section 1.5b emphasized this point; the first allowed electronic transitions of $NH(\tilde{X}^3\Sigma)$ and $O_2(\tilde{X}^3\Sigma_g^-)$ were cited as examples where $\Delta r_e \simeq 0$, and $\gg 0$ respectively. Two further examples are provided by $CS(\tilde{A}^1\Pi \leftarrow \tilde{X}^1\Sigma)$ and $ClO(\tilde{A}^2\Pi_i \leftarrow \tilde{X}^2\Pi_i)$. Their absorption spectra are shown in Figures 2.6 and 2.7 (see plate III of plate section). In CS the (0,0) band is by far the strongest; the vibrational

---

* This assumes the potential energy curve of the upper state approximates to the Morse function. If the state is 'perturbed' by a third electronic state, whose potential energy curve approaches or perhaps crosses it, the vibrational spacings in the upper state will not form a regular progression. Indeed, such irregularities indicate vibronic interaction between neighbouring electronic states.

progression based on $v'' = 0$ rapidly loses intensity and no further absorption can be detected beyond the (2,0) transition. $\Delta r_e = 0.039$ Å only. In Figure 2.6(b) the CS has been produced in a state of high vibrational excitation and many levels above $v'' = 0$ are populated. The extended sequence of bands with $\Delta v = 1$ and 2 correspond to excitation from levels with $v'' \leqslant 7$, in the ground electronic state. They are termed 'hot' bands and their intensity depends on the vibrational temperature. If the latter is high, this means that the fractional population of levels with $v'' > 0$ is large. If there is a Boltzmann distribution among the available levels, then $n_{v''}/n_{v''=0} = \exp[-(E_{v''} - E_{v''=0})/RT]$ where T is the vibrational temperature.

The ClO radicals occupy the ground vibrational level, but since $^2\Pi$ electronic states have components of both spin and orbital electronic angular momentum about the internuclear axis, they are split into two components, $^2\Pi_{\frac{3}{2}}$ and $^2\Pi_{\frac{1}{2}}$ with the first the lower. Using Mulliken's notation (see section 1.3a), the m.o.'s formed by overlap of the $2s$ and $2p$ a.o.'s of the oxygen atom with the $3s$ and $3p$ a.o.'s of the Cl atom, are filled in the order

$$\underbrace{(z\sigma)^2(y\sigma^*)^2}_{\substack{2s\text{–}3s \\ \text{overlap}}} \ \underbrace{(x\sigma)^2(w\pi)^4(v\pi^*)^3}_{\substack{2p\text{–}3p \\ \text{overlap}}}, \ \tilde{X}^2\Pi_i$$

in the ground state. The first electronic transition, $v\pi^* \leftarrow w\pi$, promotes an electron into an antibonding π-orbital and the excited $\tilde{A}^2\Pi_{\frac{3}{2}}$ state should have a greatly increased equilibrium bond length. Experimentally it is found that the ultraviolet absorption associated with the transition consists of a long progressions of bands, with a maximum intensity into $v' = 12$:[6] from an analysis of the rotational fine structure $\Delta r_e = 0.24$ Å.[7] There is also a rather weaker progression of bands, originating from the zero vibrational level of the low-lying metastable $\tilde{x}^2\Pi_{\frac{1}{2}}$ state of ClO, and populating the $\tilde{a}^2\Pi_{\frac{1}{2}}$ state.[6]

For a given electronic and vibrational (or vibronic) transition, associated with a frequency $\tilde{v}_e + \tilde{v}_v$, the change in total energy due to a rotational transition between two levels J'' and J' is

$$\tilde{v}\,(\text{cm}^{-1}) = \tilde{v}_e + \tilde{v}_v + B'J(J' + 1) - B''J''(J'' + 1)$$

where B'' and B' are the rotational constants in the initial and final states. In general these will be different, since the electronic binding, and hence the equilibrium geometry, is different in the two states (to be precise, the rotational constants are also functions of the occupied vibrational level, because of the anharmonicity of the molecular vibrations; they are even functions of the rotational energy if the effects of centrifugal distortion of the molecule are taken into account).

Electronic transitions with $\Delta\Lambda = 0$ (for example, $\Sigma \longleftarrow \Sigma$) are polarized parallel to the internuclear axis (see p. 73); for such transitions the rotational selection rule is $\Delta J = \pm 1$. The rotational fine structure has two branches, termed P ($\Delta J = -1$) and R ($\Delta J = 1$). If the transition has $\Delta\Lambda = \pm 1$, its dipole lies perpendicular to the internuclear axis and the selection rule is now $\Delta J = 0, \pm 1$: a Q-branch ($\Delta J = 0$) lies between the P and R-branches. Their frequencies are

$$\tilde{v}_P = \tilde{v}_e + \tilde{v}_v - (B' + B'')J + (B' - B'')J^2$$

$$\tilde{v}_Q = \tilde{v}_e + \tilde{v}_v + (B' - B'')J + (B' - B'')J^2$$

$$\tilde{v}_R = \tilde{v}_e + \tilde{v}_v + 2B' + (3B' - B'')J + (B' - B'')J^2$$

In the unlikely event that $B' \simeq B''$ (as is the case in the ultraviolet absorption of NH), the rotational frequencies of the P and R-branches will be regularly spaced, with a separation of $2B$, while all the lines in the Q-branch will be superimposed at the origin, to give an intense 'single' line. The P and R-branches will progress toward longer and shorter wavelengths, respectively (see Figure 2.8 (plate IV of plate section)). The absorption profile is identical to that associated with a transition promoted by infrared absorption between vibrational and rotational levels in the same electronic state. In the far more likely event that $B' < B''$, the rotational lines will not be regularly spaced. Those of the R-branch, for example, initially march in decreasing steps toward the violet, but when J reaches such a value that the negative quadratic term exceeds the positive linear term in J, they march back toward the origin. This point is best shown on a Fortrat diagram, in which the frequencies of the rotational lines are plotted against J (see Figure 2.9). The curves are all parabolas: the projections of the points on the abscissa indicate the rotational structure. Those of the R-branch form a band-head, where both the lines and the intensity pile up. The vibrational band frequencies are often measured at the heads, on the assumption that in comparison with the vibrational spacing the error so introduced is negligible. If the rotational structure is unresolved, it is not possible to locate the band origin accurately.

The general drift of the structure is toward the red, if $B' < B''$; i.e. an increase in bond length degrades the structure to the red, while a decrease degrades it to the violet. In absorption the former situation is the more common.

If either (or both) of the combining electronic states have spin or orbital electronic angular momentum, then the momenta may couple not only with each other, but also with the rotational angular momentum of the

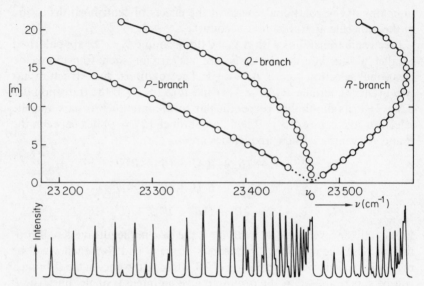

Figure 2.9 Fortrat diagram showing the P, Q and R-branches in the rotational structure of AlH (424·1 nm band—transition $^1\Pi \rightarrow {}^1\Sigma^+$). (From R. N. Dixon, *Spectroscopy and Structure*, Methuen, London, 1965.)

molecule itself. This leads to multiplet splitting of the rotational fine structure, detectable under high resolution. Possible coupling cases have been distinguished by Hund and are known as Hund's cases (a–d). The relevant nett angular momentum selection rules depend on which particular case is operative, which itself depends on the nature of the molecule and the electronic states it occupies.

Case (a) applies to molecules containing at least one heavy atom, when the splitting of energy levels due to spin–orbit coupling, is large compared to the separation between rotational energy levels. The components of spin and orbital angular momentum add vectorially to give a resultant Ω, which combines with the rotational angular momentum components to give nett angular momenta $J = \Omega, \Omega + 1 \ldots$. The rotational energy is then given by the equation

$$F(J,\Omega) \simeq B\{J(J + 1) - \Omega^2\}$$

(Note that J now includes the electronic angular momentum as well; the molecular rotational quantum number is symbolized by N (nuclear) under these circumstances.) It cannot apply in singlet ($S = 0$) or Σ, ($\Lambda = 0$) states, because spin–orbit interaction is impossible.

Case (b) applies when the spin–orbit coupling is very small. The electronic orbital angular momentum now couples with the molecular rotation to give resultants symbolized by the quantum number $K = \Lambda$,

$\Lambda + 1 \ldots$. The total angular momentum J is the resultant of K and the spin S, $(K + S) \geqslant J \geqslant (K - S)$. This case applies to $\Pi, \Delta \ldots$ states of molecules composed of light atoms, and in contrast to case (a) to all molecules in Σ, $(\Lambda = 0)$ states.

Case (c) is analogous to (j, j) coupling in atoms, but this and the final case are relatively unimportant, and the reader is referred to Herzberg for the finer details of the coupling between the molecular and electronic rotational motion.[8]

The relative intensities of the vibrational bands and of their rotational structure depend on the relative populations of the vibrational and rotational levels in the absorbing (or emitting) molecules, and on the transition probabilities between the levels of the upper and lower states. At thermal equilibrium the populations are determined by the Boltzmann distribution corresponding to the temperature of the surrounding molecules. The higher the temperature the higher the quantum numbers of populated levels. Since transitions into high rotational levels of the final state are possible only from high rotational levels of the initial state (ΔJ cannot exceed ± 1), the rotational structure is most strongly developed when the temperature is greatest. For example, the emission spectra of radicals in flames show far more rotational structure than in absorption at room temperature (see Figure 2.10 (plate IV of plate section)). In such an environment, the distribution may not correspond to an equilibrium Boltzmann distribution, and if diatomic species are excited in a chemical reaction, rather than by energy transfer in collisions, a distribution according to the Boltzmann equation will border on the miraculous. The same is true of the vibrational level populations; spectral analysis of chemiluminescence or transient absorption of excited molecules or radicals produced in chemical reactions gives a deep insight into the distribution of the energy liberated in elementary exothermic steps.

While the spacing of the lower vibrational levels is usually $\sim RT$ at room temperature, the lower rotational levels are separated by very much smaller energies. When there is a Boltzmann distribution, the majority of molecules occupy the ground vibrational level. The relative rotational populations are given by

$$\frac{n_J}{n_{J=0}} = (2J + 1) \exp\left[-\frac{hcBJ}{RT}(J + 1) \right]$$

where $2J + 1$ is the statistical weight of the J'th level. Since the lower rotational levels are so closely spaced, the increase in the statistical weight initially outweighs the decrease in the exponential factor with increasing J, and the maximum population might lie at $J \sim 5$, say. Schematic distributions are represented in Figure 2.11.

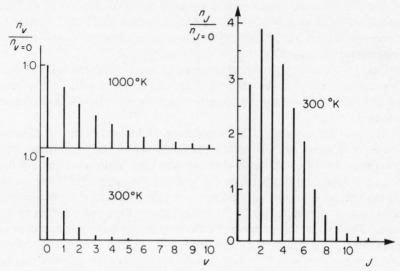

Figure 2.11　Vibrational and rotational Boltzmann distributions.

The transition probabilities for vibrational bands are determined by the Franck–Condon Principle, and have already been discussed. In the case of rotation, the transition probabilities are restricted by the selection rule for ΔJ, but within that restriction, they are effectively constant being independent of both ΔJ and J, provided the molecule is not homonuclear. If it is (i.e. if it has a centre of symmetry), the total wave function of the molecule must be either symmetric or antisymmetric with respect to exchange of the two nuclei (i.e. inversion through the centre of symmetry). When this operation is performed account must be taken of the spin of the nuclei; the property of spin is possessed not only by electrons but also by the protons and neutrons of the nucleus, all of which possess it in units of $\frac{1}{2}\hbar$.* They are all Fermi particles and their wave functions, like those of the electron, are antisymmetric with respect to particle exchange (Pauli Principle). On the other hand, when the nett nuclear spin only has components which are *integral* multiples of $\hbar$, the nucleus behaves like a Bose particle, whose wave function is symmetric to particle exchange (i.e. inversion through the centre of symmetry). Thus the proton, with spin $\frac{1}{2}\hbar$* is a Fermi particle, ^{16}O and ^{2}H with zero and unit nuclear spins respectively, are Bose particles.

The nett molecular wave function is given by the product $\psi_e\psi_v\psi_r\psi_n$, where ψ_n is the nuclear spin wave function. If we take the case of the H_2

---

* Along any given direction.

molecule, whose nuclei are Fermi particles, the nett wave function must be antisymmetric with respect to exchange of the two protons. Since they both have spin $\frac{1}{2}$, they may combine to produce either a singlet nuclear spin state (nett spin zero) or a triplet spin state (nett spin unity). The first two states are respectively antisymmetric and symmetric toward exchange of nuclear coordinates (inversion through the centre of symmetry) (see p. 27). If the *nett* wave function is to be antisymmetric toward exchange of the two nuclei, as Nature requires, the singlet state can combine only with symmetric, and the triplet state only with antisymmetric rotational levels. These turn out to be those with J even and J odd respectively; thus (provided the electronic and vibrational parts of the wave function are symmetric) rotational levels with J even have unit nuclear statistical weight, while those with J odd have a weight of three and the intensities of successive rotational lines will alternate in the ratio $3:1$.

2.2b Continuous spectra

If a diatomic molecule is excited into an electronic state where it has a potential energy sufficient for it to dissociate into its component atoms, the associated absorption spectrum will be continuous. If the upper state is attractive, the vibrational bands converge with decreasing wavelength and merge into a continuum at the dissociation limit (cf. the ionization limit). If it is repulsive, the absorption spectrum is continuous at all wavelengths. Since the dissociating molecule flies apart during the first 'vibration' there can be no vibrational or rotational structure: the molecule has no time to execute either motion and its potential energy is converted into the translational kinetic energy of the separating atoms.*

These possibilities are well illustrated by electronic transition in the halogens and their hydrides. Four variations on the theme are shown in Figures 2.12 (a–d). In the first three, the upper state is attractive, but its equilibrium bond length is much greater than in the ground state. Because of the Franck–Condon Principle the larger the extension, the greater the potential energy in the upper state, and the more likely is it to exceed the dissociation limit. The less the wave function of the vibrational level $v'' = 0$ overlaps those of the bound vibrational levels in the upper state, the less will be the intensity and extent of the banded part of the spectrum. In the final figure, the absorption is entirely continuous.

If all the absorbing molecules occupy the ground vibrational level, the relative intensities of the vibrational bands $(v', 0)$, depend on the Franck–

---

* At sufficiently high energies, the molecule may be excited into an electronic state lying above the ionization limit. A series of Rydberg bands converge at the ionization limit. An alternative possibility at high energies is the dissociation $AB + h\nu \rightarrow A^+ + B^-$.

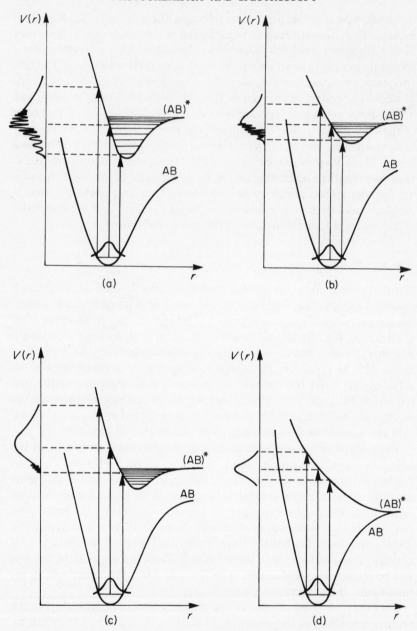

Figure 2.12 Schematic potential energy curves for electronic transitions where $\Delta r_e > 0$ or $\Delta r_e = \infty$. (a), (b), (c) appropriate for $^3\Pi_{0u}^+ \leftarrow {}^1\Sigma_g^+$ transitions in I_2, Br_2, Cl_2; (d) appropriate for $^1\Pi_1 \leftarrow {}^1\Sigma^+$ transitions in HI or HBr.

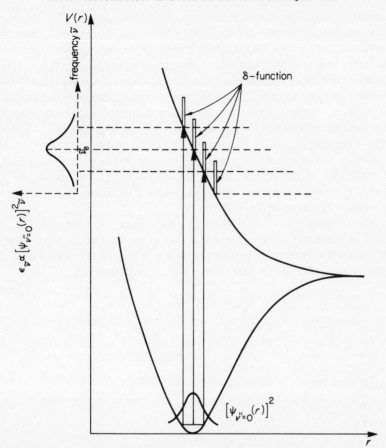

Figure 2.13 Construction of repulsive potential curve assuming its 'vibrational' wave functions to be represented by a series of 'δ-functions'. (N.B., the δ-function ought to be infinitely narrow; in practice it has been treated as a step function with a constant amplitude over a narrow interval Δr, say 10^{-3} nm.)

Condon factors $[\int \psi_v^* \psi_{v''=0} \, dr]^2$. When the upper state is continuous, it has been found practicable to represent the 'vibrational' wave functions $\psi_{v'}$ by a δ-function,[9] which has an amplitude only at the internuclear separation corresponding to the classical turning point (see p. 61 and Figure 2.13). If the function is normalized, it will have the same value for each level in the continuum; for a transition into any given level in the upper state the vibrational overlap integral reduces to a value proportional to $\psi_{v''}$ measured at the internuclear distance r_v corresponding to the absorbed frequency v. The Franck–Condon factor reduces to $[\psi_{v''=0}(r_v)]^2$, and the extinction coefficient at frequency v is (see p. 76)

$$\varepsilon_v \propto v[\psi_{v''=0}(r_v)]^2$$

The absorption profile is effectively a reflection of $[\psi_{v''=0}]^2$ from the upper potential energy curve, slightly weighted by the absorption frequency. If the absorption spectrum is entirely, or very largely continuous, the 'reflecting' section of the curve is usually close to linear. In this situation the absorbed energy $h\nu$ decreases linearly with the internuclear separation of the absorbing molecule, and over a moderate frequency range ($< 5000\ cm^{-1}$) closely approximates the bell-shaped, Gaussian profile of the ground vibrational wave function; ($\psi_{v''=0} \propto \exp -\{\beta(r - r_e)^2\} \propto \exp -\{constant(\tilde{\nu}_e - \tilde{\nu})^2\}$ where $\beta(cm^{-1})$ is a constant, and $\tilde{\nu}_e$ is the frequency at the absorption maximum, corresponding to the excitation of molecules at the equilibrium bond length r_e). This is shown for the near ultraviolet absorption of Cl_2 in Figure 2.14, where $\log \varepsilon_\nu$ is seen to be a linear function of $(\tilde{\nu}_e - \tilde{\nu})^2$.

When higher vibrational levels are populated, the absorption in the wings is increased at the expense of that at the maximum. This reflects the transfer of molecules from $v'' = 0 \rightarrow 1$, where $(\psi_{v''})^2$ is zero near the equilibrium position. In general, the nett extinction coefficient is

$$\varepsilon_\nu = \sum_{v''=0}^{v''} N_{v''}\varepsilon_{v''}$$

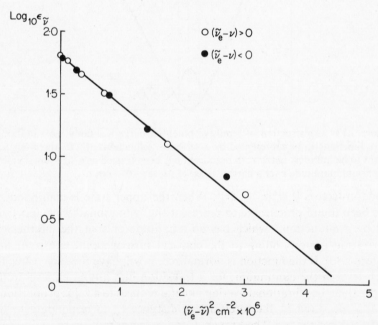

Figure 2.14 Quadratic dependence of $\log \varepsilon_{\tilde{\nu}}$ on $(\tilde{\nu}_e - \tilde{\nu})$ for the $^1\Pi_1 \leftarrow \tilde{X}^1\Sigma_g^+$ transition in Cl_2. ($\tilde{\nu} = 30,500\ nm^{-1}$.)

where $N_{v''}$ and $\varepsilon_{v''}$ are the fractional populations and extinction coefficients of molecules in vibrational levels v''. An increase in temperature has the effect, both of spreading, and 'flattening' the absorption continuum.

The frequency range of a continuum depends on the slope of the reflecting section of the repulsive upper potential energy curve. If it is very shallow, the continuum has a small spread, and the long wavelength limit of the absorption is fairly sharp, allowing a reasonable estimate of the minimum energy required to dissociate the molecule in its upper state. A steep curve generates a broad spectrum: the long wavelength limit is ill-defined and only permits an estimate of the upper limit of the dissociation energy. It is also possible that such a curve actually has a shallow minimum, which is inaccessible by excitation from the ground vibrational level, because of the Franck–Condon restriction. In general, it is impossible to locate the $(0, 0)$ transition with any certainty from an absorption spectrum which is largely or entirely structureless.

So far the discussion has assumed an attractive lower state; however, continuous absorption will also result from an electronic transition between a repulsive ground state and an attractive (or repulsive) upper state. 'Ah', you say, 'how can there be a repulsive ground state?' The answer is that we have already encountered one, in the discussion of the collisional broadening of spectral lines (p. 92). For example, if the pressure of mercury or cadmium vapour is increased sufficiently, absorption continua develop which are associated with the excitation of collision complexes of the atoms in their ground electronic states.[8] The ns and np atomic orbitals overlap to form bonding and antibonding combinations, and if the two atoms occupy the n^1S_0 ground state, the dimer has the electronic configuration $(\sigma_g ns)^2(\sigma_u^* ns)^2, {}^1\Sigma_g^+$; the bonding and antibonding orbitals are both filled and there is no nett bonding so that the ground state is repulsive. On the other hand, the first excited state can be formed by combination of one normal (n^1S_0) and one excited (n^3P_1) atom, which produces an excited dimer with the electronic configuration $(\sigma_g ns)^2(\sigma_u^* ns)^1(\sigma_g np)^1, {}^3\Sigma_u^+$; this has only one electron in an antibonding orbital and should lead to a nett attraction.

Many emission continua are known which are associated with the decay of excited dimers into non-bonding ground states, both in the vapours of metal and inert gas atoms, and in their mixtures. Those of the inert gases provide useful white light sources in the vacuum ultraviolet stretching from the limits of 60 nm (He_2^*) to 210 nm (Xe_2^*). Excited dimers are also of importance in the photochemistry of fluorescent polyatomic molecules in solution, where the term 'excimer' has gained popular acceptance (see p. 207).

2.2c Diffuse spectra: predissociation

It sometimes happens that the potential energy curve of an electronic state, into which a molecule is excited by light absorption, is crossed by that of another; if the excited molecule vibrates through the region of the crossing point, it may transfer into the second electronic state in a radiationless transition. If the absorbed energy exceeds the dissociation limit of the second state, the molecule will fly apart.

The phenomenon was discovered by Henri in his study of the banded ultraviolet absorption spectrum of $S_2(\tilde{X}^3\Sigma_g^-)$ vapour.[10] There are slight perturbations in the rotational structure at the long wavelength end of the $(v', 0)$ progression, and around the $(10, 0)$ band at 280 nm, the rotational lines broaden and partially merge, although the vibrational band structure remains clear. Resolved rotational structure reappears beyond the $(13, 0)$ band but disappears completely when $v' > 17$. Henri suggested that the 'rotational blurring' was caused by delayed dissociation in the upper state, which set in when the molecule was excited into the appropriate vibrational levels. He termed it *predissociation*; the phenomenon is best explained by reference to the potential energy diagram (Figure 2.15 (see plate V of plate section)).

The primary transition $(\tilde{B}^3\Sigma_u^- \leftarrow \tilde{X}^3\Sigma_g^-)$ populates a stable upper state. At some point on its potential energy curve, probably near to the terminus of the seventeenth vibrational level, the curve is crossed by a weakly attractive $^3\Pi_u$ state, into which a radiationless transition can occur.[12] If the transition occurs in $\sim 10^{-11}$ sec, i.e. in a time of the order of the rotational period, the rotational structure will be ill-defined, although the vibrational structure will remain clear. The normal (radiative) lifetime of the molecule is $\sim 10^{-7}$ sec so that in the absence of the competing radiationless transition, both rotational and vibrational energies will be quantized. The natural line width in this situation will be

$$\Delta\tilde{v} = \frac{1}{2\pi c\tau} = 5 \times 10^{-5}\,\text{cm}^{-1}$$

This is about one thousandth the average Doppler width of the spectral lines, and broadening through predissociation will not be evident until it occurs in times $\leqslant 10^{-10}$ sec. If it occurs in 10^{-12} sec (after ~ 10 vibrations in the initially populated state), the natural line width is ~ 5 cm^{-1}, which is far greater than the rotational constants of S_2 in either state (< 0.3 cm^{-1}), and the rotational structure disappears. When the specific rate of predissociation is $< 10^{10}$ sec^{-1}, it cannot be detected by line broadening; its diagnosis requires other methods, such as the detection of atomic reactions or the sudden disappearance of fluorescent emission beyond a particular vibrational band. Fluorescence lifetimes are typically $\sim 10^{-8}$ sec.

The probability of predissociation depends on

(i) the selection rules for radiationless transitions: Kronig's rules,

(ii) the Franck–Condon Principle, and

(iii) the energy of the crossing point in relation to the ultimate dissociation limit.

The selection rules require that the total spin and orbital symmetries (or angular momenta) of the intersecting states be unchanged in the course of the radiationless transition. In consequence $+ \leftrightarrow -$ transitions, and $u \leftrightarrow g$ transitions in centro-symmetric molecules are forbidden (in contrast to radiative transitions where $u \leftrightarrow g$ is allowed), as are those with ΔS and $\Delta \Lambda \neq 0$. However, conservation of spin angular momentum is not essential if there is strong spin–orbit interaction (heavy atoms, Hund's case (c)), and intersystem crossings with $\Delta S = \pm 1$ may occur. Similarly conservation of the orbital angular momentum is relaxed by coupling of the molecular and electronic rotational motion. The greater the molecular rotational quantum number, the more probable are the transitions in which $\Delta \Lambda = \pm 1$ (in polyatomic molecules the rigorous electronic selection rules can be relaxed by vibronic coupling: this is impossible in diatomic molecules, since the vibrations cannot be other than totally symmetric and each vibronic state must have the same symmetry as the parent electronic state). On the basis of Kronig's rules, the spontaneous transition in S_2 at $v' \geqslant 17$ is allowed (except in the unlikely event $J' = 0$).

Assuming the radiationless transition is allowed by the selection rules, it is still necessary that the molecule vibrates into the neighbourhood of the crossing point if the transition is to occur. It cannot readily transfer from one state to the other, unless the vibrational overlap integral between the two states is large, as required by the Franck–Condon Principle.

Since the vibrational wave functions extend some distance outside the boundary of the classical potential energy curve there will still be some overlap when the vibrational energy lies a little above or below the classical crossing point, though the transition probability will fall in both directions. Although the potential energy of the molecule may lie a little below that required classically, it may still transfer to the other state by 'tunnelling' through the potential barrier. With these thoughts in mind, we can return to the discussion of the predissociation in S_2.

Fair and Thrush[11] have suggested that the weaker predissociation maximum which develops around $v' = 10$ corresponds to an appreciable vibrational overlap of the secondary maxima which lie *within* the terminal loops of the vibrational wave functions (see Figure 2.16). This avoids the

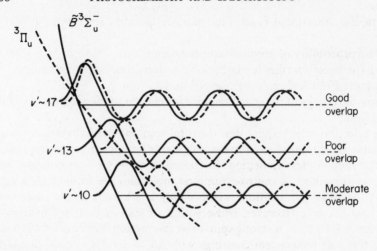

Figure 2.16   Schematic representation of primary and secondary overlap maxima in the predissociation of S_2 (cf. Figure 2.39).

postulate of a relatively inefficient, spin-forbidden predissociation into a $^1\Sigma_u^-$ state, originally postulated by Lochte–Holtegreven on experimental evidence now known to be of doubtful validity.[12] On this basis, it is suggested that the predissociating $^3\Pi_u$ state crosses the inner limb of the $\tilde{B}^3\Sigma_u^-$ state around $v' = 18$, with the two curves running closely together down to $v' \simeq 9$. Since the $^3\Pi_u$ state correlates with two normal sulphur atoms (3^3P), while the $\tilde{B}$ state correlates with one normal and one excited (3^1D) atom (see section 2.3c), the two curves must cross again (see Figure 2.15). This would account for the perturbations near the origin of the $(v', 0)$ progression. Herzberg has distinguished type Ia, Ib and Ic predissociations,[8] depending on whether the predissociation limit lies near, above or below the crossing point, on the right hand limb (see Figure 2.17). The predissociation limit in S_2 clearly lies near $v' = 9$, and the construction of Figure 2.15 shows that the predissociation of S_2 falls into category Ib.

The wave mechanical notions can be reformulated in a crude classical analogy. Imagine you are a little demon, oscillating back and forth in the potential well of the initially excited state. As you shoot up the slope you encounter a spring-loaded trap-door, leading to another slope. If the spring is weak the door opens readily (transition allowed) but if it is stiff it will take some time to open (transition forbidden). The longer you spend sitting on it, the more chance you have of falling through (Franck–Condon factor). If you never reach the trap-door, you will never get through (unless you dig a hole in the wall—the tunnel effect), and if you shoot past it too quickly, you will not give the door sufficient time to open.

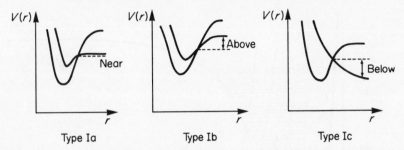

Figure 2.17 Possible types of predissociation in diatomic molecules through intersection of potential energy curves.

But if the door marks the limit of your oscillation you will sit on it for some time and have the best chance of making the transition.

When radiationless transition is forbidden by the electronic selection rules, the rate at which it occurs may be increased by collisions with surrounding molecules. If dissociation is promoted in this way it is termed *induced predissociation*. For example, the banded absorption in iodine, which gives the vapour its violet colour is associated with the transition

$$I_2(\tilde{B}^3 \Pi_{0,u}^+)_{v',J'} \longleftarrow I_2(\tilde{X}^1\Sigma_g^+)_{v'',J''}$$

At low pressures, the reverse transitions give rise to fluorescence but some of the excited molecules predissociate, since iodine atoms can be detected through their optical[13] and electron spin resonance[14] spectra; the two processes are in competition. If increasing pressures of foreign gases are added, or if the iodine pressure itself is increased, the fluorescence originating from $v' \geqslant 7$ is reduced: it is said to be *quenched*. The potential energy curve of the $\tilde{B}^3 \Pi_{0,u}^+$ state is crossed at vibrational levels near and above $v' = 7$, by those of other electronic states which are repulsive, but into which radiationless transitions are symmetry-forbidden (see Figure 2.18). The restrictions are relaxed by the van der Waals interaction with colliding molecules, which introduces a perturbation into the Hamiltonian energy operator and modifies the molecular wave functions of the excited molecule.

Many years ago it was reported that the application of a magnetic field also quenched the $\tilde{B} \longrightarrow \tilde{X}$ fluorescence of I_2 and it was concluded that the selection rules for predissociation were relaxed by magnetic interactions.[15a] Recent quantitative measurements have confirmed the effect,[15b] which requires fields in the range $\geqslant 10$ kG. The magnetic field couples the $\tilde{B}$ state $((O_u^+)$ in the symbolism appropriate to j,j coupling) with a repulsive state (O_u^-), which dissociates into two unexcited $I(5^2P_{\frac{3}{2}})$ atoms. The coupling must be small since increasing magnetic fields do not

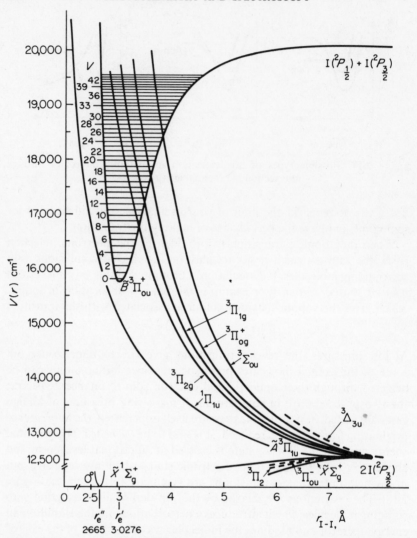

Figure 2.18 Electronic states of iodine, showing location of repulsive states responsible for quenching by induced predissociation, relative to the excited bound state. (Taken from J. I. Steinfeld, *J. Chem. Phys.*, **44**, 2740, (1968).)

cause any measurable increase in the rotational line widths of the $\tilde{B} \leftarrow \tilde{X}$ absorption bands.[15c]

2.3 DECAY AND DEACTIVATION OF EXCITED ATOMS AND DIATOMIC MOLECULES

Photochemistry is concerned with the chemical reactions which follow absorption of light by atoms, molecules, ions or complexes. The stages of the reaction can be divided into primary and secondary processes, with the former relating to all the elementary steps which involve the primarily excited species, and the latter to all the chemistry which subsequently follows. If the light is absorbed by an atom the excited atom may

(i) re-emit some or all of its energy as fluorescence or phosphorescence,

(ii) transfer some or all of its energy to other atoms or molecules with which it may collide,

(iii) extract an atom from, or form a complex with a colliding molecule, or

(iv) ionize.

The first possibility represents partial or complete 'wastage' of the absorbed energy, from the point of view of any subsequent chemical change. On the other hand, if one wants to find out what the excited atom is 'up to', and in particular, how efficiently it is doing it, it is very useful to have it signalling its whereabouts without having to look for it. Any 'chemical' primary process which competes with re-emission will reduce its intensity and also the lifetime of the excited atom. Thus the measurement of the quenching of the emission as the pressure of a foreign substrate is increased, allows the efficiency of step (ii) (which may lead to chemistry), and step (iii) (which already is chemistry) to be estimated.

The same is also true of diatomic molecules, but in their case dissociation, predissociation, induced predissociation and radiationless transition back into the ground electronic state provide alternative primary processes; in principle, fluorescence may compete with all but the first of these. The primary processes can be separated into those which are physical, and only involve a redistribution of the energy in the system as the atoms or molecules within it change their quantum states, and those which are chemical and involve a re-grouping of the atoms and molecules in the system. Each is in competition with all the others, and a study of the one requires a study of them all.

2.3a Fluorescence and phosphorescence in the gas phase

The simplest route for the decay of an electronically excited state is spontaneous re-emission of the absorbed energy. If this involves a reverse transition back to the level from which it was excited, then the emitted frequency will be identical with that absorbed: it is resonance radiation. If the excited atom or molecule can lose part of its energy within its radiative lifetime and then make a radiative transition into the ground electronic state, the emitted radiation is termed fluorescence. When the emission is associated with a spin-forbidden transition the interval between absorption and re-emission will be extended and it is called phosphorescence. Resonance radiation is only excited if the pressure is low, and the excited atom or molecule remains isolated during its radiative lifetime. If it suffers collisions in this period, some of its energy may be transferred and the resonance emission is quenched. Consider the resonance emission in mercury vapour. The spectral line at 253·7 nm is associated with the transition $6^3P_1 \rightarrow 6^1S_0$ (allowed under (j,j) coupling). If the mercury vapour pressure is increased the emission is rapidly quenched, since the resonant radiation is 'imprisoned' by reabsorption; the apparent life of the 6^3P_1 state is increased and, with it, the chance of collisional deactivation into the metastable 6^3P_0 state. It has a mean radiative lifetime ($\sim 10^{-2}$–10^{-3} sec) and unless it is deactivated by collision with the walls of the containing vessel, its decay into the ground electronic state is associated with a phosphorescent emission at 265·4 nm. A similar effect occurs when foreign gases such as N_2 are added to mercury vapour; it will be discussed in the next section.

Following the Franck–Condon Principle, a diatomic molecule excited in an electronic transition will often be endowed with vibrational energy also. This is represented in Figure 2.19 where the equilibrium internuclear bond length has increased in the upper state. The most probable electronic transitions in absorption populate excited vibrational levels in the upper state. Its radiative lifetime is $\sim 10^{-8}$ sec if the electronic transition back to the ground state is fully allowed, and longer if it is not. Since the radiative lifetime for a transition into a lower *vibrational* level within the excited electronic state is typically $\sim 10^{-3}$ sec, the isolated molecule will remain in the same vibronic level during its fluorescent lifetime. It may be that the internuclear separation at the instant of emission will also be the same as it was at the instant of absorption, but since it will have performed at least 10^5 oscillations in that time, there is a fair chance that it is not! The chance is given by the square of the amplitude of the vibrational wave function in the level it occupies; in upper vibrational levels this has large maxima near both the classical turning points (see Figure 2.19) which allows the possibility of both resonant and fluorescent emission, even in an isolated molecule at low pressure. In these circumstances, fluorescent decay populates upper vibrational levels of the ground electronic state.

At a pressure of one atmosphere, the mean interval between collisions for any given molecule is typically $\sim 10^{-10}$ sec, and the excited molecule will suffer many of them during its radiative lifetime, particularly if this exceeds 10^{-8} sec. Any vibrational or rotational quanta with which it was initially endowed can be transferred to the molecules with which it collides (and perhaps be transformed into translation and rotation in the process): if the excited molecules suffer enough collisions before they make up their

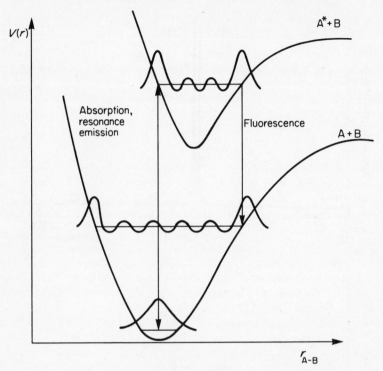

Figure 2.19 Fluorescence of an isolated diatomic molecule excited by monochromatic light.

minds to return home, there may be none of them left in the initially
populated vibrational (and rotational) level, and no light will be emitted
at the resonant frequency. The situation differs from the radiative decay
of atoms in that there are several vibrational levels in the ground electronic
state which can be populated, rather than a unique ground level (unless
the atomic levels are split by spin–orbit interaction as in the alkali metals).
Necessarily, the fluorescence will lie to the red of the absorption.

These possibilities are well illustrated by the fluorescent decay of iodine.
If the vapour is exposed to the green spectral line emitted by mercury at
546·075 nm, it is selectively excited into the thirty-fourth rotational and
twenty-fifth vibrational level of the $\tilde{B}^3\,\Pi_{0,u}^+$ electronic state.[16,17]* At very
low pressures ($\leqslant 0\cdot2$ torr), the excited molecule returns to the rotational
levels $J'' = 33, 35$, and vibrational levels $v'' = 0, 1, 2, \ldots$, of the ground

---

* For many years it was thought to be the twenty-sixth vibrational level, but the vibrational
numbering in the upper state is now known to have been in error.[17]

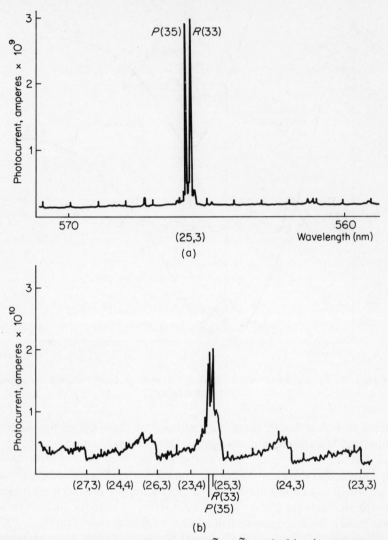

Figure 2.20 Fluorescent emission from I_2 ($\tilde{B} \rightarrow \tilde{X}$), excited by the green mercury spectral line at 546 nm. (a) Resonance fluorescence from 0·2 torr of I_2 vapour, showing emission from $v' = 25$ and $J' = 34$ only. (b) Fluorescence spectrum in the presence of 3·10 torr of hydrogen, showing the development of vibrational bands through energy transfer. (From J. I. Steinfeld and W. Klemperer, *J. Chem. Phys.*, **42,** 3745, 1965).)

electronic state (see Figure 2.20) (transitions with $\Delta J = 0$ are forbidden in the resonance spectrum in order to conserve rotational symmetry: see ref. 8, p. 254). The mean radiative lifetime of the upper state is $7·2 \times 10^{-7}$ sec[18] and if some inert foreign gas is introduced at a partial

PLATE I

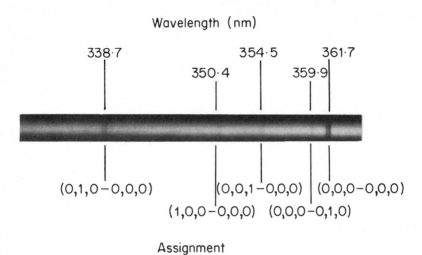

Figure 1.34 Vibrational structure in the $\tilde{A} \leftarrow \tilde{X}$ absorption spectrum of $(\dot{N}{=}CF_2 \cdot (v'_1, v'_2, v'_3 - v''_1, v''_2, v''_3)$

(*Facing page 114*)

PLATE II

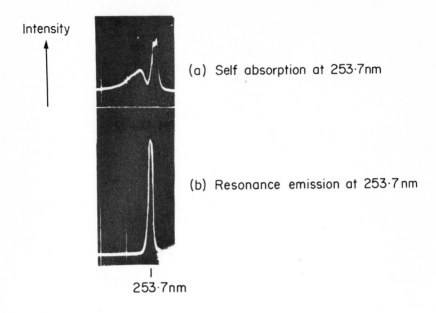

Intensity

(a) Self absorption at 253·7nm

(b) Resonance emission at 253·7 nm

253·7nm

Figure 2.4 Self-absorption of the Hg $(6^3P_1 \rightarrow 6^1S_0)$ resonance line in a medium pressure Hg arc lamp. (a) 'Reversed' emission spectrum (medium pressure lamp); (b) resonance emission (low pressure lamp)

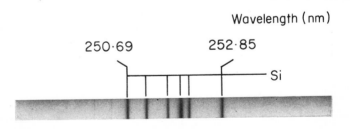

Wavelength (nm)

250·69   252·85

Si

Figure 2.5 'Reversed' spectral lines of silicon, appearing in absorption against the continuum emitted by a silica capillary flash discharge lamp

PLATE XII

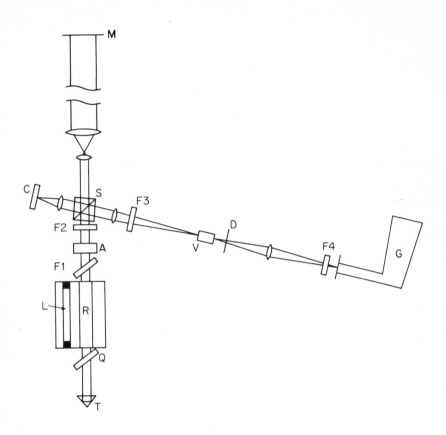

A	Frequency doubling crystal (Ammonium dihydrogen phosphate)	F_4	Biphenylene filter (opaque at 347 nm)
		G	Spectrograph
C	Fluorescent solution providing background continuum	L	Flash lamp
		M	Mirror
D	Aperture stop	Q	Q-switch
F_1	Filter (transmits > 630 nm)	R	Ruby
F_2	Filter (attenuation at 694 nm)	S	Beam splitter
F_3	Biphenylene filter (attenuation at 347 nm)	T	Prism
		V	Reaction cell

Figure 3.12 (a) Nanosecond laser flash photolysis—spectrographic apparatus (After G. Porter and M. R. Topp, *Proc. Roy. Soc. A*, **315**, 163 (1970))

PLATE XI

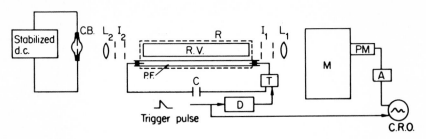

C	High voltage capacitor	$L_1 L_2$	Lenses
T	Flash lamp trigger	$I_1 I_2$	Iris diaphragms
D	Trigger delay	R	Reflector
P.F.	Photolysis flash tube	M	Monochromator
R.V.	Reaction vessel	P.M.	Photomultiplier
C.B.	Continuous background lamp	A	Amplifier

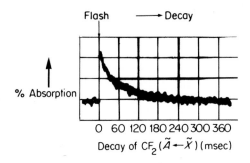

Figure 3.1 (b) Schematic diagram of flash photolysis system—photoelectric mode

PLATE X

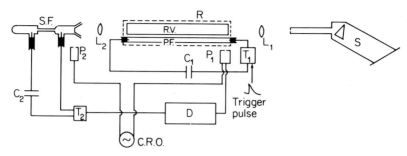

C_1, C_2	High voltage capacitors		P.F.	Photolysis flash tube
			R.V.	Reaction vessel
T_1, T_2	Flash lamp triggers		R	Reflector
P_1, P_2	Photocells		L_1, L_2	Lenses
S.F.	Background spectro-		D	Electronic delay unit
	scopic flash tube		S	Spectrograph

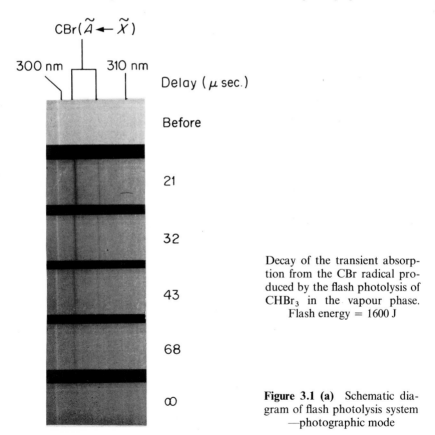

$$CBr(\widetilde{A} \leftarrow \widetilde{X})$$

300 nm   310 nm

Delay (μ sec.)

Before

21

32

43

68

∞

Decay of the transient absorption from the CBr radical produced by the flash photolysis of $CHBr_3$ in the vapour phase. Flash energy = 1600 J

Figure 3.1 (a) Schematic diagram of flash photolysis system —photographic mode

PLATE IX

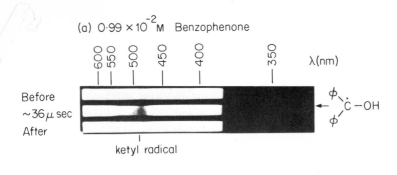

(a) 0.99×10^{-2} M Benzophenone

λ(nm)

600 550 500 450 400 350

Before
~36 μ sec
After

ketyl radical

$\underset{\phi}{\overset{\phi}{>}} \dot{C} - OH$

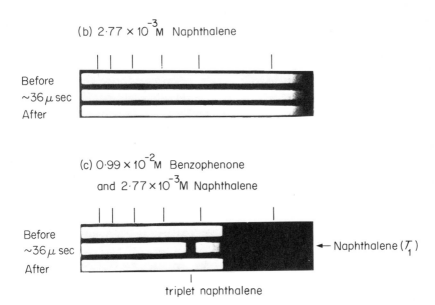

(b) 2.77×10^{-3} M Naphthalene

Before
~36 μ sec
After

(c) 0.99×10^{-2} M Benzophenone
and 2.77×10^{-3} M Naphthalene

Before
~36 μ sec
After

← Naphthalene (T_1)

triplet naphthalene

Figure 2.62 Flash excitation of benzophenone: naphthalene in benzene through a Pyrex filter (which prevents direct excitation of the naphthalene) (From G. Porter and F. Wilkinson Ref. 141)

PLATE VIII

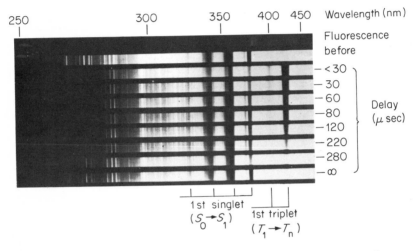

Figure 2.51 Series of spectra showing the transient absorption of triplet anthracene in hexane solution (Note the returning intensity of the singlet absorption bands as the triplet absorption fades) (Taken from G. Porter and M. Windsor, *Discussions Faraday Soc.*, **17**, 178 (1964))

PLATE VII

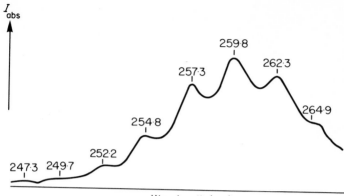

Figure 2.38 (c) Progression of diffuse vibrational bands in the $\tilde{A} \leftarrow \tilde{X}$ transition of $\cdot NF_2$ (produced through the thermal dissociation of N_2F_4: $N_2F_4 \rightleftharpoons 2NF_2$) (After P. L. Goodfriend and H. P. Woods, *J. Mol. Spec.*, **13**, 63 (1964))

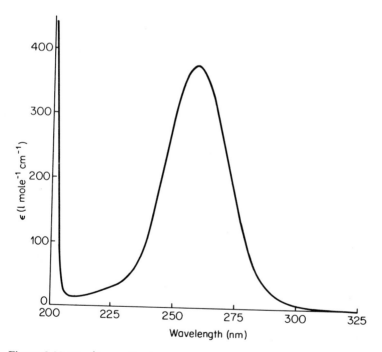

Figure 2.38 (d) $^1(n \rightarrow \sigma^*)$ absorption band of CH_3I vapour, showing complete absence of vibrational structure

PLATE VI

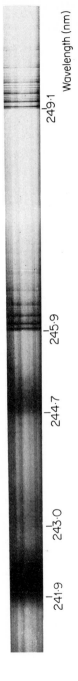

Figure 2.38 (a) The onset of predissociation in the $\tilde{B} \leftarrow \tilde{X}$ absorption band system of NO_2 (From Herzberg, *Electronic Spectra and Electronic Structure of Polyatomic Molecules*, Vol. III.

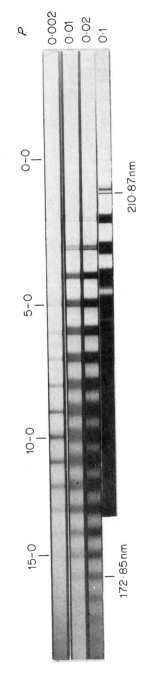

Figure 2.38 (b) The $\tilde{A} \leftarrow \tilde{X}$ absorption bands of ND_3 showing the diffuseness introduced by predissociation. (Photographed at four different pressures, *P*) (From Herzberg, *Electronic Spectra and Electronic Structure of Polyatomic Molecules*, Vol. III, Copyright 1966, by Litton Educational Publishing Inc., with permission

PLATE V

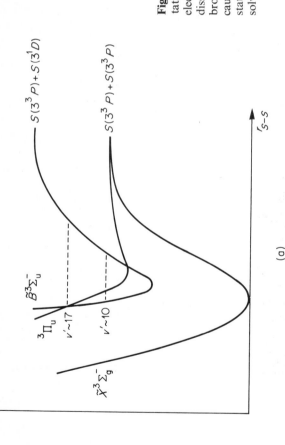

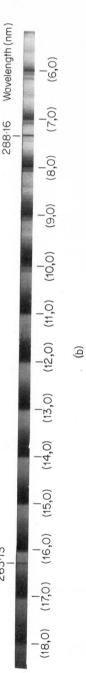

Figure 2.15 (a) Schematic representation of potential energy curves of electronic states involved in the predissociation of S_2. (b) Onset of the line broadening in S_2 ($\tilde{B}^3\Sigma_u^- \leftarrow \tilde{X}^3\Sigma_g^-$) caused by predissociation of the $\tilde{B}$ state (Note the partial return of resolved rotational structure between the (14, 0) and (17, 0) bands)

(a)

$S(3^3P) + S(3^1D)$

$S(3^3P) + S(3^3P)$

$\tilde{B}^3\Sigma_u^-$

$^3\Pi_u$

$v' \sim 17$

$v' \sim 10$

$\tilde{X}^3\Sigma_g^-$

$V(r)$

r_{s-s}

(b)

Wavelength (nm)

288·16

263·13

(18,0) (17,0) (16,0) (15,0) (14,0) (13,0) (12,0) (11,0) (10,0) (9,0) (8,0) (7,0) (6,0)

PLATE IV

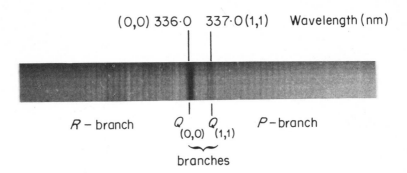

(0,0) 336·0   337·0 (1,1)   Wavelength (nm)

R – branch   Q
(0,0) Q
(1,1)   *P* – branch

branches

Figure 2.8 Absorption bands of NH showing the regular *P*, *Q* and *R* rotational structure (From D. Husain and R. G. W. Norrish, *Proc. Roy. Soc. A*, **273**, 145 (1963))

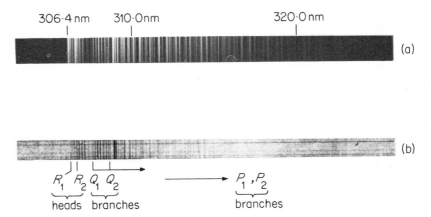

306·4 nm   310·0 nm   320·0 nm

(a)

(b)

R_1 R_2 Q_1 Q_2 P_1, P_2

heads branches   branches

Figure 2.10 The (0, 0) band of OH at 306·4 nm photographed (a) in emission from and O_2–coal gas flame, and (b) in absorption at room temperature following flash photolysis of an O_3–NH_3 mixture. Note the much more highly developed rotational structure in the flame spectrum. (The two sets of bands are associated with the $^2\Pi_{\frac{3}{2}}$ and $^2\Pi_{\frac{1}{2}}$ states of OH ($\tilde{X}\ ^2\Pi_i$). cf. ClO where the spin–orbit interaction is much larger and the levels are much more widely separated) (Figure 2.10(b) taken from N. Basco and R. G. W. Norrish, *Proc. Roy. Soc. A*, **260**, 293 (1960))

PLATE III

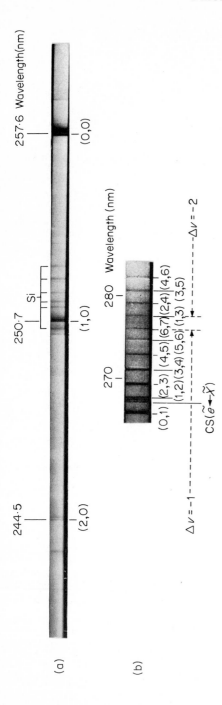

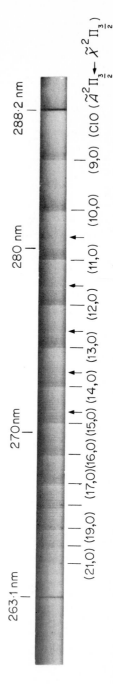

Figure 2.6 Vibrational structure associated with the $\tilde{A} \leftarrow \tilde{X}$ electronic transition of CS (produced in the flash photodissociation of CS_2 vapour). (a) The $(v', 0)$ progression (note the rapid decrease in relative intensity); (b) the $\Delta v = -1$ and -2 sequences in vibrationally excited CS (overlapping another electronic transition $\tilde{e} \leftarrow \tilde{X}$) (Figure 2.6(b) taken from A. B. Callear, *Proc. Roy. Soc. A*, **276**, 401 (1963))

Figure 2.7 Vibrational progression in the electronic absorption spectrum of ClO. The more intense progression is associated with absorption from the ground state ($\tilde{X}^2\Pi_{\frac{3}{2}}$), while the bands marked with a small arrow arise from the $^2\Pi_{\frac{1}{2}}$ state which is separated from the ground state by weak spin-orbit interaction

PLATE XIII

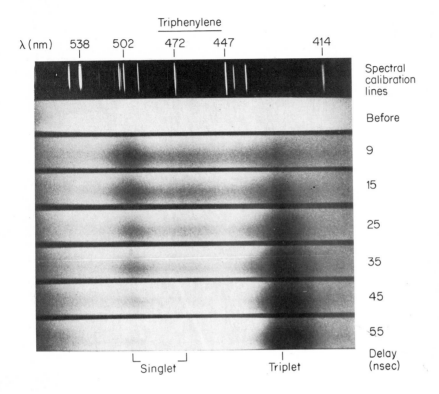

Figure 3.12 (a) (*continued*)   Decay of excited singlet triphenylene following laser flash excitation at 347 nm

PLATE XIV

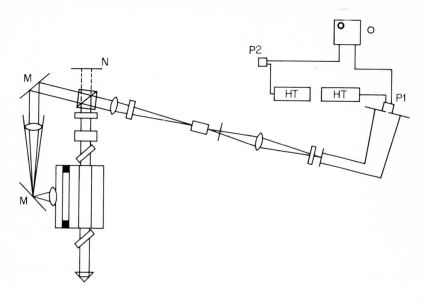

H.T. High voltage power supplies for photomultipliers P_1 and P_2
M   Mirrors
N   Safety screen
O   Oscilloscope

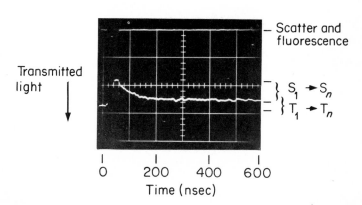

Transient absorption in a solution of triphenylene at 505 nm following laser
flash photolysis at 347 nm

Figure 3.12 (b) Nanosecond flash photolysis—photoelectric apparatus (After
G. Porter and M. R. Topp, *Proc. Roy. Soc. A*, **315**, 163 (1970))

pressure of 1 torr or so, or if the iodine pressure itself is increased, the excited iodine can lose some of its vibrational and rotational energy in collisions. This allows fluorescence bands from levels with $v' < 25$ and $J' < 34$ to develop and the emission spreads over a broader spectral range (see Figure 2.20). It is also possible for a particularly energetic collision to transfer energy from translation into vibration, producing iodine molecules in $v' = 26$. When these fluoresce, they emit to the blue of the exciting light; their emission is known as anti-Stokes fluorescence, in contrast to the Stokes fluorescence which lies to the red.

Measurement of the relative intensities of the emitted spectral lines, as a function of the pressure and nature of the foreign gas, provides particularly detailed information on the elementary processes by which energy is transferred from the excited molecule. Collisions can also induce pre-dissociation of the iodine so that there is a subtle interplay between all the competing processes in the excited molecule. Similar experiments have been performed on NO[19] and S_2,[20] using monochromatic light to excite the molecules into a specific resolved quantum state and then charting their course out of that state under the influence of collisions. The mechanisms of energy transfer will be discussed in the following section.

2.3b Energy transfer

An electronically excited atom or diatomic molecule can lose some or all of its excess energy in the gas phase by transferring it to other atoms or molecules with which it collides. When the transfer changes the electronic state of the donor, its fluorescence (or phosphorescence) will be quenched. Such a change is inevitable when the donor is an atom, but a photoexcited diatomic molecule may lose excess vibrational (and rotational) energy while remaining in the same electronic state if the transfer of its electronic excitation is the less probable process. The acceptor may be excited into higher translational, rotational, vibrational or electronic states (or combinations thereof). If as a consequence, the acceptor emits radiation or some chemical change follows, the process is said to be *photosensitized*. One observes emission from, or reaction involving the transparent acceptor rather than the absorbing donor.

The bimolecular process of energy transfer has to compete against 'natural' unimolecular decay processes such as fluorescence or phosphorescence, radiationless electronic transition, and predissociation (but not direct dissociation since this occurs within the first vibration and wins in any contest). In the case of electronically excited atoms, only the first of these can occur (at energies below the ionization threshold), and their photochemistry always involves energy transfer to, or chemical reaction with a substrate. It should cause no surprise therefore to learn that the study of photosensitization by atoms has exercised the minds of photo-chemists since the 1920's, and will exercise ours throughout section 3.1. In principle, energy transfer from electronically excited diatomic mole-cules can also promote chemical change, but this has rarely been observed in practice; most diatomic molecules either dissociate on excitation, or else the energy transferred is too small to be effective.

(i) *Atoms*

In the absence of any substrate, the excitation and decay of an excited atom A* can be expressed by the scheme

	Process	Rate
$A + h\nu \rightarrow A^*$	excitation	I_{abs}
$A^* \quad \rightarrow A + h\nu$	emission	$k_1[A^*]$

Add a substrate Q, and the additional steps

	Process	Rate
	energy transfer	$k_2[A^*][Q]$
$A^* + Q \rightarrow A \quad + Q^*$	elec $\rightarrow$ elec	
$\rightarrow A \quad + (Q)_v$	elec $\rightarrow$ vibl	
$\rightarrow A \quad + (Q)_t$	elec $\rightarrow$ transl	
$\rightarrow AX + P$	atom transfer	$k_3[A^*][Q]$

may occur, $(Q \equiv PX)$

In the first case

$$\frac{d[A^*]}{dt} = I_{abs} - k_1[A^*]$$

and in the second

$$\frac{d[A^*]'}{dt} = I_{abs} - k_1[A^*]' - (k_2 + k_3)[A^*]'[Q]$$

Under steady illumination both rates can be set to zero, and the relative steady state concentrations

$$\frac{[A^*]}{[A^*]'} = 1 + \frac{k_2 + k_3}{k_1}[Q]$$

The intensity of emission is proportional to the standing concentration of excited atoms, and if I and I' are the intensities of the 'natural' and quenched emission

$$\frac{I}{I'} = 1 + \frac{k_2 + k_3}{k_1}[Q]$$

A linear relation between the reciprocal of the latter and the partial pressure or concentration of the substrate, reveals the competition between step (1) and steps (2) and (3). It is known as the *Stern–Volmer relationship*. The slope of the plot gives the ratio between the probabilities of the competing steps. $k_1 = 1/\tau_0$ is the reciprocal of the mean radiative lifetime of the excited atom: if this is known, $k_2 + k_3$ can be estimated.

It is not always possible to distinguish the relative contributions of physical (energy transfer, k_2) and chemical (k_3) quenching, though experimental techniques developed by Gunning and Strausz have greatly alleviated this situation in the case of Hg (6^3P_1) atoms (see section 3.1a). Because of this the two processes are 'lumped' together to give a composite quenching constant k_Q. We have also assumed that energy transfer carries the excited atom back to its ground electronic state, i.e. we have assumed *total quenching*. However, it is commonly the case that the emission is quenched by partial deactivation into a lower-lying metastable state, for example the 6^3P_0 state of mercury. In this situation values of k_Q obtained from measurements of the fluorescence quenching will include both contributions; other techniques must be employed if they are to be separated, which allow the concentration of metastable atoms to be monitored as well as that of the fluorescent ones (see later discussion).

As an alternative to k_Q, the quenching efficiencies are commonly expressed as a cross-section, using the terminology of collision theory. The specific bimolecular collision rate between A* and Q is given by hard-sphere collision theory as

$$Z = \left(\frac{8RT}{\pi\mu}\right)^{\frac{1}{2}} \pi\sigma^2$$

where μ is the reduced mass of A and Q (gram-molecular weight units), and σ is the distance between the centres of the colliding species at the instant of collision. If quenching occurs on every 'collision', this can be equated to the specific quenching rate

$$k_2 + k_3 = \left(\frac{8RT}{\pi\mu}\right)^{\frac{1}{2}} \pi\sigma_Q^2$$

The question now arises as to what is meant by 'collision'. In the first place the cross-section $\pi\sigma_Q^2$ may include contributions from chemical as well as physical quenching; the latter may include both total and partial quenching, each of which may arise from a different type of collision and each with its own cross-section. Secondly, the hard-sphere approximation is quite unrealistic in any detailed description, and the 'average' cross-section may include many distant encounters which are effective and close ones which are not. However, if we do not know the details of the form of the potential energy of interaction between A* and Q, or the dynamics of their collision, a detailed description is not possible, and the hard-sphere approximation is the simplest. It provides a formalism through which different experimental measurements can be related. In consequence we do not expect to find any correlation between the quenching cross-sections $\pi\sigma_Q^2$, and those that would be typical of 'hard-sphere' gas–kinetic collisions, $\pi\sigma^2$. In practice the experimental quenching cross-section is defined as σ_Q^2; it may be much greater or smaller than the values expected for hard-sphere encounters.

The Stern–Volmer relationship is often reformulated, as

$$\frac{I}{I'} = 1 + p\tau_0 Z$$

where Z is the gas kinetic collision number ($\bar{Z}[Q]$), and p is defined as a probability of quenching on collision. However, in view of the complications discussed above and the uncertainty in any estimate of the collision number for electronically excited atoms, the practical value of the equation in this form is relatively limited.

Experimental methods for determining total quenching cross-sections have been discussed by Cvetanovic.[22] In addition to the obvious one of monitoring the intensity of the fluorescence as a function of the partial pressure of the quencher, it is also possible to follow the rate of decay of the fluorescence immediately after rapid extinction of the light source.[23] With $I_{abs} = 0$, integration of the equation

$$\frac{d[A^*]'}{dt} = I_{abs} - k_1[A^*]' - (k_2 + k_3)[A^*]'[Q]$$

gives

$$\ln[A^*]_t' = \ln[A^*]_0' - \{k_1 + (k_2 + k_3)[Q]\}t$$

whence

$$\ln I_t' = \ln I_0' - \{k_1 + (k_2 + k_3)[Q]\}t$$

Measurement of the fluorescent intensity I_t' for varying [Q] allows the determination of k_1 and $(k_2 + k_3)$. Both methods are hampered by the phenomenon of 'radiation imprisonment' (see section 2.1b), which reduces the observed fluorescent intensity and increases its apparent lifetime; the experimental data have to be corrected appropriately. The absolute values obtained depend on the method of correction employed.[23,24a] Yang[24b] has developed a method of eliminating the problem by extrapolating the measured fluorescence intensity back towards zero mercury vapour pressure where the problem of imprisonment does not arise. An alternative 'chemical' method has been developed by Cvetanovic for the study of energy transfer from $Hg(6^3P_1)$ atoms,[25] in which N_2O and some other substrate compete as quenchers. The excited mercury photosensitizes the decomposition of N_2O into N_2 and $O(2^3P)$, and by recording the yield of N_2 as the pressures of a series of quenchers are varied, a scale of quenching cross-sections can be derived relative to that of N_2O. Cross-sections for the partial deactivation of $Hg(6^3P_1)$ atoms by CO and N_2 have been reported by Scheer and Fine[26] who monitored the $Hg(6^3P_0)$ atoms through their ability to eject electrons from the surface of a silver electrode. Callear and Norrish[27] have measured cross-sections for deactivation of $Hg(6^3P_0)$ atoms by following the rate of decay of their absorption lines following flash excitation in the presence of a range of quenching gases.* Other techniques will be mentioned in subsequent discussion. Some values for quenching cross-sections and probabilities are collected in Tables 2.1(A) and (B).

In the discussion of their relative magnitudes and of their significance with respect to the possible mechanisms of energy transfer, we must keep in mind the following general questions:

(i) What are the possible types of potential energy curve or surface over which the quenching collision takes place?

(ii) What is the importance, if any, of the magnitude of the quantum of energy transferred from the donor, the form in which it appears in the acceptor and the correspondence in the spacings of their energy levels?

(iii) Is there any correlation between the probability of the electronic transition in the isolated excited atom (determined by the optical selection rules) and the same transition promoted through energy transfer? Is it necessary that spin angular momentum be conserved?

Some of these questions will be answered in what follows.

The main conclusion to be drawn from the existing experimental data is that the form of the interaction between the excited atom and the quencher is all-important. The stronger the attractive interaction, the greater the cross-section.

Consider first quenching by other atoms; these can accept energy in the form of electronic (E) or translational (T) excitation. If the quencher and the unexcited atom are chemically inert, their combined potential energy will rise steadily as they approach but the same may not be the

* But see Appendix I, p. 334.

TABLE 2.1(A) Total quenching cross-sections for energy transfer from excited atoms, where $\Delta E > 1$ eV, measured in $cm^2 \times 10^{16}$

Excited atom Transition Energy (eV)	As $4^2D \rightarrow 4^4S$ (1·32)	Na $3^2P \rightarrow 3^2S$ (1·95)	Cd $5^3P_1 \rightarrow 5^1S_0$ (3·78)	Hg $6^3P_1 \rightarrow 6^1S_0$ (4·86)
Argon	$2\cdot3 \times 10^{-4}$	v. small	v. small	v. small
Krypton	v. small	v. small	v. small	v. small
Xenon	0·47	v. small	v. small	v. small
H_2	1·6	7·4	3·5	6·0
D_2	0·97	—	—	8·4
N_2	0·72	14·5	0·02	0·19
CO	8·3	28·5	0·14	4·1
CO_2	0·16	—	—	2·3
N_2O	—	—	—	15·0
CS_2	—	—	—	33·0
Methane	0·28	0·11	0·012	0·06
Ethane	—	0·17	0·024	0·10
Neopentane	—	—	—	1·4
Propane	—	0·2	0·012	1·5
n-Butane	—	0·3	0·064	5·1
Isobutane	—	—	0·05	4·9
n-Pentane	—	—	—	8·6
Isopentane	—	—	—	12·0
n-Hexane	—	—	—	16·0
Ethylene	—	44	25	24
1,1 Difluoro-ethylene	—	—	—	21
Tetrafluoro-ethylene	—	—	—	7·3
Propylene	—	55	29	32
Butene-2	—	58	30·6	39
Benzene	—	75	28·4	41·9
CF_4	—	—	—	v. small
SiH_4	—	—	—	26
CH_3Cl	—	—	—	24
CH_3Br	—	—	—	28
CH_3I	—	—	—	41

case when the atom is excited into a new electronic configuration. The potential energy may rise less steeply, or if a complex is formed, pass through a minimum (cf. the excimers mentioned in section 2.2b). Fluorescent decay during the lifetime of the excited complex results in *radiative quenching* of the excited atom. In the unlikely event that the interactions are the same the curves will run parallel, otherwise they will converge

TABLE 2.1(B) Probabilities of deactivation of excited atoms per gas kinetic collision, where $\Delta E < 1$ eV: (i.e. collisionally induced spin–orbit relaxation)

Excited atom Transition Energy (eV)	Se $4^3P_0 \to 4^3P_1$ or $^3P_0$ (0.25 or 0.31)	Br $4^2P_{\frac{1}{2}} \to 4^2P_{\frac{3}{2}}$ (0.46)	I $5^2P_{\frac{1}{2}} \to 5^2P_{\frac{3}{2}}$ (0.94)
He, Xe	—	v. small	v. small
Ar	1.1×10^{-4}	v. small	v. small
H_2	0.63	7.8×10^{-3}	1.8×10^{-4}
D_2	—	1.8×10^{-2}	3.1×10^{-4}
N_2	1.2×10^{-2}	1.1×10^{-5}	3.0×10^{-6}
CO	3.1×10^{-3}	3.2×10^{-5}	6.7×10^{-6}
CO_2	0.50	—	7.7×10^{-7}
NO	—	0.22	6.7×10^{-2}
N_2O	0.48	—	—
O_2	6.5×10^{-3}	0.16	5.6×10^{-2}
H_2O	—	0.16	5.9×10^{-3}
CH_4	—	1.4×10^{-2}	2.4×10^{-4}
CF_4	—	9.5×10^{-4}	2.5×10^{-5}

Note. In general, the quencher is more efficient when the majority of the electronic energy in the atom can be taken up as vibration in the quencher. There is a rough linear correspondence between the energy deficit (i.e. the energy which has to appear in translation and/or rotation), and the logarithm of the quenching probability.

or perhaps intersect. If they do intersect, the nett probability of energy transfer will depend on the ease with which the crossing point can be reached and the probability of radiationless transition in its neighbourhood. The possible types of interaction are represented in Figure 2.21.

In the case of the rare gases, their first excited states all lie at energies far above those of the donors, and $E \to E$ transfer is impossible. The only alternatives are $E \to T$ transfer (Figure 2.21b) or radiative quenching (Figure 2.21c). If alternative (b) is appropriate, the configuration point for the motion of the approaching atoms moves along the upper curve towards the point of intersection, crosses onto the lower curve and runs down its slope as the two atoms fly apart; electronic excitation is converted into translation. This description was first given by Jablonski.[28] If the energy transferred, ΔE, is large ($> 10^3$ cm^{-1}, say), the curves will not intersect until the two atoms are very close and high relative velocities will be needed to reach the crossing point. In general, the conversion of large amounts of electronic energy into translation is very inefficient. In fact the quenching of $Hg(6^3P_1)$ by the rare gases is associated with the emission of a broad fluorescent continuum, to the red of the resonance line at 254 nm,[28b] indicating a radiative quenching mechanism rather than E → T transfer.

If ΔE were small the curves in Figure 2.21b would intersect sooner, and σ_Q^2 should rise. Mitchell and Zemansky, who wrote a classical monograph on the quenching of excited atoms in 1934,[29] observed that following the excitation of sodium atoms into the $3^2P_{\frac{3}{2}}$ state by the D line of higher frequency, the other D line could be observed in fluorescence only in the presence of inert gases. They estimated that the cross-section for the transfer of $Na(3^2P_{\frac{3}{2}}) \rightarrow Na(3^2P_{\frac{1}{2}})$ on collision with argon was comparable with the gas kinetic cross-section; ΔE is only 17 cm^{-1}. Note that in the isolated atom, the transition is forbidden by the selection rule $\Delta L = \pm 1$.

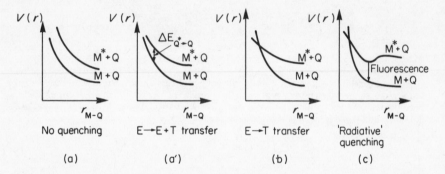

Figure 2.21   Possible potential energy curves for interaction between an excited atom M* and a quencher, Q.

If there are electronic levels in the quenching atom which lie near that of the donor, E → E transfer becomes a possibility. Cario and Franck[30] discovered that the fluorescence of thallium vapour could be sensitized by $Hg(6^3P_1)$ atoms. Emission was observed only from those electronic states of the thallium which lay near to the $Hg(6^3P_1)$ level. This type of energy transfer arises from the coupling of the transition dipole in the donor and acceptor and is known as *radiationless resonance energy transfer*. The more closely the two dipoles are 'tuned' the more efficient the process. The vibrations of two coupled pendulums (suspended from the same length of string), provides an entertaining, if naive, classical analogue. The energy remains 'electronic' and no crossing of the potential energy curves is required. If the energy match is very close ($< 10^2$ cm^{-1}) the resonance interaction may extend over many atomic diameters. However, the resonance peak is very sharp,[31] and long range interaction rapidly falls off as the energy discrepancy increases. This is illustrated very neatly, perhaps too neatly, by the observations of Beutler and Josephi[32] on the mercury-sensitized fluorescence of sodium vapour. In the absence of any added gases, $Hg(6^3P_1)$ atoms sensitize fluorescent emission corresponding to the transition $9^2S_{\frac{1}{2}} \rightarrow 3^2S_{\frac{1}{2}}$ in sodium; the energy discrepancy is $162 \, cm^{-1}$. On addition of N_2 (known to be effective in transferring $Hg(6^3P_1) \rightarrow (6^3P_0)$), the fluorescence faded and was replaced by that corresponding to the transition $7^2S_{\frac{1}{2}} \rightarrow 3^2S_{\frac{1}{2}}$. The energy of $Hg(6^3P_0)$ lies some $560 \, cm^{-1}$ below that of $Na(7^2S_{\frac{1}{2}})$; the difference must be provided by the relative energy of translation of the colliding atoms.

As well as $E \rightarrow E$ and $E \rightarrow T$ transfer, a molecular quencher can take up energy in vibration (V) and rotation, though we shall not concern ourselves with the latter possibility. There is also the possibility of chemical quenching. Which one, or which combination of these occurs, depends on the interaction between the excited atom and the molecule, and for $E \rightarrow E$ transfer, on the match between the energy stored in the donor, and that which can be taken up by the acceptor. In the latter case, efficient transfer also requires conservation of the overall spin angular momentum.

The occurrence of $E \rightarrow V$ transfer was confirmed by Polanyi and coworkers, when they were able to record the infrared emission from vibrationally excited CO and NO, photosensitized by $Hg(6^3P_1)$ atoms.[33] $E \rightarrow E$ transfer is impossible in CO since its lowest excited state lies above that of mercury. The lowest excited state in xenon carries $8.4 \, eV$, which lies well above several of the triplet states in CO, and in this system $E \rightarrow E$ transfer has been detected by observing the photosensitized phosphorescent emission from the CO.[34] Direct evidence for the chemical reaction of excited atoms is also available, principally from the experiments of Gunning and coworkers.[21]

The potential energy diagrams in Figure 2.21 now represent sections through potential energy surfaces or hypersurfaces. Vibrational excitation of the quencher would be represented by the oscillation of the configuration point along the appropriate axis in the configuration space.

The most detailed information about the mechanisms of energy transfer from excited atoms to molecules has been obtained for diatomic quenchers. It has long been known that N_2 and CO are effective in transferring $Hg(6^3P_1) \rightarrow Hg(6^3P_0)$. Callear and Norrish reported strong absorption from metastable atoms immediately following flash excitation of mercury vapour in the presence of N_2, CO, H_2O and D_2O (but not NO, H_2, O_2, CO_2, N_2O, NH_3, paraffin hydrocarbons or ethylene)[27] (see Figure 2.22).* It used to be thought that this was because the energy transferred had to

* See Appendix I, p. 334.

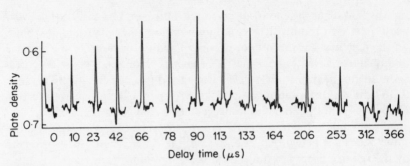

Figure 2.22 Formation and decay of Hg(6^3P_0) following flash excitation of mercury vapour in an atmosphere of nitrogen. The successive absorption maxima show the changing intensity of the atomic absorption line at 296·7 nm, associated with the transition $6^3D_1 \leftarrow 6^3P_0$. (Taken from A. B. Callear and R. G. W. Norrish, *Proc. Roy. Soc. A*, **266**, 292, (1962).)

be close to that needed to excite the quenchers into the first vibrational level; this is now known to be beside the point.[35] Scheer and Fine found that although the total quenching cross-section of CO is some ten times greater than that of N_2, the partial quenching efficiencies are about the same, suggesting that CO was far more likely to transfer the excited mercury directly into its ground electronic state.[26] This was confirmed when Polanyi and coworkers[33] detected infrared emission from CO when excited into vibrational levels $v'' \leqslant 9$ in its ground electronic state; from their observations they were able to estimate relative rate constants for $E \rightarrow V$ transfer into each level, as well as the overall cross-section for quenching the Hg(6^3P_1). Excitation into $v'' = 9$ accounts for about half the total available energy and the remainder must appear in translation. The results were discussed in terms of a mechanism, suggested earlier by Scheer and Fine[26]

$$\text{Hg}(6^3P_1) + \text{CO}(\tilde{X}^1\Sigma) \rightarrow \text{HgCO}(^3\Sigma) \text{ or } (^3\Pi) \qquad \begin{array}{l} \text{complex} \\ \text{formation} \\ \text{(assumed linear)} \end{array}$$

$$\text{HgCO}(^3\Sigma) \rightarrow \text{HgCO}(^1\Sigma) \qquad \begin{array}{l} \text{intersystem} \\ \text{crossing} \end{array}$$

$$\text{HgCO}(^1\Sigma) \rightarrow \text{Hg}(6^1S_0) + \text{CO}(\tilde{X}^1\Sigma)_{v'' \leqslant 9} \quad \text{dissociation}$$

Transfer occurs through the intermediate formation of a long-lived excited mercury carbonyl complex, in which the equilibrium C–O bond length is longer than that of CO($\tilde{X}^1\Sigma$); intersystem crossing carries the complex into its non-bonding ground state from which it dissociates

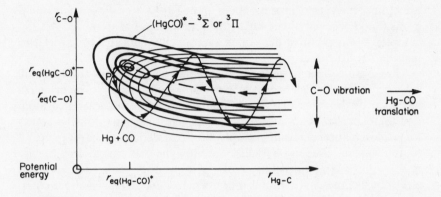

Figure 2.23 Potential energy surfaces showing interaction of CO, $Hg(6^3P_1)$ and $Hg(6^1S_0)$. Arrowed broken line represents trajectory of configuration point as $Hg(6^3P_1)$ and CO approach over the upper surface. P represents intersection where intersystem crossing takes place between triplet upper surface and singlet lower surface. Arrowed full line represents trajectory of configuration point along the lower surface as $Hg(6^1S_0)$ and CO separate.

leaving the CO vibrationally excited. The process is represented in Figure 2.23. Resonant energy transfer, in which there was no intersection of the upper and lower surfaces, would have left the CO in $v'' = 20$. Some vibrational excitation may also be promoted by repulsion between the departing mercury atom and the carbon atom to which it is attached. How much, depends on the steepness of the repulsive surface, and the 'rigidity' (force constant) of the C–O bond.

Similar results were obtained for quenching by NO, which was excited into $v'' \leqslant 16$, corresponding to about two-thirds of the available energy. The similarity suggests an analogous mechanism, but the possibility of a primary $E \rightarrow E$ transfer exciting the NO into its metastable $\tilde{a}^4\Pi$ state, which lies $4.7\,eV$ above the ground state and $0.2\,eV$ below $Hg(6^3P_1)$, could not be ruled out.[36]*

$$Hg(6^3P_1) + NO(\tilde{X}^2\Pi) \rightarrow Hg(6^1S_0) + NO(\tilde{a}^4\Pi) \quad E \rightarrow E \text{ transfer}$$

$$Hg(\tilde{a}^4\Pi) \xrightarrow{NO(\tilde{X}^2\Pi)} NO(\tilde{X}^2\Pi)_{v'' \leqslant 16} \quad \text{self-quenching}$$

Note that the overall spin is conserved in the $E \rightarrow E$ transfer, since the spin angular momentum of both partners changes by $\Delta S = 1$. The 'forbidden' transition in the one is compensated by the 'forbidden' transition in the other. The xenon-sensitized phosphorescence of CO

* The energy transfer photosensitizes the decomposition of NO. Since the energy transferred is insufficient for direct dissociation, an excited intermediate must participate at some stage.

exemplifies this type of process, but as in the $E \longrightarrow V$ transfer from mercury, it was found necessary to postulate the intermediacy of an excited $XeCO^*$ complex.[34] Competitive quenching experiments indicated a dissociative lifetime 2×10^{-6} sec.

These results have been discussed at some length because of the detail they carry, detail which has become available only recently. The principal moral to be drawn is that strong interaction between the excited atom and the quencher is all-important for efficient quenching, since it allows for a closer approach and possibly the intersection of the 'input' and 'output' potential energy surfaces. Amongst the diatomic molecules, NO and O_2 have large quenching cross-sections; they both have unpaired electrons like the excited atoms they quench. CO has a much larger cross-section than N_2 though they are isoelectronic; however, it is well known that CO can readily donate electrons to metal atoms and is thus the more likely to form complexes. The same moral had been drawn earlier and discussed by Laidler[37] and by Gunning and Strausz.[21] Inspection of Table 2.1(a) reveals trends in the relative cross-sections of series of quenchers toward the same excited atom, which parallels their electronegativity and their 'chemical reactivity'; $Hg(6^3P_1)$ shows strong electrophilic character. There is a close parallel between the quenching cross-sections, and the relative rate constants for reaction of the quencher with $O(2^2P)$ atoms.[21] The trends can be reproduced by arguments based on transition state theory.[21,37] Most strikingly, they are not too sensitive to a change in the energy content or identity of the excited atom, though the absolute values may be different (comparison between different results may be hazardous if these originate from different sources, because of variations in the methods used to obtain them).

Quenching by saturated molecules frequently leads to chemical change. Unless there is fluorine substitution, non-hydrides have larger cross-sections than hydrides. Steacie[38] and Laidler[37] suggested that quenching may conveniently be viewed as involving an incipient or real atom transfer from the quencher to the excited atom and kinetic isotope effects have been detected when paraffins are substituted with deuterium.[21] For example, quenching of $Hg(6^3P_1)$ by H_2 probably follows the course

$$Hg(6^3P_1) + H_2(\tilde{X}^1\Sigma_g^+) \longrightarrow HgH(\tilde{X}^2\Sigma) + H(1^2S_{\frac{1}{2}}) \longrightarrow Hg(6^1S_0) + 2H(1^2S_{\frac{1}{2}})$$

The reaction is 80 kJ mole^{-1} exothermic, and since much of this may be concentrated in vibration in the new bond, the HgH may not be formed in a bound vibrational level: its dissociation energy is $< 80 \text{ kJ mole}^{-1}$. A hydride must be formed in the quenching of $Cd(5^3P_1)$ since the energy available (364 kJ mole^{-1} as against $470.4 \text{ kJ mole}^{-1}$ in mercury) is not sufficient to promote direct dissociation ($D_{H-H} = 431 \text{ kJ mole}^{-1}$). In the case of $Na(3^2P)$ $(202 \, kJ \, mole^{-1})$ the deficit is too large and although its electronic energy may well be transferred into vibration in the H_2 molecule, it cannot lead to dissociation.

Unsaturated molecules, unless substituted by fluorine, are efficient quenchers of $Hg(6^3P_1)$ atoms. Some form of association involving donation of π-electrons into the vacated mercury atomic orbital seems probable. It is also probable that $E \longrightarrow E$ transfer into a triplet level in the quencher follows; although there is no direct evidence for this, the accumulated circumstantial evidence is compelling. Further discussion is deferred until section 3.1.

(ii) *Diatomic molecules*

Provided the electronically excited diatomic molecule is in a stable electronic state, does not predissociate and has insufficient vibrational energy for direct dissociation, it will survive either to fluoresce, react with, or transfer electronic, vibrational or rotational energy to the surrounding molecules until equilibrium is restored. Collisions may also induce radiationless transition into neighbouring electronic states, either by transferring the excited molecule into vibrational levels lying in the region of the intersection or by relaxing the selection rules for radiationless transition; induced predissociation may follow. A detailed kinetic scheme for the competition between fluorescence and collisional deactivation through energy transfer or induced radiationless transition is complex since it must include the possibilities of $V \longrightarrow V$, T transfer, as well as the $E \longrightarrow E$ and $E \longrightarrow V$, T transfer which occurs with excited atoms (rotational energy transfer also occurs, though it will not be discussed here). In addition, the vibrational energy may be transferred through a succession of collisions, in which several steps are necessary before the molecule reaches the foot of the 'vibrational ladder' and where the collisional efficiency falls as the ladder is descended. Where the acceptor is monatomic, only $V \longrightarrow T$ transfer is possible. Landau and Teller presented a classical mechanical description of such a process in 1936,[39] and emphasized that its probability will be a maximum when the mean duration of the collision is of the same order as the vibrational period of the excited molecule. Compared to the typical value of 10^{-13} sec most collisions at normal temperatures will be 'slow', and $V \longrightarrow T$ transfer should be relatively inefficient. At long range, the interaction potential is likely to be repulsive; suppose it has the form $V(r) \propto \exp(-r/l)$, where r is the distance between the centres of mass and l is a characteristic range in the repulsive field. A collision in which the initial relative velocity is v occurs in a time $\sim l/v$. If this is very much less than the vibrational period $1/2\pi\nu$, both atoms in the diatomic molecule have time to adjust their velocity components during the period of collision, while remaining in phase; the vibrational energy is unchanged. A fast collision (high relative velocity, steeply repulsive potential) results in the transfer of momentum between the approaching atom and one of the atoms of the diatomic molecule before the other atom can respond. The less rigid the oscillator (the longer its vibrational period), the more easily is the one atom disturbed before the other and the more readily does V $\leftrightarrow$ T transfer take place. Its probability in any given collision will be of the form

$$P = f\left(\frac{l}{v} \cdot 2\pi v\right) = f\left(\frac{l}{v} \cdot \frac{\Delta E}{\hbar}\right)$$

where $\Delta E = hv$ is the vibrational quantum transferred. Landau and Teller showed that the probability of V $\rightarrow$ T transfer from a harmonic oscillator, in a collision in which the mean of the relative velocities before and after is $\bar{v}$, is of the form

$$p \propto \exp\left\{-\frac{4\pi^2 l |\Delta E|}{\hbar \bar{v}}\right\}$$

where ΔE is the magnitude of the quantum transferred.[39] For a Boltzmann distribution of molecular velocities, the average probability of energy transfer per gas kinetic collision becomes

$$\langle p \rangle \propto \exp\left\{-3\left(\frac{\pi^2 \mu \Delta E^2 l^2}{2\hbar^2 kT}\right)^{\frac{1}{3}}\right\}$$

where μ is the reduced mass and T is the temperature. The probability should increase as μ decreases or T increases (both control the relative velocity $\bar{v}$), or as the energy transferred (ΔE) decreases: this is in agreement with observation.

The model has been reformulated in strict quantum mechanical terms, extended to three dimensional encounters and to polyatomic molecules and modified to incorporate other forms of interaction potential, which may include attractive terms. The subject has been reviewed by Cottrell and McCoubrey,[40] and more recently by Stevens.[41] The main findings which are relevant to photochemical studies are

(i) for transfer from low vibrational levels (v small), more than 10^5 collisions may be survived before V $\rightarrow$ T transfer occurs. It is further restricted by the rule $\Delta v = \pm 1$, which would be rigorous for a harmonic oscillator.

(ii) As v increases, V $\rightarrow$ T transfer becomes far more probable (for a harmonic oscillator $P_{v \rightarrow v-1} = v P_{1 \rightarrow 0}$, but as v increases so does the vibrational anharmonicity, of course). Transitions with $\Delta v = \pm 2$ or greater may occur with increasing probability (multiquantum jumps). For example, vibrational deactivation of $I_2(\tilde{B}^3\Pi_{0u}^+)$ from its twenty-fifth vibrational level, by a range of added gases occurred with gas kinetic collision efficiencies varying from 0·26 for helium to unity for SO_2.[17] Transfer with $\Delta v = -1$ was only five times as fast as with $\Delta v = -2$, though many collisions are necessary for complete deactivation.

(iii) The probability of vibrational energy transfer varies with the reduced mass; the mean collision time is $\sim l/v$ and $v \propto \mu^{\frac{1}{2}}$.

(iv) The probability is increased by a strong 'chemical' interaction between the collision partners. This may well be encouraged when the donor is also electronically excited.

(v) If the energy is transferred to another molecule rather than an atom, $V \rightarrow V$ (or $V \rightarrow R$) transfer can occur. It will be most efficient when the spacings between the vibrational levels in the collision partners are closely matched, and in this situation it will be very much faster than $V \rightarrow T$ transfer. The smaller the energy that must be converted into translation the better. Single quantum jumps are expected to be more efficient than multiquantum jumps, i.e.

$$(AB)_v + (CD)_{v'} \rightarrow (AB)_{v-1} + (CD)_{v'+1}$$

Self-relaxation may be particularly efficient because of close resonance between the vibrational levels. A tendency towards excimer formation in electronically excited molecules will also be favourable.

2.3c Dissociation: correlation rules

Given that the result of excitation of a diatomic molecule has been its dissociation, the photochemist would like to know the electronic states and velocities of the two atoms as they fly apart, and whether a radiationless transition preceded the act of dissociation. If the molecule dissociates from an electronically excited state, it is possible that at least one of the atoms will be electronically excited also, and when the absorbed energy exceeds that of the dissociation limit, the remainder will appear as kinetic energy in the separating atoms. In either case, the excess energy of the atom may increase its chemical reactivity and affect the subsequent course of the chemical change.

The electronic states in which the two atoms are produced must correlate with that of the dissociating molecule; they are limited by symmetry. The possible correlations can be found, in the main, by simple vectorial addition of the electronic angular momenta of the atoms. These give rise to sets of 'rules', known as the *Wigner–Witmer correlation rules*. Since detailed discussions have been presented by Herzberg[8] and Gaydon,[42] only a brief outline is given here. Consider first molecules in which the components of electronic orbital and spin angular momenta are separable (i.e. following Hund's coupling cases (a) or (b)). When two atoms approach, the orbital angular momenta are split into the components

$$M_{L_1} = L_1, L_1 - 1, \ldots, -L_1$$

$$M_{L_2} = L_2, L_2 - 1, \ldots, -L_2$$

by the strong internuclear electric field (Stark Effect). The resultant angular momentum about the internuclear axis can take the values

$$\Lambda = |M_{L_1} + M_{L_2}|$$

Suppose the atoms are produced in $P(L_1 = 1)$ and $S(L_2 = 0)$ states, respectively, as in the dissociation

$$HCl \rightarrow H(^2S) + Cl(^2P)$$

then $M_{L_1} = 1, 0, -1$, $M_{L_2} = 0$, and $\Lambda = 1$ (Π state) or 0 (Σ state). The HCl molecule has the electronic configuration $\sigma_{H-Cl}^2 \pi_{Cl}^4$, $^1\Sigma$ in its ground state; in its first excited state an electron is promoted out of the non-bonding π_{Cl} orbital into the H–Cl antibonding σ^*-orbital, to give the configuration $\sigma_{H-Cl}^2 \pi_{Cl}^3 \sigma_{H-Cl}^*$, $^{1 \text{ or } 3}\Pi$. All the states correlate with atoms which are configurationally unexcited (i.e. $H(^2S) + Cl(^2P)$). The Π state is doubly degenerate since only positive values of Λ are taken (the splitting of terms by the electric field is proportional to the *square* of the field strength), but the Σ states split into two components, denoted Σ^+ or Σ^- depending on whether the sum $L_1 + L_2 + \sum_i l_{1i} + \sum_j l_{2j}$ is even or odd.*
The ground state of HCl is $^1\Sigma^+$.

If both atoms are produced in P states, as in the dissociation

$$Cl_2 \rightarrow Cl(^2P) + Cl(^2P)$$

then $L_1 = L_2 = 1$, and $\Lambda = 2, 1, 0; 1, 0, -1; 0, -1, -2$ producing a Δ state, two Π states and three Σ states. Of the latter, two form a pair arising from $M_{L_1} = \pm 1$, $M_{L_2} = \mp 1$ and split into Σ^+ and Σ^- components: the third is formed from the combination $M_{L_1} = M_{L_2} = 0$, and is Σ^+ or Σ^- depending on the parity.

The spin angular momenta are unaffected by the internuclear electric field, and the components Σ along the axis are not required. The resultant spin of the molecule can be $S_1 + S_2$, $S_1 + S_2 - 1, \ldots, S_1 - S_2$. For example $O(^3P)$ and $N(^4S)$ could correlate with NO in any of the states $^6\Pi$, $^6\Sigma^+$, $^4\Pi$, $^4\Sigma^+$, $^2\Pi$ or $^2\Sigma^+$. In its ground electronic state, NO has the

---

* L_1 and L_2 are the nett orbital angular momenta of the electrons in atoms 1 and 2, and l_{1i} and l_{2j} are those of electrons i and j in atoms 1 and 2.

configuration ... $\sigma^2_{N-O}\pi^4_{N-O}\pi^{*\,1}_{N-O}$; it is the $^2\Pi$ species which is the ground state.

In molecules where Hund's case (c) is approached, and there is strong spin–orbit coupling, the total angular momenta J of the atoms can be correlated with the components Ω along the internuclear axis in similar fashion to L and Λ.

Taken alone, the correlation rules do not specify in which of the possible combinations of electronic states which satisfy the requirements of symmetry the dissociation products actually appear. However, the field is narrowed when the relative energies of the dissociation products are also taken into account. In the discussion on predissociation it was stated that radiationless transitions were only allowed between molecular electronic states of the same symmetry (which includes the spin). The potential energy curves of states of different symmetries (or spectroscopic designations) may intersect; those of the same symmetry 'avoid' the intersection (see Figure 2.24). A molecule whose vibration carries its configuration point through an intersection is likely to remain in the same electronic state and not cross into the neighbouring one. This is known as the '*non-crossing rule*'. Let us apply it to the example of NO, which has a $^2\Pi$ ground electronic state. This can be correlated not only with two normal atoms $N(2^4S) + O(2^3P)$ but also with a combination in which the nitrogen atom is excited into its 2^2D state. However, a $^2\Pi$ curve which correlates with the latter cannot cross a $^2\Pi$ curve correlating with

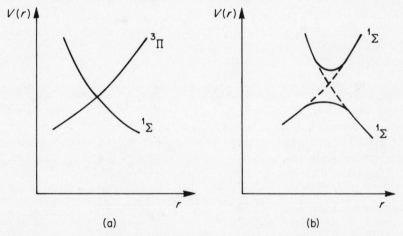

(a)   (b)

Figure 2.24 Intersection of potential energy curves. (a) different symmetries—crossing, (b) same symmetries—avoiding crossing.

two unexcited atoms; the lower molecular state must correlate with the least excited products. This is a general rule.

In some instances it is possible to determine the dissociation energies of diatomic molecules from their electronic spectra. For example, the absorption of visible light excites the electronic transition $^3\Pi_{0u}^+ \leftarrow {}^1\Sigma_g^+$ in iodine vapour, and is responsible for its violet colour. The vibrational bands converge to a limit, where normal $(5^2P_{\frac{3}{2}})$ and excited $(5^2P_{\frac{1}{2}})$ I atoms are produced. The energy of the excited atom can be found from its atomic spectrum, and hence the dissociation energy of I_2 into two normal atoms. These correlate with the ground electronic state $^1\Sigma_g^+$.

If the vibrational bands had not converged, because of Franck–Condon restrictions, an extrapolation method might have been used to estimate the series limit. However, much caution must be exercised before acceptance of the results so obtained, since the extrapolation assumes a knowledge of the shape of the upper potential energy curve. In the Birge–Sponer method, it is assumed to be described by a Morse function, so that the difference in the energies of successive vibrational levels is directly proportional to their vibrational quantum number. This assumption usually fails for high vibrational levels and the limit measured from such a linear extrapolation is too high. The potential energy curves of covalently bound molecules converge faster than the negative exponential terms in the Morse function. These, and related matters have been discussed in great detail by Gaydon,[42] to whom the reader is referred hereafter.

As well as producing atoms in excited electronic states, photodissociation may also endow them with considerable translational energy. In order to conserve momentum, the excess energy absorbed above the dissociation limit will be divided between the separating atoms in the inverse ratio of their atomic weights. In the dissociation of a molecule AB, conservation of momentum requires

$$m_A v_A = m_B v_B$$

where m and v are the atomic masses and velocities. The relative translational kinetic energies are $m_A v_A^2 / m_B v_B^2 \equiv m_B / m_A$. The classic example of 'translationally hot' atom production is the photolysis of HI vapour by light at 253·7 nm. Since the hydrogen atom $(m_A = 1)$ is so much lighter than the iodine atom $(m_B = 127)$, it carries virtually all the excess energy (127/128ths). The dissociation energy of HI is 297 kJ mole^{-1}, and at 253·7 nm, 471 kJ mole^{-1} are concentrated into the molecule. If it dissociates into atoms in their ground electronic states, the H atom carries almost 172 kJ mole^{-1} in translation; if the iodine atom is produced in the $5^2P_{\frac{1}{2}}$ state, only 84 kJ mole^{-1} are available for translational excitation. The near-ultraviolet absorption continuum of HI actually arises from several overlapping electronic transitions, all of which populate repulsive upper states. Two of these, $^1\Pi_1$ and $^3\Pi_1$ correlate with normal iodine atoms while the third, $^3\Pi_0$, correlates with $H(1^2S) + I(5^2P_{\frac{1}{2}})$.[43] Donovan and Husain have found that dissociation by polychromatic light of $\geqslant 200$ nm gives $\sim 1/5$th of the iodine atoms in the excited state.[44] The estimate is limited by the fact that the yield is likely to vary with the absorbed wavelength (see section 3.2b).

An ingenious method for probing the dynamics of photodissociation has been reported by Solomon,[45] in which the molecule is dissociated by polarized light. In the case of the halogens, the dissociation continua are associated with $^1\Pi_{1u} \leftarrow {}^1\Sigma_g^+$ electronic transitions for which the transition dipole is polarized perpendicular to the internuclear axis. In consequence, the separating atoms fly apart in the plane perpendicular to the plane of polarization of the absorbed light. They were detected by their ability to remove a tellurium mirror which had been deposited on the inside of the hemispherical containing vessel. The pressure was sufficiently low to give a mean free path greater than the vessel's diameter. The technique allows the polarization of electronic transitions producing continua to be determined. This is impossible to determine from the spectral analysis,

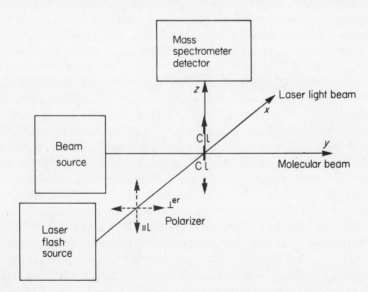

Figure 2.25 Block diagram showing principles of crossed light beam-molecular beam apparatus for studying the molecular dynamics of photodissociation. (For Cl_2, dissociation in the near-ultraviolet is associated with a perpendicular transition ($^1\Pi \leftarrow {}^1\Sigma$), and the polarizer would be set to transmit perpendicular polarized light.)

since there is no rotational fine structure. If the flight times of the separating atoms could be determined, it would then be possible to discover the manner in which the excess energy is divided between the dissociating products. This exciting aim has very recently been realized by Wilson and coworkers,[46] who have substituted a laser pulse for the continuous radiation, a small mass spectrometer detector for the tellurium mirror, and a molecular beam for the bulk absorbing gas (see Figure 2.25). This allows them to measure the velocities and angular directions of the recoiling photodissociation products when the molecule is excited with a very brief (20 nsec) pulse of monochromatic light. The results give a description of the molecular dynamics of photodissociation, including the distribution of energy among the products, and the electronic states of the dissociating molecules. The technique has been applied to polyatomic molecules as well as diatomic to give extremely detailed information; in the journalists' jargon, the technique represents a 'breakthrough' providing a new form of molecular spectroscopy for the study of dissociative electronic states.

2.4   POLYATOMIC MOLECULAR SPECTRA

Like their diatomic relations, the electronic absorption spectra of polyatomic molecules may be discrete, diffuse or continuous, but discrete spectra now become a comparative rarity, and diffuse spectra do not necessarily reflect a brief lifetime in the upper state. Fine structure may disappear simply as a consequence of molecular size and complexity. Rotational spacings are in inverse proportion to the molecular moments of inertia (see p. 65); individual spectral lines crowd more and more closely together as the molecular weights and dimensions increase until the density of rotational levels within a given energy range is such that they overlap and become unresolvable. If non-radiative intramolecular processes compete with radiative decay to reduce the natural lifetime of the excited molecule, the onset of diffuseness will be more rapid. The most common process involves radiationless transition into a neighbouring electronic state; this is favoured by increasing molecular complexity which increases the density of both electronic and vibrational levels. In addition, the selection rules for radiationless transition are far less restrictive than in diatomic molecules since they can be relaxed when the molecule is excited into vibrational modes of the appropriate symmetry. The nett vibronic symmetry can be such as to permit a transition that is electronically forbidden. Vibronic perturbations of this type are impossible in diatomic molecules (see p. 73).

If the molecule is not isolated in the vapour phase, but dissolved in solution, the perturbations caused by collisions with the solvent are sufficient to blur the rotational structure completely (cf. pressure broadening in the vapour phase), though the vibrational structure is generally left at least partially intact. In this situation, freezing the solution to a glass can often improve the resolution by reducing the thermal motion and preventing collisions. Since the larger polyatomic molecules are unlikely to be volatile, their absorption spectra and their photochemistry must perforce be studied in solutions. The field of study which has become known as 'Organic Photochemistry' relates almost exclusively to photochemical processes in condensed phases.

To summarize, the photochemist is usually concerned only with the electronic and vibrational (or vibronic) energy levels in the molecules that he studies. Discrete rotational levels are present only in relatively light molecules which are in the vapour phase; we shall be interested in them only when asking the most detailed questions about the photochemical primary process.

2.4a Types of electronic transition: notation

In simple molecules containing only a few atoms (for example, NO_2), the least tightly bound electrons occupy m.o.'s which are not localized on or between any given pair of atoms. The valence bond description is inappropriate and almost invariably electronic transitions in polyatomic molecules are discussed in terms of molecular orbital theory.[47] In more complex molecules it is often the case that the least tightly bound electrons tend to be localized on a relatively small group of neighboring atoms within the molecule, and to a good approximation low-lying electronic transitions can be treated as if localized on that group. For example, in the series of saturated, aliphatic aldehydes and ketones, the carbonyl function forms a common *chromophoric group* with its lowest lying electronic transitions set in the same region of the spectrum, regardless of the molecule in which it is bound. The longest wavelength absorption bands associated with some of the more common chromophores are collected in table 2.2. The consequences of the electronic transition may subsequently 'reverberate' around the entire molecule as it vibrates about its new equilibrium configuration, but the initial act of light absorption will tend to be localized on one part of the molecule and can be discussed in terms of localized m.o.'s. If there are two, unconjugated, non-interacting chromophoric groups within the same molecule, remaining effectively 'insulated' from each other (for example by a chain of methylene groups) the nett spectrum is the sum of each.

If the molecule has some elements of symmetry its electronic states can be expressed in the group theoretical notation discussed in section 1.3b.

TABLE 2.2 Wavelengths and extinction coefficients of some common chromophoric groups in organic molecules

Group	Molecule	$\sim \lambda_{max}$ (nm)	$\sim \varepsilon$(l mole^{-1} cm^{-1})
—Cl	CH_3Cl	173	weak
—Br	CH_3Br	204	200
—I	CH_3I	258	365
—OH	H_2O	166	100
—OCH$_3$	CH_3OH	184	150
	CH_3OCH_3	184	2,520
—SCH$_3$	CH_3SCH_3	210	1,000
—NH$_2$	CH_3NH_2	215	600
>C=C<	$CH_2=CH_2$	162	10,000
	$CH_3CH=CH_2$	175	12,000
	butadiene	210	20,000
	⬡	257	8,000
—C≡C—	$CH≡CH$	220	weak
		182	moderate
		152	strong
	$CH_3C≡CH$	187	450
—C≡N	CH_3CN	167	weak
—N$_3$	$C_2H_5N_3$	287	20
		222	150
>C=O	$H_2C=O$	295	10
	CH_3CHO	290	12
	$(CH_3)_2CO$	270	16
>C=S	$(C_6H_5)CS$	605	66
—N=N—	$CH_3N=NCH_3$	340	4·5
—N=O	$(CH_3)_3CNO$	665	15
—NO$_2$	CH_3NO_2	270	8
Aromatic ring	benzene	255	200

For example, the ground and first excited singlet states of methane can be represented by the electronic configurations (see p. 43).

$$\ldots (2a_1)^2(1t_2)^6, \tilde{X}^1A_1$$

$$\underbrace{}_{\sigma_{C–H}}$$

and

$$\ldots (2a_1)^2(1t_2)^5(3a_1)^1, \tilde{A}^1T_2$$

$$\underbrace{}_{\sigma_{C–H}} \underbrace{}_{\sigma_{C–H}^*}$$

In group theoretical notation, an electronic transition between the two is written

$$\tilde{A}^1 T_2 \leftarrow \tilde{X}^1 A_1$$

with the usual convention that the excited state is written first. It is represented in Figure 2.26. If a detailed group theoretical dscription of the m.o.'s is not feasible because of the molecular complexity, or is too detailed for the type of question being asked, then the approximate and widely used notation introduced by Kasha[48] can be employed. Under this, the m.o.'s involved in the electronic transition are described as σ- or π-bonding or antibonding (σ, σ^*, π or π^*) or non-bonding (n). A transition which transfers an electron from a σ_{C-H} into a σ^*_{C-H} orbital would be written

$$\sigma_{C-H} \rightarrow \sigma^*_{C-H}$$

or more simply, $\sigma \rightarrow \sigma^*$ populating a $^1(\sigma, \sigma^*)$ upper state. Note that now the upper state is conveniently written last! The notation is simple and convenient, but it can represent a rather severe approximation and should be used with caution. For example, in the case of methane, the $\sigma \rightarrow \sigma^*$ transition lies far down in the ultraviolet at $\leqslant 150$ nm, since the two orbitals are very strongly bonding and antibonding. The energy absorbed is of the order required to excite an electron into the $3s$ a.o. of the carbon atom, which like the σ^* orbital, is also totally symmetric (a_1). Consequently, the

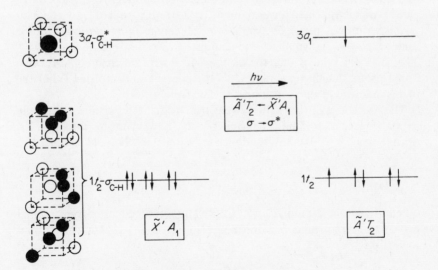

Figure 2.26   First electronic transition in CH_4.

upper state has a measure of $(3s)_C$ character. The transition is the first member of a Rydberg series, since the $(3s)_C$ orbital is unoccupied in the ground state of the atom.

The m.o.'s of formaldehyde were discussed in section 1.3c and 1.5c and the electronic configurations of its ground, and first excited singlet state, were shown to be

$$\ldots (1b_1)^2 (2b_2)^2, \tilde{X}^1 A_1$$

$$\underbrace{\qquad}_{\pi_{C-O}} \; \underbrace{\qquad}_{n_O}$$

and

$$\ldots (1b_1)^2 (2b_2)^1 (2b_1)^1, \tilde{A}^1 A_2$$

$$\underbrace{\qquad}_{\pi_{C-O}} \; \underbrace{\qquad}_{n_O} \; \underbrace{\qquad}_{\pi^*_{C-O}}$$

In group theoretical notation, the first electronic transition would be written (assuming the molecule remained planar)

$$\tilde{A}^1 A_2 \leftarrow \tilde{X}^1 A_1$$

The associated ultraviolet absorption is centred around 300 nm, but since there is no component of the transition dipole with A_2 symmetry, the transition is electronically forbidden and the maximum extinction is only $\sim 10 \, l \, \text{mole}^{-1} \, \text{cm}^{-1}$ (see p. 75). The same is true of acetaldehyde, acetone and all the other saturated aliphatic ketones, although their nett molecular symmetry is often lower than in CH_2O (point group C_{2v}); CH_3CHO possesses only a plane of symmetry on average (point group C_s), and under the lowered symmetry the analogous transition would be allowed. That it is not implies that the *local symmetry* of the chromophoric carbonyl group, which remains C_{2v}, is the controlling factor, and that the m.o.'s involved in the transition are effectively localized on it.

The absorbed photon excited an electron from the $2b_2$ orbital, essentially a non-bonding $2p$ orbital centred on the oxygen atom, into the $2b_1$ orbital which is C–O π-antibonding (see Figure 2.27). It could be described as

$$n_O \rightarrow \pi^*_{C-O}$$

or simply $n \rightarrow \pi^*$, populating a $^1(n, \pi^*)$ state. If an electron were excited out of the $1b_1$ orbital the upper electronic state would have the configuration

$$\ldots (1b_1)^1 (2b_2)^2 (2b_1)^1, \tilde{B}^1 A_1$$

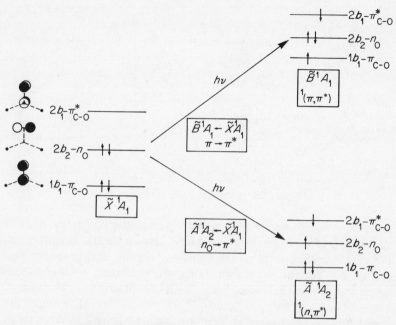

Figure 2.27 Electronic transitions in H_2CO.

with the same symmetry as the ground state; the transition

$$\tilde{B}^1 A_1 \leftarrow \tilde{X}^1 A_1$$

would be allowed for electric dipole radiation polarized along the C–O bond (A_1 symmetry). The $1b_1$ orbital is C–O π-bonding, and in Kasha's terminology the transition would be termed

$$\pi_{C-O} \rightarrow \pi^*_{C-O}$$

or $\pi \rightarrow \pi^*$ populating a $^1(\pi, \pi^*)$ state.

The near-ultraviolet absorption of CH_3I is associated with the electronic transition (see p. 44)

$$\ldots (3a_1)^2 (2e)^3 (4a_1)^1, \tilde{A}^1 E \leftarrow \ldots (3a_1)(2e)^4, \tilde{X}^1 A_1$$

$$\underbrace{}_{\sigma_{C-I}} \underbrace{}_{n_I} \underbrace{}_{\sigma^*_{C-I}} \qquad \underbrace{}_{\sigma_{C-I}} \underbrace{}_{n_I}$$

An electron is promoted out of the $2e$ orbital, effectively the doubly degenerate $\{5p_x, 5p_y\}$ orbital localized on the I atom, into the $4a_1$ orbital, mainly C–I σ-antibonding (see Figure 2.28). In Kasha's notation the

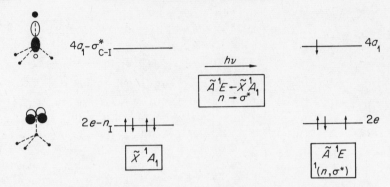

Figure 2.28   First electronic transition in CH_3I.

transition would be termed $n \rightarrow \sigma^*$ populating a $^1(n,\sigma^*)$ state. The symmetries of the two combining states allow an electric dipole transition polarized perpendicular to the C–I bond, in the (x,y) plane (symmetry species E). Yet the maximum extinction coefficient of the molecular spectrum is only $370 \, l \, mole^{-1} \, cm^{-1}$ and the integrated absorption gives an oscillator strength $f = 7 \times 10^{-3}$, very much less than unity. As in acetaldehyde, the concept of local symmetry is helpful in this case also, for if the σ^* orbital is constructed largely from the $5p_z$ orbital of the I atom, the nett electronic transition would include a large contribution from the localized transition

$$(5p_x, 5p_y)_I \rightarrow (5p_z)_I$$

Under the local spherical symmetry of the I atom, the $p \rightarrow p$ transition would be forbidden by the Laporte selection rule. In general, in molecules where the description $n \rightarrow \sigma^*$ or $n \rightarrow \pi^*$ do not represent too great approximations, the transitions are relatively weak, with maximum extinction coefficients $10^3 \, l \, mole^{-1} \, cm^{-1}$.

The π-electronic structure of benzene was discussed in section 1.3c (pp. 51–53): transfer of an electron from the highest occupied m.o. e_{1g}, into the lowest vacant m.o. e_{2u}, gives the degenerate electronic configuration

$$\ldots (a_{1u})^2 (e_{1g})^3 (e_{2u})^1$$

producing electronic states of symmetries B_{1u}, B_{2u} and E_{1u}. Of the three spin-allowed possible transitions

$$
\begin{aligned}
\tilde{A}^1 B_{2u} &\quad (\sim 250 \, nm \, ; \, \varepsilon_{max} \sim 10^2) \\
\tilde{B}^1 B_{1u} &\leftarrow \tilde{X}^1 A_{1g} \quad (\sim 200 \, nm \, ; \, \varepsilon_{max} \sim 10^4) \\
\tilde{C}^1 E_{1u} &\quad (\sim 180 \, nm \, ; \, \varepsilon_{max} \sim 10^5)
\end{aligned}
$$

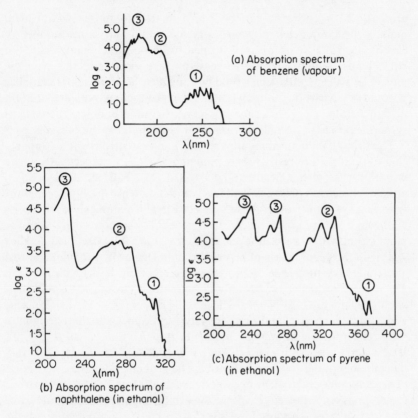

Figure 2.29   Ultraviolet absorption of aromatic hydrocarbons.

the first two are electronically forbidden, but become vibronically allowed if vibrational modes of e_{2g} symmetry are also excited (see p. 74). Each transition could be termed $\pi \rightarrow \pi^*$ though this description is one of type rather than detail.

There are strong similarities between the near ultraviolet absorption profiles presented by benzene and polycyclic aromatic hydrocarbons such as naphthalene, anthracene, pyrene and so on. It is usually possible to distinguish a band of low intensity at the longest wavelengths, followed by two, or sometimes three others of increasing intensity at shorter wavelengths (see Figure 2.29). These were originally termed the α, p and β bands by Clar,[49] and in benzene they can be correlated with transitions into the $^1B_{2u}$, $^1B_{1u}$ and $^1E_{1u}$ states. They were subsequently redesignated ^1L_a, ^1L_b and ^1B_a or ^1B_b states by Platt[50] on the basis of the free electron (FE) approximation. This assumes that the π-electron clouds of *cata*-condensed aromatic hydrocarbons* can be approximated to those of equivalent monocyclic conjugated polyenes which contain the same number of carbon atoms. The atoms are assumed to lie on the circumference of a circle whose length is equal to that of the perimeter of the aromatic molecule. Thus naphthalene would be equivalent to an imaginary circular polyene of ten carbon atoms, with ten π-electrons circulating in a uniform field in a closed ring. The cross-linking needed to form polycyclic systems is then regarded as a small perturbation of the monocyclic polyene.

The orbital angular momenta and energies of electrons circulating in a ring of radius a, are restricted to $\pm n\hbar$ and $n^2 h^2 / 2ma^2$ respectively, where m is the mass of the electron and $n = 0, 1, 2, \ldots r$. Since there are two possible directions of rotation all levels with $n > 0$ are doubly degenerate. In the case of naphthalene, the ten π-electrons will completely fill the levels with $n \leqslant 2$ in the ground electronic state and the nett orbital angular momentum will be zero. If levels up to $n = r$ are fully occupied, as they will be in the general case of an aromatic molecule ($4r + 2$ rule), the first electronic transition will promote an electron from levels with $n = r$ to $n = r + 1$, producing two doubly degenerate electronic states with nett angular momenta of $[\pm r \pm (r + 1)]\hbar$, i.e. $\pm \hbar$ or $\pm(2r + 1)\hbar$. For benzene and naphthalene $r = 1$ and 2 respectively, (see Figure 2.30) Platt designates electronic states with angular momenta $0, \pm \hbar, \pm 2\hbar, \ldots$ as $A, B, C, \ldots$ states, and those with $\pm(2r + 1)\hbar, \pm(2r + 2)\hbar, \ldots$ as $L, M, \ldots$ states, independent of the absolute value of r. The ground and first excited states of benzene and naphthalene become ^1A, ^1L and ^1B and since an electronic transition is only allowed to change the orbital angular momentum by one unit, the only one permitted is $^1B \leftarrow {}^1A$. In the FE approximation all the excited states would be doubly degenerate, but alternation of the potential field around the ring of atoms splits that of the L state into two components termed ^1L_a and ^1L_b. In molecules of lower symmetry than benzene, the cross-linking also splits the ^1B state.

Murrell has given a good discussion of the FE model, as well as alternative approximate MO treatments, and for further and more detailed information the reader is referred to his book.[47]

When heteroatoms with 'lone pair electrons' are substituted for a carbon atom in the benzene ring, both $n \rightarrow \pi^*$ and $\pi \rightarrow \pi^*$ transitions may appear in the near-ultraviolet absorption. For example, pyridine shows weak absorption to the red of the $\pi \rightarrow \pi^*$ band, associated with excitation of an electron in the non-bonding sp^2 orbital lying in the molecular plane. However, when a heteroatom is substituted for one of the hydrogen atoms, the 'lone pair orbital' is not orthogonal to the π-system and some

---

* i.e. those where all the carbon atoms lie on the perimeter of the molecule; if they do not they are described as *peri*-condensed.

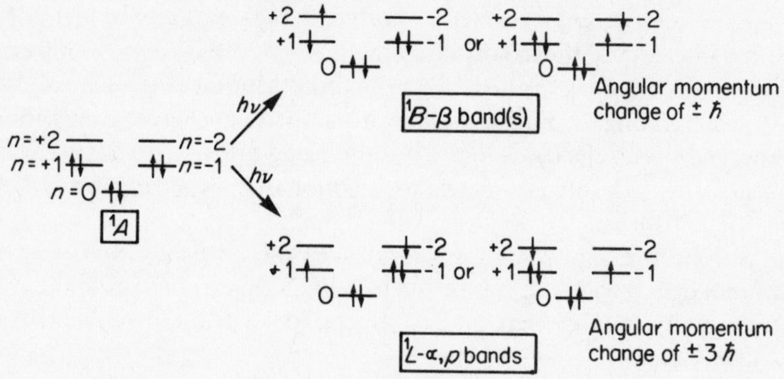

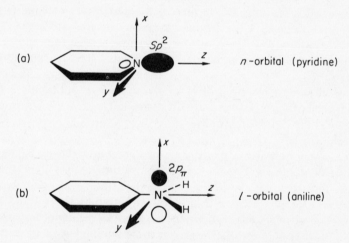

+2 —+—   ——-2   +2——   —+-2
+1—+—   —++—-1   or   +1—++   —+—-1
    0 —++—   0 —++—

Angular momentum
change of $\pm\,\hbar$

$^1B{-}\beta$ band(s)

$n=+2$ ——   ——$n=-2$
$n=+1$—++—   —++—$n=-1$
   $n=0$ —++—

^1A

+2 ——   —+—-2   +2—+—   ——-2
+1—+—   —++—-1   or   +1—++   —+—-1
    0 —++—   0 —++—

$^1L{-}\alpha,\rho$ bands

Angular momentum
change of $\pm\,3\,\hbar$

Figure 2.30   Electronic transitions in benzene on the basis of the free electron model.

(a) n-orbital (pyridine)

(b) l-orbital (aniline)

Figure 2.31   Lone pair orbitals in pyridine and aniline.

overlap may result. For example, in a planar conformation of the aniline molecule the 'lone pair' orbital on the nitrogen atom will be $2p\pi$, lying parallel to the π-system of the ring, unlike that in pyridine (see Figure 2.31). Kasha[51] has accommodated the distinction in his terminology by defining 'n' orbitals which lie in the molecular plane, and 'l' orbitals which do not. The degree of π-bonding between the nitrogen atom and the ring in an aromatic amine depends on the actual angle between the plane of the ring and that of the amino group. When they are not coplanar, the antibonding π-orbitals are termed 'a_π', and the $\pi \rightarrow \pi^*$ electronic transitions have

$l \rightarrow a_\pi$ character mixed into them. The degree of mixing can be recognized from the intensity of the absorption which is reduced by $l \rightarrow a_\pi$ character, and from the changes in the absorption spectrum measured in solution. Hydrogen bonding to, or better still protonation of the heteroatom reduces the energies of the electrons in the non-bonding orbitals. Thus solution in hydrogen-bonded solvents (such as alcohols) or in strong acids shifts ultraviolet absorption associated with $n \rightarrow \pi^*$ or $l \rightarrow a_\pi$ transitions to the blue. Solution in non-polar solvents leaves the absorption close to the position it occupied in the vapour phase. When the atom is protonated the shift is usually so large that the absorption 'disappears' from view in the near-ultraviolet.

Mulliken many years ago introduced a simple notation which is applicable to all molecules, including the diatomic ones, and which is still in use. The ground state is designated N, and excited states of types V, Q, R and T are distinguished.[52] $V \leftarrow N$ spectra are associated with intra-valency shell transitions, where there is a high degree of charge transfer within the molecule. They are intense and correlate with $\sigma \rightarrow \sigma^*$ and $\pi \rightarrow \pi^*$ transitions. $R \leftarrow N$ transitions promote electrons into extra-valency shell orbitals and populate Rydberg states. $Q \leftarrow N$ transitions are analogous to $n \rightarrow \sigma^*$ or $n \rightarrow \pi^*$ excitations, transferring electrons from orbitals effectively localized on a single atom, into antibonding intra-valency shell orbitals. They are usually of low intensity. $T \leftarrow N$ transitions populate triplet states and lead us conveniently into the final part of this discussion.

Up to the present we have tacitly assumed that both the excited and the ground electronic states have singlet spin multiplicity. Unless there is strong spin–orbit coupling of course, triplet states will not be readily accessible through direct radiative transition from the ground state. However, electronic excitation produces open shell configurations and the upper state may possess either singlet or triplet multiplicity without falling foul of the Pauli Principle. For example, the $^1A_2$ and $^1A_1$ states of formaldehyde have $^3A_2$ and $^3A_1$ counterparts, described in Kasha's notation as $^3(n, \pi^*)$ and $^3(\pi, \pi^*)$. They are the first members of a system of excited triplet states which form a 'stack' stretching away towards the ionization limits. The members of the stack are termed $T_1, T_2, T_3, \ldots, T_n$, in order of increasing energy while the excited singlet states are termed $S_1, S_2, S_3, \ldots, S_n$, with S_0 reserved for the ground state. Since this has a closed shell electronic configuration there is no T_0 state.

Although direct excitation into triplet states is spin-forbidden, they are often readily accessible through radiationless transition from neighbouring singlet states. The process is conventionally termed *intersystem crossing*. If the molecule contains a heavy atom, direct radiative transitions from the ground state become more probable, because of the increased spin–orbit coupling. They are associated with relatively weak absorption which appears to the red of the lowest energy singlet–singlet $(S_0 \rightarrow S_1)$ bands, since T_1 lies below S_1. Sometimes they may lend colour to the molecule by extending its absorption spectrum from the ultraviolet into the visible. In recent years the first electronic transitions in SO_2 and CS_2 have both been recognized as $S_0 \rightarrow T_1$, in the case of SO_2 through analysis of the rotational fine structure[53] and in both CS_2 and SO_2 by observation of the Zeeman splitting of the levels in the paramagnetic triplet state.[54] The

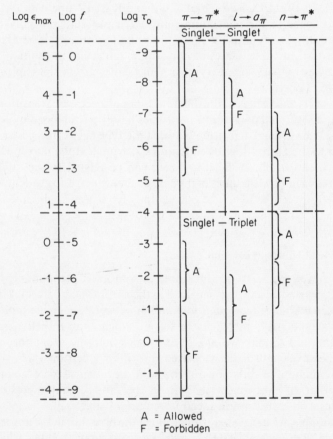

A = Allowed
F = Forbidden

Figure 2.32 Relative intensities and luminescent lifetimes associated with polyatomic molecular electronic transitions. ε_{max}—maximum extinction coefficient, f—oscillator strength, τ_0—natural radiation lifetime, sec. (From M. Kasha and H. R. Rawls, *Photochem. and Photobiol.* **7**, 561, (1968).)

transitions derive their intensity from the appreciable spin–orbit coupling induced by the relatively heavy sulphur atoms.

The spin of the lowest triplet level makes it metastable with respect to the ground singlet state and it may persist for relatively long periods before decaying. Its importance in the spectroscopy and photochemistry of the larger unsaturated, polyatomic molecules cannot be overstressed. For example, many of the photochemical reactions of electronically excited molecules which occur in solutions are those of the lowest triplet state. Indeed there are so many 'triplet' reactions now known that it is hard to realize that only 25 or so years have elapsed since Terenin[55] first suggested that the phosphorescent and chemically reactive states of eosin might be a triplet state of the molecule, and Lewis and Kasha[56] first demonstrated the equivalence of the phosphorescent and triplet states in polyatomic molecules. It was not until 1953 that Porter and Windsor[57] were first able to detect excited triplet states of polyatomic molecules in solution, yet a recent review on the chemistry of organic molecules in the triplet state includes 513 references.[58]

The relative intensities of the different types of electronic transition that occur in unsaturated polyatomic molecules have been analysed by Kasha. He finds a consistent correlation between the type of electronic transition assigned to the absorption (or emission) spectrum and the maximum and integrated extinction coefficients. These can be related to the oscillator strengths and natural lifetimes of the excited molecules, using the equations developed in section 1.5d. The correlations have been represented on a useful chart, which is reproduced in Figure 2.32.

2.4b Special effects in solution

When absorption spectra are measured in solution it is commonly found that the observed spectrum is sensitive to the nature of the solvent. There is always some kind of molecular interaction between the solute molecules and the neighbouring molecules of solvent which form a shell around it and, if the interactions are altered through electronic excitation of the solute, the wavelength of the associated absorption spectrum will be shifted. The magnitude and directions of the shift are often used to assign the character of an electronic transition in the more complex polyatomic molecules whose absorption spectra are unresolvable.

For example, we have already mentioned the blue shift in the absorption associated with $n \rightarrow \sigma^*$, $n \rightarrow \pi^*$ or $l \rightarrow a_\pi$ transitions (p. 144), when the solute is dissolved in hydroxylic solvents. The shift is particularly pronounced in these media since the hydrogen-bonding which stabilizes the ground electronic state is greatly weakened when one of the electrons is excited out of the lone pair orbital. A particularly large shift in such solvents, or on protonation can be used as a diagnostic test. The smaller blue shift that occurs in polar solvents in the absence of hydrogen-bonding arises from an alteration in the strength of the solvent–solute dipole–dipole interaction. In the ground electronic states of aliphatic ketones for example, the electronic charge distribution is weighted toward the oxygen atom because of its greater electronegativity. In the $^1(n, \pi^*)$ state however, electronic charge is transferred back towards the central carbon atom, reducing the magnitude of the nett dipole moment. If the dipole–dipole interaction is weakened the upper state is less stabilized than the ground state. In addition, the equilibrium orientation of the dipolar solvation shell which existed around the unexcited molecules becomes inappropriate in the upper state. Because of the Franck–Condon restriction, there is a considerable delay between the instant of excitation and reorientation of the solvation shell into its new equilibrium conformation. The effect has been termed orientation strain by Bayliss and McRae,[59] and we shall meet it again when discussing fluorescence. These authors have also discussed the manner in which other weaker types of molecular interaction, particularly dispersion forces, may affect absorption spectra in solution.

The absorption bands of molecules which suffer a large increase in dipole moment on excitation are usually intense, and show large *red* shifts when the polarity of the solvent is increased. The magnitude of the dipole moment in the excited state, can be estimated using an equation based on one derived by Lippert,[60] which predicts the shift in the absorption maximum $\tilde{v} - \tilde{v}_0$, when the dielectric constant D, and refractive index n, of the solvent are varied

$$\tilde{v} = \tilde{v}_0 - \frac{2\mu^2}{hca^3}\left\{\frac{D-1}{2D+1} - \frac{n^2-1}{2(2n+1)}\right\}$$

where a is the radius of the cavity in the solvent, occupied by a molecule of the solute, and μ is the dipole moment of the excited molecule.

For example, the near-ultraviolet absorption of *p*-nitroaniline shifts to the red with increasing solvent polarity, and the change in dipole moment is estimated to be ~ 10 Debyes.[61] This is very large, and the absorption is associated with an *intramolecular charge-transfer transition*. The question is, how far has the charge been transferred? Three alternatives can be considered:

(i) it could be from the lone pair orbital of the amino-group into the ring (i.e. of the type $l \rightarrow a_\pi$ as in aniline (pp. 142–144)), or

(ii) from the ring to the electron accepting nitro group, or

(iii) the whole way from the amino donor to the nitro acceptor, to give a zwitterionic structure

$$H_2\overset{\oplus}{N}\!=\!\!\left\langle\!\!\!\!\!\rule{0pt}{0pt}\right\rangle\!\!=\!\overset{\ominus}{N}O_2$$

for the excited state. (i) and (iii) would involve electrons occupying the non-bonding orbital on the amino group and the red shift would be reduced in a hydrogen-bonding solvent. This was not the behaviour observed and alternative (ii) must be the correct one. Anybody interested in the chemistry of the excited molecule would be very pleased to have information about its electronic configuration, and simple measurements of the type described can be most helpful in distinguishing alternative photochemical mechanisms.

In order for light absorption to effect charge transfer from a donating to an accepting group it is not necessary that both be bound within the same molecule. *Intermolecular charge-transfer absorption* (and emission) bands also develop within weak molecular donor–acceptor complexes. For example, tetracyanoethylene and hexamethylbenzene are both colourless solids. The one is a powerful electron acceptor and the other readily donates an electron from its π-system. Dissolved together in ethanol they produce a solution which, to the eye at least, resembles claret. The colour is associated with the transfer of an electron from the donor to the acceptor. This was first recognized by Mulliken,[62] who suggested that the ground electronic state of the neutral complex had a little ionic character mixed into it, and vice versa in the excited state. The wave functions of the two states could be written as

$$\psi_I = \psi_1(D, A) + \lambda\psi_2(D^+ - A^-); \qquad \psi_{II} = \psi_2(D^+ - A^-) + \mu\psi_1(D, A)$$

where μ and λ are $\ll 1$, and ψ_1 and ψ_2 represent the interactions of the neutral molecules, and of their radical ions. They can be improved if some interaction between excited states of the donor and/or acceptor is included, i.e. by including the possibility of electron transfer into vacant or half-filled orbitals of the acceptor from doubly or singly occupied orbitals of the donor.[63]

If the number of methyl groups substituted into the benzene ring is reduced, the absorption maximum shifts to the blue until with benzene itself only a pale yellow colour is left. To a good approximation the frequency of the central maximum is proportional to the ionization potential of the aromatic hydrocarbon.[64]

Charge-transfer absorption bands also appear when it is the ground state which is ionic or when ions are solvated or complexed. A few examples will illustrate some of the possibilities. The quaternary salts of amines can remain associated as ion pairs in solution. The pyridinium iodides absorb in the near ultraviolet and visible. Their absorption maxima shift to the blue as the polarity of the solvent is increased, reversing the trend shown in neutral molecular complexes; Kosower[65] has used the shifts to define a scale of solvent polarity at the molecular rather than the macroscopic level.* Many of the iodides present two absorption maxima separated by an interval close to that between the ground $(^2P_{\frac{3}{2}})$ and first excited $(^2P_{\frac{1}{2}})$ states of the iodine atom. They can be assigned to the alternative charge-transfer processes[66]

While in the neutral excited state, the two radical fragments may diffuse apart and dissociate, initiating a photochemical change. Kosower and Lindqvist[67] have in fact been able to detect the pyridyl radical in absorption immediately following flash excitation.

In dilute solutions of the alkali metal iodides, both ions are solvated and may separate. The ultraviolet absorption of the I^- ions is found to be dependent on the nature of the solvent, but not on the nature of the alkali metal. They also present two absorption maxima separated by the $I(5^2P_{\frac{3}{2}} - 5^2P_{\frac{1}{2}})$ interval and the absorption has been assigned to electron transfer from the anion to the surrounding solvation shell; this is commonly termed 'c.t.t.s.' (charge transfer to solvent).[68] In aqueous solutions it has been shown that the electron circulates in a hydration shell, and is associated with an absorption band centred around 700 nm. The band has been detected during flash excitation of aqueous solutions containing I^- and $Fe(CN)_6^{4-}$ as well as aqueous solutions of the phenate ion.[69] The iodine atom produced in the iodide solutions is rapidly scavenged to produce the I_2^- radical ion, which can be detected as a transient species in absorption, and eventually the stable I_3^- ion.[70]

* The energies in kcal mole^{-1} corresponding to the frequencies of the absorption maxima, define a scale of solvent 'z-values'.

Dissociation may also follow light absorption by transient metal complexes, where electron transfer between the ligand and the central ion can produce free atoms or radical ions. For example, the ultraviolet light absorption of the ferric ion complexes of the type $Fe^{3+}X^-$ is associated with the process

$$Fe^{3+}X^- + hv \rightarrow (Fe^{2+}X)$$

where X^- could be halide, hydroxide, azide and so on.[71] At longer wavelengths, the transition metal ions absorb more weakly due to the so-called '$d \rightarrow d$' transitions localized on the central ion. These cannot occur in the spherically symmetrical free ion since the d orbitals are degenerate, and even if they were not, the transitions would violate the Laporte selection rule. Under the lower symmetry of the ligand field which surrounds the complexed ion, the degeneracy is split, and the transitions become allowed.

We have travelled a long way from the absorption spectra of isolated molecules in the gas phase with which the section began, and we have only discussed the changes in electronic energy which may occur in polyatomic molecules, and latterly in ions. The absorption of many polyatomic molecules is structured, because changes in vibrational energy accompany their electronic excitation. If the molecules are isolated in the vapour phase and are not too complex, they may also exhibit rotational fine structure under very high resolution. These matters are discussed in the following section.

2.4c Rotational and vibrational structure

(i) *Rotational structure*

The appearance of the rotational structure in a molecular electronic absorption spectrum is determined by the molecular complexity, the lifetime of the upper state, the relative equilibrium geometries of both electronic states and by their symmetries, which control the permitted changes in the rotational quantum numbers. Assuming you have the use of a spectrograph with a resolving power sufficient to record faithfully the profile of a Doppler broadened spectral line, the chance of actually seeing any rotational fine structure depends on the spacings between the rotational levels and on the actual line widths. In a molecule of the size of benzene the moments of inertia about the principal axes are small enough to crowd the rotational levels together, and so the average spacings in a given electronic or vibronic band are less than the Doppler width and the individual lines are unresolved. Any process that competes with radiative decay from the upper state makes matters worse by increasing the line width. Completely resolved structure is restricted to molecules with not more than three 'heavy atoms' (like carbon, nitrogen or oxygen) along their spine, possibly combined with 'light' atoms (such as hydrogen or deuterium) which do not markedly increase the moments of inertia.

The orientation of the transition moment strongly influences its general appearance by restricting the allowed changes in rotational quantum numbers. For example in a symmetric or slightly asymmetric rotor such as NO_2, $\Delta K = 0$ when the transition moment is directed parallel to the top axis, but $\Delta K = \pm 1$ when the transition moment is in a perpendicular direction. In either case $\Delta J = 0 \pm 1$ (except for the $K = 0 \rightarrow 0$ sub-band where $\Delta J = \pm 1$ only). Thus a vibronic band associated with a perpendicular transition consists of a progression of 'K sub-bands', each having its 'J-structure' separated into P, Q, and R-branches. When the change in molecular dimensions is small, the Q-branches are prominent and the profile resembles that of an infrared vibrational band (see Figure 2.33). When there is an appreciable increase in size, as is usually the case when an electron is excited into an antibonding m.o., the J-rotational structure degrades to the blue, and the R-branches develop band heads, which may remain prominent even under relatively poor resolution. The 'K-structure' may march to the red or blue depending on the detailed changes in the moments of inertia on excitation.

Analysis of rotational fine structure recorded under high resolution allows estimates of the molecular moments of inertia and identification of the symmetries of the upper and lower states. In stable molecules the ground state is usually totally symmetric. Provided there is sufficient experimental information to match the geometrical unknowns, it then becomes possible to determine equilibrium bond angles and bond lengths in the excited molecule. Unfortunately this technique can seldom be applied to give detailed information about the photochemical primary process, since the electronic transitions which lead to photochemical change are rarely associated with resolved rotational structure. This is frustrating because changes in molecular structure which follow electronic excitation may be of profound importance in the subsequent photochemistry. Ideally, one would like to chart the dynamics of the photoexcited molecule as it dissociates, or vibrates through its new equilibrium configuration, or into that of some neighbouring electronic state or isomer. Our present knowledge of these matters is still vestigial. However, there are other methods of estimating the changes in equilibrium geometry in the absence of completely resolved structure. One is to 'synthesize' the unresolved rotational band contours by assuming trial sets of rotational constants and selection rules and using high speed digital computers, calculate the frequencies and

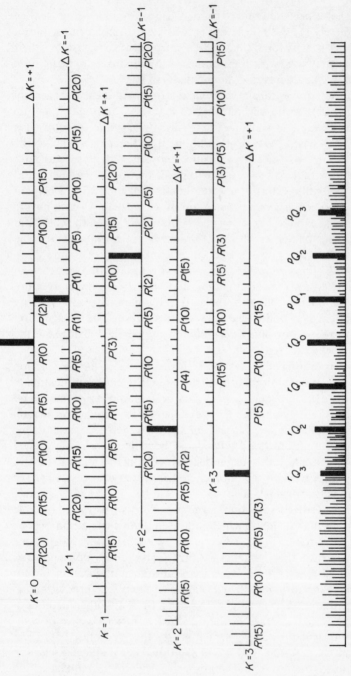

Figure 2.33 Calculated rotational structure of a perpendicular band of a symmetric rotar, where B′ ≃ B″ and A′ ≃ A″ (From Herzberg, *Electronic spectra and Electronic Structure of Polyatomic Molecules*, Vol. III, Copyright 1966, by Litton Educational Publishing, Inc., with permission). In the upper part seven sub-sub-bands are shown separately. At bottom the sub-bands are superimposed. The Q-branches in which many lines coincide are shown as black rectangles. The height of these rectangles and of the lines representing the P and R-branches indicate the intensities. No intensity alternation is indicated.

relative intensities of the spectral lines until there is agreement with experiment. This technique is particularly useful in 'heavy' molecules where the rotational structure is compressed, or broadened by pre-dissociation, etc. Alternatively, attention can be focused on the vibrational structure.

(ii) Vibrational structure

Unless the lifetime of the photoexcited molecule is shorter than $\sim 10^{-12}$ sec, it will have time to execute a few vibrations and resolved vibrational structure will appear in its absorption spectrum. The spacings will be those within the excited state, though if the temperature is sufficient some 'hot bands' may also appear to the red end of the spectral origin, due to excitation out of excited vibrational levels of the ground state. The vibrational progressions that appear will reflect the changes in equilibrium molecular geometry, but will be restricted by the requirements of symmetry (see section 1.5c). In allowed electronic transitions provided the two states belong to the same point group, bands based on the ground vibrational level of the lower state can only be associated with totally symmetric vibrations in the upper state. Other vibrations may subsequently be excited during the lifetime of the excited molecule, since anharmonicity mixes the normal modes and permits intramolecular energy transfer, but this will not be reflected in the absorption spectrum. The relative intensities of the bands that do appear are governed by the Franck–Condon Principle; the longer the progression, the greater the change in the appropriate normal coordinate. If the change is not too large so that the harmonic approximation is not too severe, the changes in the normal coordinates can be estimated from the relative intensities.[72]

For example the $\tilde{A}^1 B_{2u} \leftarrow \tilde{X}^1 A_{1g}$ absorption band of benzene shows a long progression in the totally symmetric ring 'breathing' frequency of the upper state, with spacings $\sim 925\ \mathrm{cm}^{-1}$, (see Figure 1.36). Because the transition is only allowed under vibronic perturbation (section 1.5c) each band also includes one quantum of an e_{2g} vibration of the upper state. In fluorescence, the spacings increase to $990\ \mathrm{cm}^{-1}$ which is the 'breathing' frequency of the ground state, (see Figure 2.34) (the vibration is also observable in the Raman spectrum). It can be deduced that $^1(\pi \rightarrow \pi^*)$ excitation has preserved the regular hexagonal shape, but slightly reduced the force constant of the breathing mode and increased the equilibrium C–C bond length by an estimated 0.02 Å.[73] The $n_O \rightarrow \pi^*_{C-O}$ transition in carbonyl compounds reduces the 'chemical' bond order from 2 to 1.5, increasing the equilibrium C–O bond length and giving rise to progressions in the C–O stretching frequency of the upper state. In CH_2O the bands are clearly resolved, with spacings $\sim 1180\ \mathrm{cm}^{-1}$: the frequency in the ground

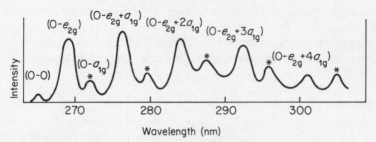

Figure 2.34 The fluorescence spectrum of benzene in an ethanol matrix at 77°K, showing the progressions in the a_{1g} 'breathing' mode of the ground state. (After Y. Murakami and Y. Kanda, *Bull. Chem. Soc. Jap.*, **41**, 2599 (1968).) (The more intense progression carries one quantum of the e_{2g} ring deformation mode of the ground state. The weaker system of bands (*) that begins at the spectral origin would be forbidden in an isolated molecule, but derives its intensity from the lowered symmetry induced by solvent perturbation.)

state is 1750 cm^{-1}. In more complex aldehydes and ketones the structure is often heavily blurred both by predissociation and the crowding of rotational levels, though progressions may sometimes still be discerned as broad maxima or shoulders superimposed on a continuous background (see Figure 2.35 and 2.36).

It is quite common for the molecular point group to alter on excitation because of a change in the overall shape. The first electronic transitions in CO_2, C_2H_2 and CH_2O all lower the molecular symmetry, from linear to bent in CO_2 and C_2H_2 and from planar to pyramidal in CH_2O (see section 1.3c). The latter change is reflected in the short progressions in the bending vibration which accompany the more prominent C–O stretching

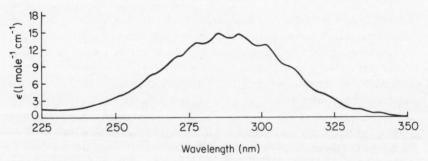

Figure 2.35 Absorption profile of the $^1(n_O \rightarrow \pi^*_{C-O})$ transition in C_2H_5CHO (vapour), showing progression of diffuse bands associated with C–O stretching vibration.

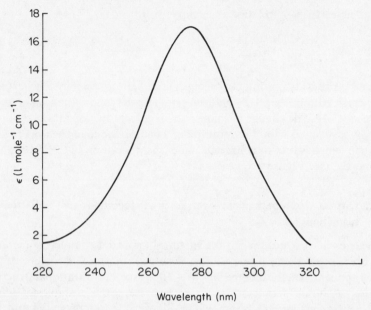

Figure 2.36 Absorption profile of acetone vapour $^1(n_O \rightarrow \pi^*_{C-O})$ transition.

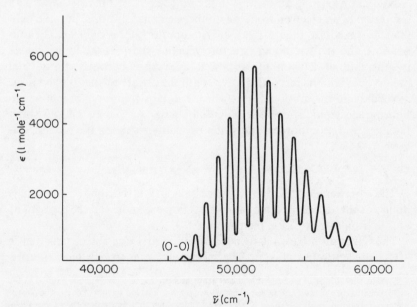

Figure 2.37 Ultraviolet absorption of ammonia ($\tilde{A} \leftarrow \tilde{X}$ transition): showing long progression in deformation frequency of upper state.

frequencies. In NH_3 the first electronic transition

$$\ldots \underbrace{(3a_1)^1}_{n_N} \; \underbrace{(4a_1)^1}_{\sigma^*_{N-H}}, \; \tilde{A}^1 A_1 \leftarrow \ldots \underbrace{(3a_1)^2}_{n_N}, \tilde{X}^1 A_1$$

actually raises the symmetry from pyramidal to planar. The excited molecule adopts the D_{3h} conformation and its upper state should be termed $\tilde{A}^1 A_2''$. The change in shape is evident from the long progression of bands, associated with the out-of-plane bending mode of the upper state, a mode which is totally symmetric under the operations of the lower point group C_{3v} (see Figure 2.37).

2.4d Diffuse and continuous spectra: predissociation and radiationless transitions

Processes that reduce the natural lifetime of an excited molecule increase the absorption spectral line widths. Resolved rotational structure begins to disappear when the lifetime falls to $\sim 10^{-11}$ sec and vibrational structure begins to blur at lifetimes $\leqslant 10^{-12}$ sec. Representative examples of diffuse and continuous spectra are shown in Figure 2.38 (see plates VI and VII of plate section). Let us consider them in turn.

NO_2 shows strong absorption in the range 258–235 nm associated with excitation of an electron from the doubly occupied $4b_2$ m.o. into the half-filled $6a_1$ m.o. (see Figure 1.21).[74] The upper orbital is only weakly anti-bonding and the rotational structure remains sharp at the longer wavelengths, but at 248 nm the rotational structure becomes diffuse and predissociation sets in. At this wavelength the absorbed energy carries the molecule into a vibrational level which overlaps a continuum in a neighbouring electronic state. The potential energy surface of the two states intersect and radiationless transition probably results in the predissociation*

$$NO_2 + h\nu \xrightarrow{\;<248\text{ nm}\;} NO + O(2^3P)$$

The rotational structure in successive bands becomes increasingly diffuse as the accessible region of intersection expands, and the probability of radiationless transition increases (cf. S_2).

The absorption bands associated with the first electronic transition in NH_3, are entirely diffuse.[74] The electronic state from which the pre-

---

* It was thought until recently that the O atom was produced electronically excited in the 2^1D level, but careful study of the spectrum shows predissociation setting in at energies below the necessary minimum threshold.[74] Herzberg[74] has suggested that either the O atom is in its ground state, 2^3P or possibly the dissociation products are $N(2^4S) + O_2(\tilde{X}^3\Sigma_g^-)$. Chemical evidence indicates that $O(2^1D)$ atoms *are* produced by absorption at shorter wavelengths.

dissociation

$$NH_3 + h\nu \rightarrow NH_2 + H$$

occurs is accessible from all of the vibrational levels which can be populated by light absorption from the ground state. Unlike NO_2, there is now no discernible predissociation limit, although in ND_3 the bands which populate the ground and first vibrational levels of the upper state do have resolvable rotational structure. However, the half-widths of the lines in the (0,0) band are $\sim 2 \cdot 5 \, \text{cm}^{-1}$ compared to an expected Doppler width of only $\sim 0 \cdot 09 \, \text{cm}^{-1}$. Since the mean lifetime of the upper state is related to the half-width as $\tau \sim 1/2\pi\Delta\nu$, the rate of predissociation is $\sim 5 \times 10^{11} \, \text{sec}^{-1}$.

The NF_2 radical exists in thermal equilibrium with its dimer perfluorohydrazine in the vapour phase, analogous to the $NO_2 - N_2O_4$ system.[75] NF_2 absorbs in the near-ultraviolet, probably due to excitation of an electron from the filled $6a_1$ m.o., largely localized on the nitrogen atom, into the singly occupied $2b_1$ m.o. which is weakly N–F π-antibonding[76] (see Figure 1.21). The corresponding transition in CF_2 does not lead to predissociation but in NF_2 it is barely possible to distinguish even any vibrational structure. The rate of predissociation into $NF + F$ must approach $10^{13} \, \text{sec}^{-1}$.

There is no trace of structure of any kind in the near-ultraviolet absorption profile of CH_3I. The $n_1 \rightarrow \sigma^*_{C-I}$ transition promotes an electron into an orbital which is strongly C–I antibonding and the excited molecule is known to dissociate at the C–I bond

$$CH_3I + h\nu \rightarrow CH_3 + I$$

The absence of even vibrational structure indicates a lifetime $\sim 10^{-13} \, \text{sec}$ and it probably dissociates on a repulsive potential energy surface within a single vibrational period. The absorption profile is bell-shaped, and the extinction coefficient is a Gaussian function of the absorption frequency, closely similar to the dissociation continua of the diatomic halogens.[77] Unlike them however, CH_3I is polyatomic with nine normal vibrational modes; the potential energy surface of the $^1(n,\sigma^*)$ state would be expected to be steeply repulsive along only one of the normal coordinates, i.e. that which stretches the C–I bond. On the basis of the Franck–Condon Principle this corresponds, of course, with just the vibrational mode that would be excited and in the act of excitation it would be 'selected' from all the others. The Gaussian profile is a 'reflexion' of the vibrational probability function of the level $v'' = 0$ in the steeply repulsive part of the upper state (cf. section 2.2b).

The suggested mechanism of dissociation in CH_3I differs from that in the other three examples since we have not found it necessary to include a radiationless transition along the dissociation path. Of course there may be one which has unit probability and cannot be detected, in which case the point is rather academic and the dissociation can be termed 'direct'. However, there is a difference between dissocation in polyatomic and diatomic molecules. In the latter we were able to distinguish direct dissociation, which occurred in the initially excited state during the first 'vibration', from predissociation which was delayed by an intervening radiationless transition. In a polyatomic molecule the distinction has to be modified since it possesses more than one vibrational mode. In order for it to dissociate it is essential that sufficient energy be concentrated in those vibrational modes which carry it along a dissociation path. It is quite possible for the molecule to be excited into vibrational modes which do not do this, although its total vibrational energy may lie above a dissociation threshold. Until sufficient can be redistributed into 'dissociative' modes the molecule will remain intact while its atoms 'dance a Lissajous jig' about the equilibrium conformation. Because of the anharmonicity of the molecular vibrations the redistribution and eventual dissociation does occur of course, but only after some delay;* on this basis it can be termed a predissociation, although it is impossible in diatomic molecules. Herzberg[74] has termed it a type II predissociation to distinguish it from the type I predissociations (p. 108) which do require a radiationless transition. It is also conceivable that a molecule could fly apart if it acquired sufficient rotational energy; if it did it would be following a type III mechanism of predissociation.

The selection rules for type I predissociation are similar to those in diatomic molecules, but because of the possibility of vibronic interactions they are far less restrictive. Radiationless transitions will have a high probability when the vibronic symmetries of the interacting states are the same, when their spins are the same and when the Franck–Condon vibrational overlap integrals are large. Conservation of symmetry becomes less and less restrictive as the number of symmetry elements in the molecular framework is reduced, i.e. when there is less symmetry to conserve! In the absence of spin–orbit coupling, however, the spin coordinates are not related to the molecular framework and spin conservation must be maintained.

At first sight, one would expect the spin selection rule to prevent any appreciable population of triplet states in molecules which are composed

---

* Prediction of the delay takes us into the realm of unimolecular reaction rate theory, of which we will have something to say in section 2.5a.

of light atoms and which have singlet ground states. In practice, inter-system crossing between excited singlet and triplet states is often efficient, particularly in large, unsaturated polyatomic molecules which can resist rapid photodissociation. The efficiency usually depends on the relative rates of radiationless transition and radiative (fluorescent) decay. For a fully allowed transition, the latter cannot occur at a rate greater than $\sim 10^9$ sec^{-1}, compared with a maximum rate for a fully allowed radiation-less transition of $\sim 10^{13}$ sec^{-1}. It is possible, therefore, to tolerate a reduction by a factor of at least 10^4 before radiative decay could begin to compete. The degree of inefficiency commonly introduced in singlet–triplet transitions falls in the range 10^4–10^6 and since triplet states *are* observed to be populated, it follows that the rates of radiationless transition must be near the maximum possible, particularly in the more complex polyatomic molecules. In their case the most important factors controlling the rates of radiationless transitions (aside from spin conservation), seem to be the Franck–Condon overlap integrals and the spacings of the vibronic levels which are being populated.[78] When these are closely packed they form a quasi-continuum and the effective factor becomes the density of vibrational levels within the interaction energy of the intersecting electronic states. This increases with the molecular size and complexity and should make radiationless transitions increasingly probable, in accord with observation (see further discussion in section 2.5c).

Since the amplitudes of vibrational wave functions do not immediately fall to zero on either side of the classical turning points, but oscillate as the normal coordinates vary between the two extremes, it is possible for radiationless transitions to occur at a significant rate, *even when there is no* '*intersection*' of the unperturbed potential energy surfaces. The vibrational overlap integral may be appreciable inside the confines of the lower surface (see Figure 2.39) allowing the molecule to 'tunnel' its way through to the neighbouring electronic state. The probability of tunnelling becomes less as the length of the 'tunnel' increases. If the potential energy curve (b) in Figure 2.39 were displaced slightly to the right, so that the curves (a) and (b) ran more nearly parallel, the vibrational overlap would become very small. It follows that radiationless transitions between stable molecular electronic states with parallel and widely separated potential energy surfaces, can generally be discounted.

The probability of tunnelling is also sensitive to isotopic substitution since the vibrational spacings and wave functions are compressed by an increase in the reduced mass of the oscillator. For example, after deutera-tion a hydrocarbon will require many more vibrational quanta to reach a given vibrational energy in a C–H stretching mode. The higher the vibrational quantum number, the smaller the amplitude of the wave

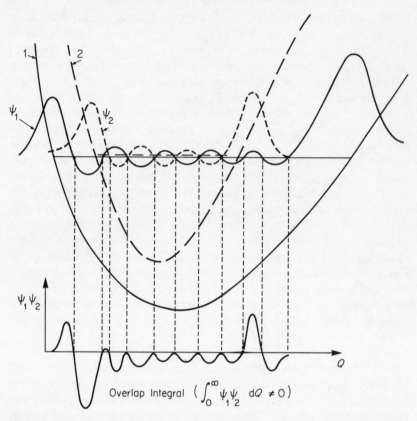

Figure 2.39 Vibrational overlap in non-intersecting surfaces (schematic). In the situation shown, the area under the product function $\int_0^\infty \psi_1 \psi_2 \, dQ$, although small, would not be zero.

function between the classical turning points of the normal mode, and the smaller will be any vibrational overlap inside the confines of the potential energy surface. Deuteration of aromatic hydrocarbons greatly reduces the rates of radiationless transition between the first triplet and ground singlet states. Robinson and Frosch[79a] attributed the change to the reduced overlap between the zero vibrational level of the triplet state and the highly excited vibrational levels of the ground singlet state. The result also shows that C–H stretching modes are active in the radiationless transition.

Ting[79b] has revised an earlier model proposed by Franck and Sponer, and suggested that radiationless decay follows intramolecular resonant transfer of the electronic energy in the triplet state, into iso-energetic vibrational levels of the ground state. Assuming these to be C–H stretching

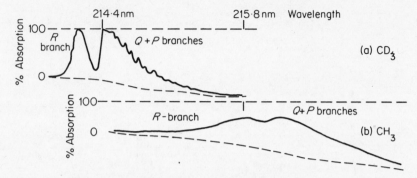

Figure 2.40 $\tilde{B} \leftarrow \tilde{X}$ absorption bands of CD_3 and CH_3, showing the partial resolution of rotational structure promoted by deuteration.

modes, he finds a remarkably impressive correlation between the predictions of his model and the experimental data presently available.

In lighter polyatomic molecules, it is often the case that deuterium substitution reduces the rate of predissociation, and sharpens the rotational fine structure in an absorption spectrum. This is shown particularly clearly in the ultraviolet absorption of CH_3 (see Figure 2.40), NH_3 and CH_2.

2.5 DECAY AND DEACTIVATION OF EXCITED POLYATOMIC MOLECULES

Once the molecule has absorbed a photon and occupies an electronically excited state, it will leave that state by the quickest available route. The nett probability of decay, summed over each exit route from the initially excited state must be unity (Einstein's Law of Photochemical Equivalence), and each individual probability represents the quantum efficiency for that particular process. As a simple illustration suppose the exit routes for a molecule promoted into its first excited singlet level were thought to be restricted to fluorescence, radiationless transition into a triplet level (intersystem crossing) or perhaps into the ground state (internal conversion); the nett quantum efficiency would be

$$\phi_{fl} + \phi_{ISC} + \phi_{IC} = 1$$

If experimental measurements showed that

$$\phi_{fl} + \phi_{ISC} = 1$$

internal conversion would be excluded. In general, a polyatomic molecule may have many options to choose from, and the most important are

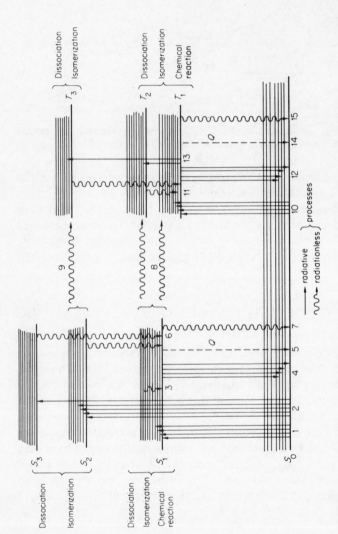

Figure 2.41 Jablonski diagram showing principal decay processes in polyatomic molecules (specific unimolecular rate, sec^{-1}).

(1) Absorption (10^{15}) $S_0 \rightarrow S_1$.
(2) Absorption (10^{15}) $S_0 \rightarrow S_2, S_3, \ldots$.
(3) Vibrational relaxation (10^{12}) S_1.
(4) Fluorescence (10^9–10^6) $S_1 \rightarrow S_0$.
(5) Bimolecular quenching ($k_Q[Q]$)$S_1 \rightarrow S_0$.
(6) Internal conversion and vibrational relaxation (10^{12}) $S_n \rightarrow S_1$.
(7) Internal conversion and vibrational relaxation (?) $S_1 \rightarrow S_0$.
(8) Intersystem crossing (10^7) $S_1 \rightarrow T_2, T_1$.

(9) Intersystem crossing (?) $S_n \rightarrow T_n$ (probably too slow to compete against (6)).
(10) Absorption (10^{15}) $S_0 \rightarrow T_1$.
(11) Internal conversion and vibrational relaxation (10^{12}) $T_n \rightarrow T_1$.
(12) Phosphorescence (10^3–10^{-2}) $T_1 \rightarrow S_0$.
(13) Absorption (10^{15}) $T_1 \rightarrow T_n$.
(14) Bimolecular quenching ($k_Q[Q]$)$T_1 \rightarrow S_0$.
(15) Intersystem crossing and vibrational relaxation (10^4–10^{-2}) $T_1 \rightarrow S_0$.

represented on the quantum state diagram, Figure 2.41 (often termed a Jablonski[80] diagram after its innovator). In order to set the scene, we shall briefly summarize the possibilities and then discuss each in more detail in the subsequent sections of this chapter.

The first spin-allowed electronic transition will excite the molecule into vibrational (and rotational) levels in the singlet state S_1. If the transition alters the equilibrium geometry the majority of molecules will be vibrationally, as well as electronically excited, and the vibrational energy may well be sufficient to promote dissociation. However this might well be delayed while the vibrational energy migrates around the molecule, and in the interval the energy may be transferred in collisions. This will be particularly rapid in solution, and if there is a significant potential energy barrier along the dissociation path the quantum efficiency of this process will fall to zero.

Molecules which do not dissociate may live long enough to re-emit a photon and fluoresce, populating upper vibrational levels in the ground state, S_0. Alternatively a radiationless transition may produce the same ultimate result, or it may populate vibrationally excited levels in a neighbouring triplet state or in an isomeric electronic state. There is evidence that in some molecules (for example, benzene), internal conversion may proceed through the intermediacy of some isomeric conformation.[81]

If the molecule reaches a triplet state it may internally convert to a lower triplet level, if there is one, where it too may dissociate, isomerize, transfer its vibrational energy, or perhaps transfer to the ground state in a radiationless transition. It might also cross back into the S_1 state if it remained isolated in the vapour phase, and nothing happened in the meantime, or if transfer back to S_1 were encouraged by thermal activation of the triplet molecule (the latter process produces the phenomenon of delayed fluorescence, see p. 197). A molecule that survives all these fates may live long enough to phosphoresce and decay into excited vibrational levels of the ground state. A long residence time in the lowest triplet level may allow absorption of a second photon which excites the molecules into a higher triplet state. Biphotonic processes are quite common in frozen solutions, where diffusional quenching of the excited molecules is prevented, and during flash excitation where the absorption of high light intensities can produce a high temporary population of the triplet state.

Bimolecular processes that compete with fluorescent or phosphorescent decay, quenching the luminescence of the excited molecule, may include electron transfer, atom transfer, complex formation or association, intermolecular energy transfer and so on. Electronic energy transfer may sensitize fluorescence, phosphorescence or chemical reaction of the acceptor. Collision with molecules containing heavy atoms, or the formation of association complexes in the excited state can alter the rates of intersystem crossing and allow more efficient transfer from S_1 to T_1 (as well as T_1 to S_0).

Polyatomic molecules excited into singlet states lying above S_1 are usually so susceptible to radiationless decay, that a fluorescent return to the ground state is extremely improbable. Predissociation, isomerization or internal conversion into lower excited singlet states, can all occur within the hypothetical fluorescent lifetime, and in a complex polyatomic molecule the rate of internal conversion is probably $\sim 10^{12}$ sec^{-1}. Internal conversion or intersystem crossing from S_1 or T_1 into the ground electronic state is generally far slower; this is generally attributed to the greater energy gap. In consequence, if the molecule does fluoresce or phosphoresce it will be from the lowest excited singlet or triplet level, regardless of the level excited initially. The only known exceptions to this rule are azulene and some of its derivatives (see p. 181). By the same token electronic energy transfer from upper singlet and triplet levels can probably be discounted in the great majority of molecules, though there is evidence for energy transfer from the second triplet level in anthracene.[82]

2.5a Dissociation

We have seen that a polyatomic molecule, excited into a vibronic level above one of its dissociation limits, need not necessarily dissociate within a single vibrational period (section 2.4d). Access to the dissociation path may require a redistribution of the vibrational energy among the normal modes of the electronically excited state (cf. unimolecular decomposition) or a radiationless transition into some other electronic state (cf. predissociation in diatomic molecules). The process of unimolecular decomposition can be visualized in terms of the motion of a configuration point on a potential energy hypersurface.

Consider the excitation of a triatomic molecule ABC from the lowest vibrational level of its ground state into a more weakly attractive excited state. We shall assume A and C have very different masses, so that the bond stretching modes approximate to A–B, and B–C stretch. Suppose that the AB–C bond is the weakest, and that excitation alters the equilibrium bond lengths but not the bond angle, so that the electronic transition can be represented on a two-dimensional contour diagram such as Figure 2.42. In order of increasing energy, possible dissociation routes are

$$\text{critical energy}$$

$$(ABC)_v^* \longrightarrow \begin{array}{lll} AB + C & (1) & E_1^* \\ A + BC & (2) & E_2^* \\ A + B + C & (3) & E_3^* \end{array}$$

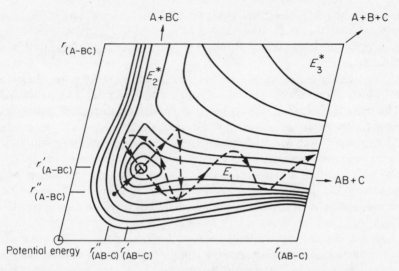

Figure 2.42   Potential energy surface representing the photodissociation of a triatomic molecule from a bound excited state.

where $(ABC)_v^*$ represents the vibrationally excited molecule in the upper electronic state, and E_1^*, E_2^* and E_3^* are the minimum threshold energies for each dissociation path.† At the instant of excitation molecules which occupied the lowest vibrational level in the ground state will be compressed relative to the new equilibrium conformation (Franck–Condon Principle) and the initial motion in the upper state will excite both stretching modes. The configuration point travels down towards the bottom of the well and up the opposite slope, carrying the molecule along the dissociation route (3). However, this is the least favoured exit since complete atomization requires the most energy. It is unlikely ever to compete with (1) or (2), and in the molecule chosen we have assumed that the vibrational energy E of the excited molecule lies within the limits E_1^*, $< E < E_3^*$. If sufficient could be transferred into the B–C stretching mode (Figure 2.42), the molecule would be able to dissociate by route (1). Transfer is impossible when the vibrations are truly harmonic, but fortunately (from the chemists' point of view) the exit valleys introduce a measure of anharmonicity into the potential energy function and the normal modes can mix. The configuration point does not retrace its original path when it falls back

† The electronic states of the products are not specified; though photodissociation commonly produces electronically excited fragments (see p. 170). We are also neglecting the dissociation path $(ABC)_v^* \rightarrow AC + B$ which would require excitation of the molecular bending mode.

towards the central well, but swings round in a series of loops. The contribution of the B–C stretching mode gradually increases until the configuration point escapes through the exit valley (1) and the molecule dissociates.

The average time that elapses before this happy result is the problem to which the theories of unimolecular reaction rates are addressed. There are two types of theory, based either on a statistical or a dynamic model. The first developed by Hinshelwood,[83] Kassel,[84] Rice and Ramsperger,[85] and Marcus[86] assumes that the normal modes of vibration in the energized molecules are anharmonic, so that free energy flow from one normal mode to another may occur. On this assumption it is only necessary that the molecule contain just sufficient energy for dissociation, no matter how it is distributed, and it will find its way into the appropriate vibration. The other approach developed by Slater[87] is conceptually more attractive since it focuses attention on the dynamic motion of the molecule. Unfortunately it assumes harmonic vibrations, which maintains a mathematical elegance but is chemically unrealistic. There is no energy flow between the modes and the vibrational motion of the molecule is the sum of all normal modes. The reaction coordinate is taken to be some linear combination of the normal modes (in the triatomic molecule ABC, a combination of the A–B and B–C stretching modes, which actually stretch both bonds), and when the total vibrational energy is such that the critical coordinate can be stretched to its breaking point, the molecule will dissociate. This requires that more energy be concentrated into the molecule than the bare minimum required in the HKRRM model. In both cases the rate of decomposition of the energized molecule k_E, is given by an equation of the form

$$k_E = A\left(1 - \frac{E^*}{E}\right)^{n-1} \qquad E \geqslant E^*$$

where E and E^* are the vibrational energy content of the molecule and its critical energy for decomposition. Marcus and Rice have suggested that E and E^* be measured from the zero point level.[86]

In the statistical (HKRRM) model, all the n oscillators are assumed to have a common frequency, which can be identified with the A factor.† In the dynamic (Slater) model it is a weighted root-mean-square frequency of the n normal modes which are able to contribute to stretching the normal coordinate. In Slater's model n must be less than the total $3N - 6$ modes of the molecule, whereas the HKRRM model includes no such

† The model has been modified to accommodate the reality of different vibrational frequencies (see the discussion in Johnston's[88] book on reaction rate theory).

restriction. In practice n is commonly found to be $\sim \frac{1}{2}(3N - 6)$. In both cases A is typically $\sim 10^{13} \sec^{-1}$, and on either theory $k_E \simeq 10^{13} \sec^{-1}$ only when $E^* = 0$ (i.e. the electronic state of the molecule is entirely repulsive), or $E \gg E^*$. It follows that a potentially dissociating molecule may be deactivated through intermolecular vibrational or electronic energy transfer in collisions, if the specific rate at which this occurs is much in excess of k_E. Once the vibrational energy content of the excited molecule is reduced to $E \leqslant E^*$, dissociation is impossible. As a result of this behaviour the quantum efficiency of decomposition in polyatomic molecules is often pressure sensitive, decreasing as the pressure increases. Indeed, this is diagnostic of an 'excited molecule mechanism', in which the excited molecule must be regarded as a kinetically identifiable intermediate susceptible to non-chemical deactivation. Such pressure dependence will only be apparent when the 'lifetime' of the energized molecule $1/k_E$ is of the same order as the intervals between the molecular collisions which are effective in causing deactivation. These are typically $\sim 10^{-10} \sec$ at atmospheric pressure and normal temperature; under these conditions when $k_E \gg 10^{10} \sec^{-1}$, the participation of an 'excited' molecule need not be considered and there is no experimental difference between the dissociation of diatomic and polyatomic molecules. The rate of dissociation decreases rapidly as the number of contributing normal modes is increased; as a rough working rule, a molecule with $E^*/E \sim 0.75$ should be at least tetratomic before the factor $(1 - E^*/E)^{n-1}$ becomes small enough to bring k_E down to a value $\sim 10^{10} \sec^{-1}$.

Before leaving the subject of unimolecular decomposition, we must emphasize an important distinction between thermal dissociation (pyrolysis) and photodissociation (photolysis). Thermal activation energizes the molecule 'from the bottom up', through intermolecular vibrational energy transfer in collisions. Its total energy increases step-by-step as it ascends the vibrational ladder toward a dissociation threshold. As soon as the energy is just a little higher than the floor of the exit valley, the molecule may dissociate, but the products will have very little excess energy and will be produced in thermal equilibrium with the surrounding molecules. In contrast, light absorption excites the molecule into upper vibrational levels in an excited electronic state in a single step. If it can find a dissociation path, it starts well up on the vibrational ladder, and it is almost invariably the case that more energy has been absorbed than the bare minimum necessary for dissociation. If the dissociating state does not correlate with ground state products, either or both of them may be electronically excited. Some of the remaining excess energy may be concentrated into vibration (or rotation), while the rest will appear as translational energy, divided between the products in inverse proportion to their masses (see p. 132). In general, the products of photodissociation are not produced in thermal equilibrium with the surroundings but are introduced into the system as 'hot' species, often chemically more reactive than their thermalized counterparts.

The division of the excess energy between vibration, rotation and translation is controlled by the molecular dynamics of the dissociation process. For example, when dissociation occurs within a single vibration in a triatomic molecule ABC, vibrational excitation of AB can be expected when the A–B bond length in $(ABC)^* > ABC, AB$. On the other hand, if the bond length in $(ABC)^* \simeq ABC, AB$ almost all the excess energy will appear as translation.[89] These two situations can be represented on a potential energy diagram, assuming no change in bond angle, as in Figure 2.43. Case (a) is appropriate for the photodissociation of NOCl at wavelengths ~ 200 nm, where the NO is found to carry at least 11 vibrational quanta (about half the total excess energy), while case (b) is appropriate to the photodissociation of the cyanogen halides, also at wavelengths ~ 200 nm, where the CN carries very little vibration. When there is a change in bond angle on excitation, some of the excess energy would be expected in rotation in the diatomic fragment.

In larger polyatomic molecules, the excess vibrational energy in a polyatomic fragment can be sufficient to allow secondary unimolecular dissociation or isomerization. For example, the photolysis of polyhalomethanes such as $CHBr_3$ at wavelengths $\geqslant 200$ nm, produces halomethyl radicals with high vibrational energy contents.[90]

$$CHBr_3 + h\nu \longrightarrow (\cdot CHBr_2)_v + Br\cdot$$

Some of these have sufficient energy to split out a molecule of HBr,

$$(\cdot CHBr_2)_{E > E^*} \longrightarrow \cdot CBr + HBr$$

while others may be deactivated by transferring some of their vibrational energy in collisions

$$(\cdot CHBr_2)_{E > E^*} \xrightarrow{+M} (\cdot CHBr_2)_{E < E^*}$$

Photolysis of cyclobutanone produces CO and a very distorted and therefore highly vibrating molecule of cyclopropane.[91]

$$H_2C \underset{CH_2}{\overset{CH_2}{<}} C = O + h\nu \longrightarrow \left\{ H_2C \underset{CH_2}{\overset{CH_2}{<}} \right\}_v + CO$$

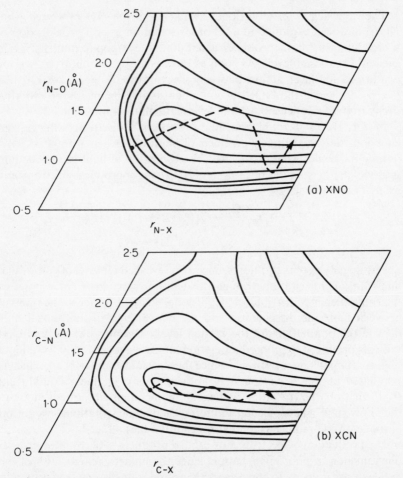

Figure 2.43 Potential energy surfaces for photodissociation of XNO and XCN. Trajectories begin from the equilibrium configuration of the *ground* electronic state (Franck–Condon Principle).

Those with sufficient energy may isomerize to propylene, or be deactivated to levels below the critical energy through collisional energy transfer.

The high degree of vibrational excitation in the NO produced from NOCl has been exploited in a 'photodissociation laser'.[92] Absorption of a high intensity flash produces a temporary overpopulation of NO in excited vibrational levels. As soon as there is a population inversion, the conditions for laser action between the vibrational levels are satisfied (see p. 79), and if the NOCl is contained in a resonant cavity, radiative decay from the upper vibrational levels stimulates infrared emission.

In the 1930's Terenin and his coworkers[93] discovered fluorescent emission from the primary products of photolysis when a range of poly-atomic molecules were dissociated by absorption of light in the vacuum ultraviolet. For example, photolysis of H_2O vapour produces electronically excited OH radicals

$$H_2O + hv \xrightarrow{136\,nm} OH(\tilde{A}^2\Sigma^+) + H(1^2S)$$
$$\downarrow$$
$$OH(\tilde{X}^2\Pi) + hv_{fl}$$

Spectral analysis of their fluorescence also reveals their level of vibrational and rotational excitation, though this will not necessarily correspond to their initial energy distribution. Collisional energy transfer in the interval between molecule dissociation and the fluorescence of the radical may deplete the populations of the excited levels. Despite this, the hydroxyl radicals produced by photodissociation at 123·6 nm are found to retain a high rotational 'temperature'.[94] Since the electronic state of H_2O, which is populated at this wavelength, is known to have a linear equilibrium conformation, the transition will excite the bending vibrational mode in the upper state. The molecular motion should favour rotational excitation of the hydroxyl radical as the hydrogen atom flies off.

The production of electronically excited products is, of course, a general phenomenon in photodissociation, since the higher excited states of the molecule are unlikely to correlate with ground state dissociation products. Many examples could be quoted and a few are collected in Table 2.3. Although absorption in the vacuum ultraviolet increases the chances of electronic excitation in the products, the phenomenon is by no means restricted to this region of the spectrum, and two examples of particular note are the formation of $O(2^1D)$ and $S(3^1D)$ from O_3 and COS. Since the two atoms have ^3P ground states their radiative decay is both spin and symmetry-forbidden and the excited atoms are metastable. As energetic, chemically reactive intermediates they play an important role in the reactions which follow photodissociation. Their pattern of chemical reactivity differs from that of the unexcited atoms.

The degeneracy of the ^2P electronic states of the halogen atoms is split by spin–orbit coupling. In fluorine, the splitting is very small, but in the

TABLE 2.3 The production of electronically excited products (underlined) in the photodissociation of polyatomic molecules

Molecule	Absorbed wavelength (nm)	Products
H_2O	123·6	$\underline{OH(\tilde{A}^2\Sigma^+)} + H(1^2S)$
NH_3	123·6	$\underline{NH(\tilde{c}^1\Pi)} + H_2(\tilde{X}^1\Sigma_g^+)$
COS	200	$CO(\tilde{X}^1\Sigma^+) + \underline{S(3^1D)}$
	150	$CO(\tilde{X}^1\Sigma^+) + \underline{S(3^1S)}$
NO_2	129·5	$\underline{NO(\tilde{A}^2\Sigma^+)} + O(2^3P)$
		$\underline{NO(\tilde{B}^2\Pi)} + O(2^3P)$
NOCl	160	$\underline{NO(\tilde{A}^2\Sigma^+)} + Cl(3^2P)$
O_3	300	$\underline{O_2(\tilde{a}^1\Delta_g)} + O(2^1D)$
CF_3I	254	$CF_3(\tilde{X}^2A_1) + \underline{I(5^2P_{\frac{1}{2}})}$

heavy iodine atom the $5^2P_{\frac{1}{2}}$ state lies $7{,}600\ cm^{-1}$ ($\equiv 91\cdot1\ kJ\ mole^{-1}$) above the $5^2P_{\frac{3}{2}}$ ground state. The excited atom is metastable since decay on the ground state is symmetry-forbidden. The majority (and possibly all of the I atoms produced through photolysis of CF_3I in the near-ultraviolet ($n_I \rightarrow \sigma^*_{C-I}$ transition), are in the $5^2P_{\frac{1}{2}}$ state and they can be detected in absorption immediately following flash excitation of the parent molecules.[95] The population inversion was demonstrated very prettily by Kasper and Pimentel[96] in the photodissociation laser. Radiative decay of an excited I atom in an optically resonant cavity stimulates the decay of other excited atoms until the population inversion is removed.

$$CF_3I + h\nu \xrightarrow{\sim 250\ nm} CF_3\cdot + I(5^2P_{\frac{1}{2}})$$

$$I(5^2P_{\frac{1}{2}}) \longrightarrow I(5^2P_{\frac{3}{2}}) + h\nu_{7,600\ cm^{-1}}$$

$$I(5^2P_{\frac{1}{2}}) + h\nu_{7,600\ cm^{-1}} \longrightarrow I(5^2P_{\frac{3}{2}}) + 2h\nu_{7,600\ cm^{-1}}$$

etc.

The emitted frequency $7{,}600\ cm^{-1}$ lies in the near infrared.

The correlation of the electronic states of a molecule with those of its dissociation products is governed by consideration of their symmetry and relative energies and by the non-crossing rule (see p. 131). If spin–orbit coupling cannot be neglected the symmetry of the spin wave functions must be included in the overall symmetry and the correlation rules based on spin alone are relaxed. In the absence of spin–orbit coupling, the spin correlation rules are those discussed in section 2.3c. For example, the ground state of N_2O is a singlet $\tilde{X}^1\Sigma^+$ and it cannot correlate with the ground states of $N_2(\tilde{X}^1\Sigma_g^+)$ and $O(2^3P)$. If the oxygen atom is produced in the 2^1D state all is well. Photodissociation of diazomethane, CH_2N_2 from an upper singlet state cannot produce ground state methylene, since the ground state is a triplet, $\tilde{X}^3\Sigma_g^-$. The primary products of photolysis in the near ultraviolet are singlet methylene (which is bent) and $N_2(\tilde{X}^1\Sigma_g^+)$.

$$CH_2N_2 \text{ (singlet)} \rightarrow CH_2(\tilde{a}^1A_1) + N_2(\tilde{X}^1\Sigma_g^+)$$

A polyatomic molecule which is on the way to dissociating into two fragments will be distorted out of its equilibrium conformation and some of its symmetry elements will be lost. The symmetry correlation rules are based on the behaviour of the electronic wave functions under the operations of the point group of the distorted molecule. For example, dissociation of CH_4 at one of the C–H bonds must initially distort the molecule from a tetrahedral conformation (point group T_d), into one of threefold symmetry as in CH_3F (point group C_{3v}), (see Figure 2.44). The possible electronic states of its dissociation products can only be found by use of group theoretical methods; all that can be done here is to indicate the approach, though the application of the group theoretical concepts is not difficult. The enthusiast will find a detailed account in Herzberg's Volume III.[74]

The first $\sigma \rightarrow \sigma^*$ transition in CH_4 lies at wavelengths ~ 150 nm and excites the molecule into the $\tilde{A}^1T_2$ state (p. 137). Under the operations of the C_{3v} point group, its threefold degeneracy is split and it can correlate with electronic states of symmetry $^1A_1$ and ^1E. The dissociation products would be a methyl radical and a hydrogen atom. At 150 nm the excess energy will be insufficient for electronic excitation of the atom: it will be produced in the ground state 1^2S_g. In the distorted molecule the H atom will lie on the threefold symmetry axis; in this situation its $1s$ wave function will be unchanged by any of the symmetry operations of the C_{3v} point group: it will belong to the totally symmetric representation A_1. The possible symmetry species Γ of the CH_3 radical are found by equating the

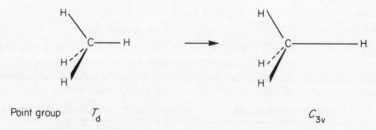

Point group T_d C_{3v}

Figure 2.44 Lowered symmetry following distortion of CH_4.

symmetry species of the direct product $A_1 \times \Gamma$ to that of the dissociating molecule, A_1 or E.

$$A_1 \times \Gamma = A_1, \qquad \therefore \Gamma = A_1$$
$$A_1 \times \Gamma' = E, \qquad \therefore \Gamma' = E$$

To satisfy the requirements of symmetry the radical can only be formed in $^2A_1$ or ^2E states. Herzberg has shown that in its ground electronic state $\cdot CH_3$ is actually planar.[74] In this conformation the symmetry species A_1 and E (point group C_{3v}) go over the species A_2'' and E' of the point group D_{3h}, and the ground state of $\cdot CH_3$ is in fact $\tilde{X}^2A_2''$ (i.e. it has the same symmetry as the singly occupied $2p\pi$ orbital localized on the central carbon atom). While the ground state of CH_4 must correlate with ground state products, its energy lies far below that of the excited $\tilde{A}^1T_2$ level. Any $\cdot CH_3$ radicals produced by dissociation of $CH_4(\tilde{A}^1T_2)$ are likely to be electronically excited.

$$CH_4(\tilde{X}^1A_1) \rightarrow CH_4(\tilde{A}^1T_2) \rightarrow CH_3(\tilde{A}^2E') + H(1^2S_g)$$

At this point it is a little cruel perhaps, to reveal that the photoexcited molecules much prefer to eliminate molecular hydrogen! The distortion leading to $CH_2 + H_2$, alters the molecular shape from tetrahedral into one of two-fold symmetry as in CH_2F_2, (point group C_{2v}). Again the H_2 must be produced in its ground electronic state, $\tilde{X}^1\Sigma_g^+$ and application of the same type of argument shows that the methylene could be produced in $^1A_1$, $^1B_1$ or $^1B_2$ states; it could not be produced in its triplet ground state. The least excited singlet has A_1 symmetry, but again this would correlate with the ground electronic state of CH_4. The dissociation path of lowest energy may well produce CH_2 in an excited singlet state

$$CH_4(\tilde{X}^1A_1) \rightarrow CH_4(\tilde{A}^1T_2) \rightarrow CH_2(\tilde{b}^1B_1) + H_2(\tilde{X}^1\Sigma_g^+)$$

The alternative dissociation

$$CH_4(\tilde{A}^1T_2) \rightarrow CH_2(\tilde{X}^3\Sigma_g^-) + H_2(\tilde{a}^3\Sigma_u^+)$$

also satisfies the requirements of symmetry and spin, but it can be excluded on chemical grounds since the $\tilde{a}^3\Sigma_u^+$ state of H_2 is repulsive. The final products would be $CH_2(\tilde{X}^3\Sigma_g^-) + 2H(1^2S_g)$, which conflicts with experimental observation. Where there is a range of possible products, each of which satisfy the correlation rules, the choice may be limited by their relative energies and by the non-crossing rule. This was discussed in section 2.3c. As the number of symmetry elements in the dissociating molecule is reduced, the symmetry requirements become less restrictive until the most important factors become the magnitude of the excess energy and the total spin.

2.5b Fluorescence

If the polyatomic molecule is excited into a stable electronic state and remains there for long enough, it may fluoresce and return to the ground state. The time available depends on the magnitude of the electronic transition moment, which is directly proportional to either the oscillator strength or the integrated extinction of the absorption spectrum (see p. 76). It cannot be less than 10^{-9} sec, and if the transition is forbidden by symmetry, or if there is very little overlap between the vacated and populated orbital, the radiative lifetime may be in the microsecond region.

In the vapour phase at pressures supporting no more than a few centimetres of mercury, the interval between collisions is $\sim 10^{-8}-10^{-9}$ sec and collisions during the radiative lifetime are almost inevitable. If the decay is forbidden there will be many collisions. These will spread the internal energy of the excited molecule among the lower vibrational and rotational levels and just as in I_2, the line-like resonance spectrum will be replaced by a broad fluorescence spectrum which lies to the red of the absorption. Eventually the internal energy will relax into the equilibrium Boltzmann distribution and, if the vibrational spacings in the upper and lower states are comparable, the absorption and fluorescence spectra may present a 'mirror image' relationship (see Figure 2.45). When the fluorescent molecule is dissolved in solution it reaches thermal equilibrium very quickly and all the emission will be from the lowest vibrational levels of the excited state.

The quantum efficiency of fluorescence is found by recording the total number of photons emitted by the fluorescent molecule throughout its fluorescence spectrum and comparing it with the number of photons absorbed at a given frequency. Thus if I_{fl} is the nett fluorescence intensity and I_{abs} is the intensity of the absorbed light, the fluorescence efficiency is

$$\phi_{fl} = \frac{I_{fl}}{I_{abs}}$$

If the optical density O.D. $= \varepsilon_v cl \ll 1$ at the exciting frequency then $I_{abs} \simeq I_0 \varepsilon_v cl$, where I_0 is the incident light intensity. For low enough concentrations or path lengths the fluorescence intensity is

$$I_{fl} \simeq \phi_{fl} I_0 \varepsilon_v cl$$

When the fluorescence is recorded as a function of the frequency of the exciting light, it should vary in the same way as ε_v, the extinction

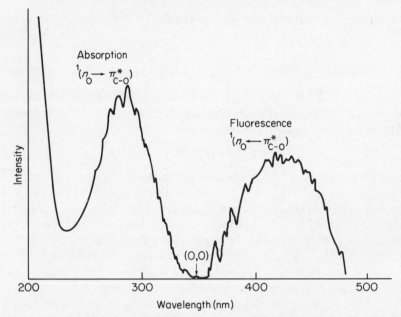

Figure 2.45   Absorption and fluorescence spectra of acetaldehyde. (After K. K. Innes and L. E. Giddings, Jr., *J. Mol. Spec.*, 7, 435, (1961)).

coefficient of the absorption spectrum; the absorption and excitation spectra should coincide, provided the fluorescence efficiency remains constant throughout the excitation spectrum.

The quantum efficiency of fluorescence is rarely unity, even when the molecule is isolated in the vapour phase,* or held in a rigid, inert matrix. Radiative decay is a relatively slow process which has to compete against a variety of possible radiationless processes. Thus saturated polyatomic molecules are most unlikely to be fluorescent. The first electronic transition is either $n \rightarrow \sigma^*$ or $\sigma \rightarrow \sigma^*$ promoting an electron into an antibonding σ-orbital. The absorption spectrum is continuous and the excited molecule usually dissociates.

Although the quantum efficiency of fluorescence from unsaturated molecules is generally independent of the wavelength of the exciting light, there are cases where the efficiency falls as the wavelength decreases. Two examples are NO_2 and benzene. NO_2 is an orange gas which absorbs through most of the visible and near-ultraviolet regions of the spectrum.

* Measurement of the absolute fluorescence intensity at very low pressures in the vapour phase presents severe experimental problems. The absolute intensity of a polychromatic fluorescence emitted in all directions at low intensity, has to be compared with that absorbed in a banded spectrum from a monochromatic, collimated beam.

The colour is associated with the electronic transition $\pi^*_{N-O} \rightarrow \pi^*_{N-O}$)

$$\ldots (4b_2)^2(2b_1)^1, \tilde{A}^2 B_1 \longleftarrow \ldots (4b_2)^2(6a_1)^1, \tilde{X}^2 A_1$$

$$\underbrace{\quad}_{\sigma^*_{O-O}} \underbrace{\quad}_{\pi^*_{N-O}} \qquad \underbrace{\quad}_{\sigma^*_{O-O}} \underbrace{\quad}_{\pi^*_{N-O}}$$

(see Figure 1.21). At wavelengths to the red of 397·9 nm the spectrum is sharp and absorption in this region excites the fluorescence of the molecule. At shorter wavelengths it predissociates

$$NO_2(\tilde{A}^2 B_1) \rightarrow NO(\tilde{X}^2 \Pi) + O(2^3 P)$$

possibly via an excited $^2 A_1$ state, and the rotational lines are broadened.[74] Many years ago, Norrish[97] discovered that when the molecule was excited in this region its fluorescence efficiency fell to zero. The radiative lifetime is $4·4 \times 10^{-5}$ sec[98] (which is remarkably long), and fluorescent decay has no chance against the more rapid predissociation.

Absorption of mercury radiation at 253·7 nm transfers benzene into a low lying vibronic level of the first excited singlet state (see p. 146). In the vapour phase at pressures less than 1 mm mercury, approximately 40 per cent of the excited molecules fluoresce back to the ground state.[99] Most of the remainder cross over to an iso-energetic level of the neighbouring triplet state, while a few transfer into an isomeric conformation:[100]

benzene   benzvalene

When higher vibronic levels are populated by absorption at shorter wavelengths, the fluorescence efficiency decreases and at < 240 nm no emission can be detected at all.[81] Neither is there any evidence of inter-system crossing, though stable isomers can still be isolated. All the available evidence suggests that an acceleration in the rate of isomerization in the higher vibronic levels excludes the competing decay processes.[81] Initially the isomers must be endowed with a great deal of vibrational energy. If they can transfer enough to neighbouring molecules they are stabilized but most revert back to benzene in its ground electronic state (see Figure 2.46).

$$\text{Benzene } (S_0) \xrightarrow{h\nu} \text{Benzene } (S_1)_{vib} \xrightarrow{I.S.C.} \text{Benzene } (T_1)_{vib}$$

$$\big\uparrow\text{+M} \qquad\qquad\qquad\qquad\qquad\qquad\swarrow$$

$$\text{Benzene } (S_0)_{vib} \xleftarrow{\text{+M}} (\text{Isomers})_{vib} \xrightarrow{\text{+M}} \text{Isomers}$$

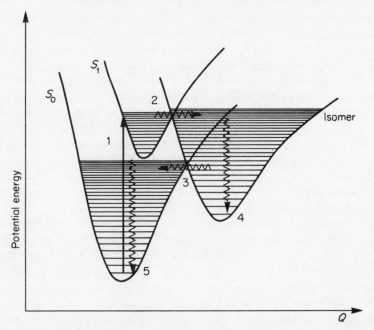

Figure 2.46 Hypothetical cross-section through potential energy surfaces in benzene: Q is a normal mode which couples the interacting electronic states (see ref. 81). (1) excitation; (2) isomerization; (3) reverse isomerization; (4, 5) vibrational relaxation in collisions.

If this interpretation is correct, it shows how a radiationless transition into an isomeric conformation can provide a convenient 'stepping stone' for the internal conversion of the energy in the excited molecule.

The rates of the intramolecular processes which compete with fluorescent decay depend on the nature of the molecule and of its first excited singlet state. Generalizations are dangerous, but the following are useful:

(i) Molecules which have high fluorescence efficiencies commonly decay from $^1(\pi,\pi^*)$ states. For example, many polycyclic aromatic hydrocarbons fluoresce strongly. The potential energy hypersurfaces of their ground and first excited singlet states run nearly parallel and the Franck–Condon overlap factors between the ground vibronic levels of the upper states and the isoenergetic vibronic levels of the ground states are very small. Unless there is some 'isomeric stepping stone', the probability of internal conversion is likely to be negligible.

(ii) Molecules where the lowest excited singlet state is $^1(n,\pi^*)$ usually have very low fluorescence efficiencies. For example, no fluorescence has yet been detected from benzophenone. If the molecule is frozen in a rigid glass to eliminate the possibility of diffusional quenching of its excited state, the $n_O \rightarrow \pi^*_{C-O}$ transition excites an intense phosphorescence, with a quantum efficiency of 0·71. The lack of fluorescence is due to efficient intersystem crossing into a neighbouring triplet state, (see p. 189).

(iii) In contrast, fluorescence from $^1(l,a_\pi)$ states may be quite strong, as in the cases of dimethyl aniline and anisole.

(iv) When the aromatic ring is substituted with both electron donor and electron acceptor groups, the first electronic transition populates an intramolecular charge transfer state (for example, p-nitroaniline). Again fluorescent decay can be detected particularly at low temperatures. A particularly striking example is 'Michler's ketone', 4,4'tetramethyl-diaminobenzophenone, which fluoresces strongly in polar solution at low temperature, in contrast to the unsubstituted parent molecule.

The choice of a polar solvent in the last example was not fortuitous, since the relative energies of the molecular electronic states are sensitive to the solvent polarity. In particular, those which have large dipole moments are shifted to lower energies in solvents of increasing polarity. In Michler's ketone both $n \rightarrow \pi^*$ and charge-transfer transitions occur in the violet and near-ultraviolet, and the excited singlet states have comparable energies. In a polar solvent, the first transition moves to the blue, while the second moves to the red. The first excited singlet state becomes the charge-transfer state, and the molecule is able to fluoresce.

(v) When fluorescence spectra are recorded in solution, it is usually found that the position of the (0,0) band in any vibrational progression, does not match that in the corresponding absorption spectrum. The gap between the two 'origins' is due to the different solvation in the two states. At the instant of absorption, the excited molecule finds itself with a solvation shell appropriate to the ground state. Changes in the electron distribution and the molecular geometry impose a new structure on the solvation shell, but the molecular motions are relatively slow and cannot respond quickly enough (Franck–Condon Principle). However, the fluorescent decay is delayed for at least 10^{-9} sec, which leaves ample time for the relaxation of the solvation shell and the stabilization of the excited molecule. At the instant of emission the new solvation shell is inappropriate to the unexcited molecule and the ground state is de-stabilized. As a result, the origins of the absorption and fluorescence spectra do not coincide. In a polar solvent the magnitude of the gap can be used to estimate the dipole moment of the excited molecule. In rigid solutions at low temperatures, the gap disappears. When the molecular motion is frozen the solvation shell can no longer relax into a new con-formation around its fluorescent molecule.

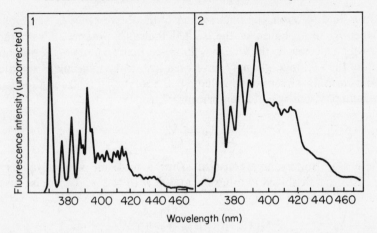

Figure 2.47   Prompt fluorescence emission spectra of 2×10^{-5} M pyrene, showing enhanced resolution at low temperatures 1, in ether–pentane–ethanol at 77°K; 2, in ethanol at room temperature; excitation at 313 nm; half-band width of silica prism analysing monochromator was 0·38 nm at 400 nm. (Taken from C. A. Parker, *Photoluminescence of Solutions*, Elsevier, Amsterdam (1968)).

(vi) Freezing the solution usually improves the resolution of any vibronic structure in the fluorescence spectrum as well as increasing its intensity (see Figure 2.47). The resolution of the absorption and excitation spectra are similarly affected. The lack of collisions in a rigid solvent medium prevents both collisional perturbation and diffusional quenching of the excited molecule.

(vii) The fluorescence emitted by any given molecule is polarized in a plane which contains the axis of the transition dipole. If we let $I_{\parallel}$ and $I_{\perp}$ represent the intensities of the components of fluorescence that are polarized parallel and perpendicular to the plane, then the *degree of polarization p* is defined by the expression

$$p = \frac{I_{\parallel} - I_{\perp}}{I_{\parallel} + I_{\perp}}$$

Ideally $I_{\perp} = 0$ and $p = 1$. Natural light has $I_{\parallel} = I_{\perp}$ and $p = 0$. When there are a large number of fluorescent molecules randomly oriented in space the ideal will not be realized of course. However, if plane polarized light is used to excite the fluorescence of molecules which are rigidly held in a solid matrix, it will only be absorbed by those molecules which are appropriately orientated. In this situation it can be shown that the average polarization is $p = \frac{1}{2}$, (assuming that the fluorescence and absorption are both associated with the same electronic transition).

In a fluid solution, the polarization of the fluorescence is 'scrambled' further by the rotation of the excited molecules associated with their Brownian motion. By the time they fluoresce, their memory of the polarization of the exciting light may have been lost. If the molecule is spherical and its rotation sweeps out a volume V, the rotational relaxation time in a medium of viscosity η at a temperature T, is

$$\rho = \frac{3V\eta}{kT}$$

where k is Boltzmann's constant. Thus a molecule with a radius of 5×10^{-8} cm dissolved in n-hexane at 300°K ($\eta \simeq 3 \times 10^{-3}$ poise) has a rotational relaxation time of

$$\rho \simeq \frac{3 \times \frac{4}{3}\pi(5 \times 10^{-8})^3 \times 3 \times 10^{-3}}{1\cdot4 \times 10^{-16} \times 300} \simeq 10^{-10}\ \text{sec}$$

This figure is smaller than the minimum radiative lifetime and the fluorescence would be strongly depolarized. Perrin[101] has shown that if the depolarization is solely due to rotation $1/p$ should vary linearly with T/η. If the molecular volume is known, the measurement of p as a function of T or η allows an estimate of the radiative lifetime. Migration of the electronic energy from molecule to molecule before the photon is finally emitted also causes depolarization (see the discussion on resonance energy transfer in section 2.5f). Unless the transition dipoles of each molecule are accurately aligned, the orientation of the one which finally emits will bear little relation to that which was initially excited. This effect can usually be neglected in dilute solutions since the probability of energy transfer and the average number of molecules in the 'migration chain' fall as the concentration is reduced.

2.5c Radiationless conversion

Measurement of the fluorescence efficiencies of aromatic hydrocarbons at very low pressures in the vapour phase has established that internal conversion and intersystem crossing are genuine unimolecular processes that can occur spontaneously in an isolated molecule. For example, the fluorescence efficiency of isolated benzene molecules, excited by light at 253·7 nm is 0·39.[99] Some of the excited molecules isomerize (see section 2.5b) while the majority cross into the triplet state. The presence of triplet benzene has been inferred by its ability to sensitize cis–trans isomerization in butene-2, or the green phosphorescence of biacetyl, through intermolecular triplet energy transfer. In toluene, the fluorescence efficiency from the ground vibronic level in the first excited singlet state of the isolated molecule is only 0·3, and experiments with added butene-2 suggest that within experimental error the quantum deficit is made up by intersystem crossing.

If the radiative lifetime of the isolated molecule calculated from its integrated absorption coefficient is τ_0 and the natural fluorescence decay time is $\tau < \tau_0$, then

$$\frac{1}{\tau} = \frac{1}{\tau_0} + \sum_i k_i$$

where k_i is the first order rate constant of the ith radiationless decay process. In the past few years, the quantitative prediction and experimental measurement of the values of k_i has exercised many minds more agile than the author's,[78] and the results of their labours are of fundamental importance in photochemistry. The following generalizations can be made

(i) It has been a general rule that the observable luminescence of a polyatomic molecule is almost always associated with a transition from the lowest excited state of a given multiplicity to the ground state. Rapid radiationless transitions reduce the lifetimes and quantum yields of emission from the higher excited states. The 'rule' is of course a relative one, since what is 'observable' depends on the sensitivity of the observer's detection system. With a sufficiently sensitive fluorimeter it should be possible, in principle, to observe any residual luminescence. One should modify the rule to state that luminescence from the lowest excited state is almost always far more intense than any residual luminescence from higher states. Very recently, Hojtink[104] has been able to record the very weak fluorescence emitted from the second excited singlet states of several aromatic hydrocarbons.

Rules are made to be broken, and a much-quoted exception is azulene, where the rule is inverted. The molecule absorbs weakly in the red ($S_0 \rightarrow S_1$), and strongly in the near-ultraviolet ($S_0 \rightarrow S_2$). Absorption in the latter region excites a violet fluorescence with a spectral profile which provides a fair reflection of the second absorption band (see Figure 2.48). After many attempts, and several false trails, very weak emission has at last been detected from the first singlet and triplet states following excitation with picosecond (10^{-12} sec) laser light pulses.[103] Fluorescent emission the S_1 state has a quantum efficiency of only $\sim 10^{-7}$. It is plausible to assume that the exceptional behaviour of azulene is associated with the low energy and intensity of the $S_0 \rightarrow S_1$ transition, which favours internal conversion and opposes radiative decay. The $S_0 \rightarrow S_2$ transition requires

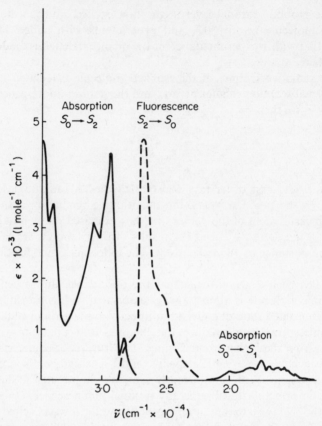

Figure 2.48   Absorption (—) and fluorescence (– – –) of azulene in ethanol at 20°C. (After C. A. Parker, *Photoluminescence of solutions*, ref. 124).

twice as much energy and is more intense by two or three orders of magnitude.

However, this is a very rare example and its behaviour is the reverse of that of the great majority of polyatomic molecules, where the excited states lie much closer in energy to their neighbours than to the ground state. Polyatomic molecules excited into levels above the first quickly undergo radiationless transition and, in a dense medium, rapidly transfer their vibrational energy to reach the lowest excited state with a Boltzmann energy distribution. The photochemical behaviour of large polyatomic molecules in solution is almost always dictated by the chemistry of their first excited states.

(ii) A kinetic description of radiationless conversion splits the process into two sequential steps. The first, which is thought to be rate determining, is the radiationless transition between nearly iso-energetic levels in neighbouring electronic states. It is followed by rapid dissipation of the vibrational energy of the newly populated state, which prevents the reverse transition from taking place (see Figure 2.41). In the pioneering theory developed by Robinson and Frosch[79] it was assumed that the excited molecules were bathed in a condensed medium, which acted as heat sink for the vibrational energy.

In the vapour phase, the rate of energy transfer would be greatly reduced. At a pressure of 1 mm mercury, the interval between molecular collisions is $\sim 10^{-7}$ sec, considerably longer than the time required for internal conversion. At first sight there seems no reason why the reverse transition into the original excited state should not occur in the interim. However, we have seen that internal conversion of the excitation in polyatomic molecules still proceeds under these conditions. This must be due to efficient *intra*molecular energy dissipation by a mechanism well-known in the theories of unimolecular reaction rates (see section 2.5a). If the molecule is sufficiently complex, the vibrational energy can be 'diluted' among a large number of normal modes in the lower excited state so that the molecule rarely re-attains the nuclear conformation in which the radiationless transition occurred. In effect, some of the electronic potential energy is converted into vibrational potential energy, in an intramolecular process.

(iii) Under the Born–Oppenheimer approximation the electronic states of the molecule are independent of the nuclear displacements; the electronic and vibrational wave functions are separable. In a molecule undergoing radiationless transition the approximation has broken down. Its electronic wave functions are also functions of the normal coordinates and neighbouring electronic states are coupled by a vibronic perturbation. The number of effective normal modes is restricted by symmetry; to promote internal conversion they must belong to the same symmetry species as the direct product $\phi_i \times \phi_f$, where ϕ_i and ϕ_f are the unperturbed electronic wave functions of the initial and final states.* If neither is degenerate all the effective modes will have to belong to the same symmetry species.

(iv) Using time dependent perturbation theory it can be shown that the probability of radiationless transition in an isolated molecule takes the form[78,79,102]

$$k = \frac{4\pi^2}{h} \cdot V^2 \rho$$

* When there is intersystem crossing, the electronic wave functions are also perturbed by spin–orbit coupling and the symmetries of the spin wave functions have also to be included.

where ρ is the density of vibronic levels in the final state (i.e. the number of vibrational levels in a unit energy interval at the energy of the initially populated state) and V is the interaction or resonance energy between the two states. It is summed over the manifold of vibronic levels in the final state, which lie close to the level initially populated. It can be factored into the product of an electronic term, whose value is inversely proportional to the energy gap between the zero point levels of the two electronic states, and a Franck–Condon vibrational overlap term. Internal conversion is favoured by a small energy gap, a high density of vibronic states, and appreciable vibrational overlap. The first criterion accounts for the lack of internal conversion from the second excited singlet level in azulene, and the last explains the increase in the luminescence efficiencies of aromatic hydrocarbons (including azulene) on deuteration (see section 2.4c).

(v) The high rate of internal conversion in complex polyatomic molecules is associated with the high density of vibronic states. Bixon and Jortner[102] have estimated the density of states in an exemplary molecule in an electronic state which has an average vibration frequency of $1{,}000\ \mathrm{cm}^{-1}$, and has absorbed an energy equivalent to $8{,}000\ \mathrm{cm}^{-1}$. Increasing the number of atoms from 3 to 4 to 5 to 6 to 10, raises the density of vibronic states per cm^{-1} from 0·06 to 4 to 50 to 400 to 400,000! These figures explain why radiationless conversion in a large polyatomic molecule from a low vibronic level of the initial state into high vibronic levels of the final state is so much faster than the reverse process (cf. para. (ii)). In the forward direction the molecule 'sees' a quasi-continuum of exit channels but in the reverse direction it 'sees' very few (see Figure 2.49).

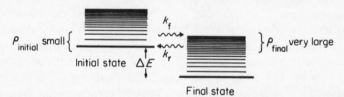

Figure 2.49 Radiationless conversion in a complex polyatomic molecule. When $\rho_{\text{final}} \gg \rho_{\text{initial}}$, $k_f \gg k_r$.

The statistics favour radiationless transition into a lower electronic state.* On the other hand, the rate of radiationless transition is also a function of the interaction energy V, which varies inversely with the energy

* If the molecule is symmetrical, coupling with many of the vibronic states may be symmetry-forbidden, so that the 'effective' density of states may represent only a small fraction of the total.

gap ΔE, between the two states, (Figure 2.49). Although a large energy gap increases the density of vibronic states it decreases V and radiationless transitions between two widely separated states are inefficient, not withstanding the large value of ρ. This is clearly shown in the rates of intersystem crossing from the first excited singlet states of benzene and anthracene. S_1 lies much nearer to T_2 than T_1 in these molecules (see Figure 2.50), and despite the greater density of states in T_1, intersystem crossing transfers the molecules into vibronic levels in T_2.

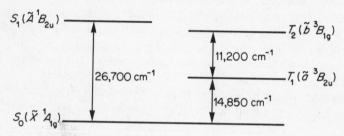

Figure 2.50 Electronic energy levels in anthracene.

(vi) The role played by the medium, in the probability of radiationless conversion, is not entirely clear at the present time. It can certainly act as a heat sink, and solvents containing heavy atoms induce external spin–orbit perturbations which accelerate the rates of intersystem crossing. However, in the absence of specific solute–solvent interactions the medium probably has little effect on the rates of radiationless transitions in the larger polyatomic molecules. In the pure state, there may be strong inter-molecular coupling of the vibronic states of individual molecules, and their behaviour may differ from that in solution. The relative spacings of the electronic levels can be altered by a crystal field, and in the crystalline state the rates of radiationless conversion may be very different from those in dilute solution.

2.5d Phosphorescence and the triplet state

(i) General discussion

The long-lived phosphorescence of many substances in the solid phase or in a glass was explained for many years in terms of the formation of a metastable tautomeric form of the molecule, and it was not until 1941 that Lewis and his coworkers first suggested that the metastable phos-phorescent state of fluorescein might be identified with a triplet state.[105a] The same suggestion was made by Terenin[55] in 1943, with respect to eosin, and the further important observation was made that oxygen, with its triplet ground state, might be able to induce the formation of triplet eosin through intermolecular perturbation. Most important of all, Terenin pointed out that if triplet eosin were formed in solution, it would probably act as a photochemical intermediate. A year later in 1944, Lewis and Kasha[56] published a classic paper in which they identified the phosphorescence of a large range of aromatic molecules with the $T_1 \rightarrow S_0$ transition, although the arguments they used were not entirely rigorous. The first unequivocal demonstration of the equivalence of phosphorescent and triplet states was made by Lewis and Calvin in 1945.[105b] If the phosphorescent state of fluorescein is a triplet it should have a paramagnetic susceptibility corresponding to two unpaired electrons; it was found that the phosphorescent state (in a boric acid glass) was indeed paramagnetic, and some years later Lewis, Calvin and Kasha showed that the susceptibility had the expected value.[105c] In 1955, Evans[106] described a magnetic balance with which he was able to follow the decay of the paramagnetic susceptibility, and found that it closely matched that of the phosphorescence.

Since that time very detailed information about the triplet states of organic molecules has been obtained from their paramagnetic resonance absorption spectra, principally by Hutchison[107] who was the first to observe the absorption, and by van der Waals and their coworkers.[108] The degeneracy of the three components of the triplet state may be split both by the application of an external magnetic field (typically $\sim 3,000\ G$) and the internal magnetic interactions which involve the spinning electrons. Magnetic dipole transitions between the sub-levels can be detected at microwave frequencies (see section 2.5d (ii)).

Long-lived phosphorescence can only be detected when the excited molecules are embedded in a solid or viscous matrix. In the liquid or vapour phases the triplet state has a much shorter lifetime and, until the advent of high intensity, synchronized flash techniques in the early 1950's, there was no direct evidence for the population of triplet states in solution. This was remedied by Porter and Windsor[57] when they discovered the transient absorption spectra of a range of triplet aromatic hydrocarbons, immediately following flash excitation of dilute solutions at room temperature. Triplet-to-triplet absorption had previously been detected, only in rigid glasses at low temperature. The classic plate (Figure 2.51 (see plate VIII of plate section)) shows the temporary depletion of the ground singlet state of anthracene, as approximately 30 per cent of the molecules are transferred into the triplet state, through intersystem crossing from S_1. The transient absorption bands correspond to the excitation of molecules in the first triplet level into higher triplet levels. Since that time, further absorption bands have been detected at longer wavelengths and the absorption labelled '1st triplet' actually corresponds to the transition $T_1 \rightarrow T_5$. T_2 lies a little below S_1 (see Figure 2.50).

The rapid decay of the triplet state restores the $S_0 \rightarrow S_1$ absorption bands to their original intensity, but since no phosphorescence can be detected this must be achieved through a radiationless process. When the triplet concentration is low the decay follows first order kinetics, which appears consistent with the transition $T_1 \rightsquigarrow S_0$.

However, the first order rate 'constant' was 'time dependent' in that its value depended on the year in which the experiment was done![109] It was also viscosity dependent since the lifetime increased in viscous solvents (hence the detection of phosphorescence in glassy solutions). It soon became clear that the rapid decay in solution was due to the presence of adventitious impurities in the solutions, particularly oxygen, which were able to quench the triplet molecules at rates limited only by their diffusion through the solvent (see section 2.5f). Because of its long radiative lifetime, the triplet state may still be quenched when the impurity levels are as low as 10^{-8} M. As more and more care was taken to remove trace contaminants so the lifetime increased, and the viscosity dependence was evidently associated with the change in second order quenching constant k_{Q_i}. Provided the solutions are very thoroughly deoxygenated and chemically inert, it is possible to observe phosphorescence in fluid solutions of low viscosity at room temperature.[110] The observed decay times of aromatic hydrocarbons commonly lie in the range 1–10 msecs, still several orders of magnitude less than in a glass. Presumably the discrepancy is due to quenching by other impurities, though the possibility of the first order rate constants for radiationless conversion being viscosity or temperature dependent cannot be excluded, particularly when the viscosity is high.

The 'observed first order' rate constant in contaminated solutions has the form $(k_{ph} + k_{T_1 \rightarrow S_0} + \sum_i k_{Q_i}[Q_i])$, where $(k_{ph} + k_{T_1 \rightarrow S_0})$ is the combined rate constant for intersystem crossing and phosphorescent decay into the ground state. At high triplet concentrations $[T]$, a second-order term $k_2[T]^2$ also contributes to the decay. In the gas phase, it becomes the only term of any significance at high absorbed intensities and the decay is rapid.[111]

The quantum yield of triplet formation ϕ_T, is given by the rate of intersystem crossing from the excited singlet state rate constant k_T relative to the nett rate of depopulation of the singlet state. If the only other processes which operate are fluorescence and internal conversion, then

$$\phi_T = \frac{k_T}{k_{fl} + k_{IC} + k_T} \equiv \quad \begin{array}{l}\text{fraction of molecules which} \\ \text{transfer from the singlet to} \\ \text{the triplet state.}\end{array}$$

The lifetime of the excited singlet state, is given by

$$\tau_{fl} = \frac{1}{k_{fl} + k_{IC} + k_T}$$

and under these conditions we also have

$$\phi_T = k_T \tau_{fl}$$

so that intersystem crossing is favoured by a long fluorescent lifetime. An accurate value of the parameter ϕ_T is essential in any kinetic study of the chemical reactions of a triplet molecule, and much effort has been directed towards its measurement. The original method relied on the fact that several aromatic compounds dissolved in glassy media at low temperature have a nett luminescence yield close to unity. This implies that fluorescence and phosphorescence provide the sole decay routes, so that $\phi_{fl} + \phi_{ph} \simeq 1$. In this situation $\phi_{ph} \simeq 1 - \phi_{fl} \simeq \phi_T$, and ϕ_T can readily be found. However, relatively few molecules are so well-behaved, and the values of ϕ_T may vary both with the temperature and the nature of the solvent. Since the majority of photochemical reactions of the triplet state have been studied at room temperature, other methods must be devised. The most direct involves measurement of the depletion in the concentration of molecules in the ground singlet state, when they are excited into the lowest triplet level by absorption of a high intensity flash.[111b]

If the fluorescence lifetime is known (either from direct measurements using pulsed or rapidly extinguished light sources, or by measurement of the phase shift between the fluorescence and exciting light when the latter is modulated at a known frequency), the rate constant for intersystem crossing k_T can be derived from ϕ_T. Intersystem crossing is particularly efficient when the molecule contains a heavy atom which increases the spin–orbit coupling and mixes the singlet and triplet wave functions (see section 2.1a), and when the singlet–triplet energy gap is small. Estimated values of k_T usually lie in the range 10^6–10^{10} sec^{-1}.

The separation between the singlet and triplet states of a given electronic configuration arises from the repulsion experienced by neighbouring electrons. If light absorption excites an electronic transition between the molecular orbitals φ and φ^*

$$\ldots (\varphi)^1 (\varphi^*)^1, \; {}^1\Phi^* \text{ or } {}^3\Phi^* \leftarrow \ldots (\varphi)^2, \; {}^1\Phi$$

the separation between the singlet and triplet states is given by twice the electron exchange integral

$$K = \int \int \varphi \varphi^* \frac{e^2}{r_{ij}} \varphi^* \varphi \, d\tau_i \, d\tau_j$$

where r_{ij} is the distance between the two electrons. This is physically equivalent to the repulsion energy between two identical charge densities, $\varphi\varphi^*e$; if φ and φ^* do not overlap then $K = 0$. In transitions of the type $n \longrightarrow \pi^*$, $n \equiv \varphi$ and $\pi^* \equiv \varphi^*$. The two orbitals barely overlap since they lie in orthogonal planes and both the charge densities $n\pi^*e$ and the singlet–triplet splittings are small. In contrast, the splittings in (π,π^*) configurations are relatively large, since the overlap is now considerable. For example, the separation between the lowest (n,π^*) levels in benzophenone is $1,750 \text{ cm}^{-1}$ compared with a gap of $21,000 \text{ cm}^{-1}$ between the lowest (π,π^*) levels in naphthalene. On this information alone, one would expect intersystem crossing to be much faster in benzophenone, and in fact $\Phi_T = 0.99$ in a rigid glass at $77°\text{K}$, compared with 0.71 for naphthalene. However, the comparison is misleading, since the rate of intersystem crossing is a function of the energy gap between S_1 and the neighbouring triplet state, which may well be T_2. In addition, when the molecule contains a heteroatom, it is possible that T_2 belongs to a different type of electronic configuration. Some possible dispositions of the lower (n,π^*) and (π,π^*) levels in such molecules are shown in Figure 2.52. El Sayed[112] has pointed out that intersystem crossing is favoured when the neighbouring singlet and triplet states belong to different electronic configurations. When spin–orbit coupling is treated as a simple first order perturbation, it does *not* mix the wave functions of the singlet and triplet states which have the same electronic configurations for example both (π,π^*) or (n,π^*). The efficient intersystem crossing in benzophenone is very probably due to a radiationless transition from $S_1-^1(n,\pi^*)$ into $T_2-^3(\pi,\pi^*)$. Presumably the same must be true of many other aromatic ketones which are non-fluorescent but phosphoresce strongly at low temperatures. Mercifully, the intersystem crossing from T_1 to S_0 does not suffer from such subtleties, since there are no triplet (or singlet) levels between them!

The relative levels of the excited states are sensitive to changes in the polarity of the solvent, hydrogen bonding, protonation in the excited state and so on (see section 2.4a), as well as the nature of any substituent groups or atoms introduced into the molecule. Slight changes in the relative spacings disturb the delicate balance between the competing decay processes, both in singlet and triplet levels, and it is not surprising to find the fluorescence and triplet yields varying from solvent to solvent. When the excited states are closely spaced, inter- or intramolecular perturbations may actually reverse their order and completely change both the photophysical and photochemical behaviour of the excited molecule. For example, in cyclohexane the lowest triplet level of p-aminobenzophenone is $^3(n,\pi^*)$. Substitution of the amino group introduces an intramolecular charge transfer band into the absorption spectrum (see p. 147), and a

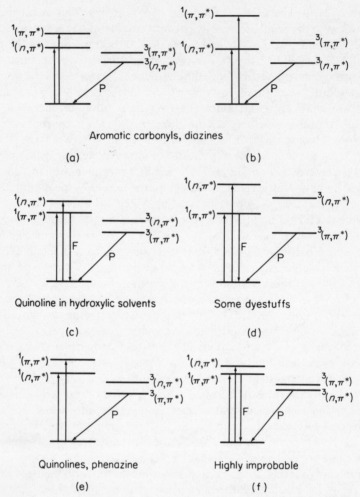

Aromatic carbonyls, diazines

(a)   (b)

Quinoline in hydroxylic solvents   Some dyestuffs

(c)   (d)

Quinolines, phenazine   Highly improbable

(e)   (f)

Figure 2.52 Possible arrangements of the lower (π,π^*) and (n,π^*) energy levels in polyatomic molecules (from F. Wilkinson, *Luminescence*, Ed. E. J. Bowen, Van Nostrand, London, 1968. Chap. 8). (N.B. since ΔE_{S-T} in $(\pi,\pi^*) > \Delta E_{S-T}$, in (n,π^*) the situation can arrive in which S_1 is (n,π^*) while T_1 is $^3(\pi,\pi^*)$.)

triplet charge-transfer level $^3(\text{CT})$ lies just above the $^3(n,\pi^*)$ state. In iso-propanol their order is reversed since the CT states are stabilized in the polar solvent, while the (n,π^*) levels shift to higher energies. The excited molecule is able to abstract a hydrogen atom from cyclohexane but not isopropanol.[113]

TABLE 2.4 Triplet state parameters

	ϕ_{ph}/ϕ_{fl}	$\tau_{ph}(sec)$	$k_{ph}(sec^{-1})$	$k_{T_1 \to S_0}(sec^{-1})$
Benzophenone	>1,000	0·0047	160	50
Naphthalene	0·09	2·3	0·016	0·42
1-Br-naphthalene	164	0·018	7	43
1-I-naphthalene	>1,000	0·002	100	400

If the triplet molecule survives all fates other than decay by phosphorescence or intersystem crossing into the ground state, its lifetime will be

$$\tau_{ph} = \frac{1}{k_{ph} + k_{T_1 \to S_0}}$$

Such a situation is likely to be realized when the molecule is isolated in a rigid, inert matrix at low temperature. Under these conditions, the quantum yield of phosphorescence is

$$\phi_{ph} = \underset{\substack{\text{quantum yield of} \\ \text{triplet formation}}}{\phi_T} \times \underset{\substack{\text{phosphorescent} \\ \text{fraction}}}{\left(\frac{k_{ph}}{k_{ph} + k_{T_1 \to S_0}} \right)} = \phi_T k_{ph} \tau_{ph}$$

and the rate constants k_{ph} and $k_{T_1 \to S_0}$ can be found from ϕ_{ph}, ϕ_T and τ_{ph}.* Unfortunately, accurate measurement of the phosphorescence yield ϕ_{ph} is difficult and relatively few reliable values are available. The results derived from some measurements in an alcohol–ether glass at 77°K[114] are collected in Table 2.4. The figures reflect three general observations:

(a) phosphorescent decay is much faster from $^3(n,\pi^*)$ than $^3(\pi,\pi^*)$ states;

(b) intersystem crossing into the ground state competes with radiative decay, even in a rigid glass at 77°K;

(c) substitution by heavy atoms accelerates intersystem crossing into, and out of the lowest triplet state.

Unfortunately, nearly all the data refer to frozen solutions at 77°K and it is not at all clear what effect, if any, the viscosity of the medium may have on the rates of intersystem crossing. The phosphorescent lifetimes of all aromatic compounds are reduced in fluid media and unless great efforts have been made to exclude oxygen and other contaminants, no phosphorescence can be detected at all. Attempts to detect their phosphorescence in the vapour phase have been unsuccessful.

* If k_{ph} is assumed to be independent of temperature, measurement of φ_{ph}/τ_{ph} at different temperatures, reflects the behaviour of ϕ_T.

On the other hand, biacetyl phosphoresces strongly in all three phases and the vapour phase phosphorescence of SO_2, for example, is well known. One cannot generalize; it is quite possible that the lack of phosphorescence in aromatic compounds is due to very efficient quenching by adventitious impurities or by photochemical products. For example, Kaplan and Wilzbach[100] have demonstrated that triplet benzene in the vapour phase readily transfers its energy to its isomer, benzvalene. Steven's attempts to measure the triplet lifetime of aromatic hydrocarbon vapours were plagued by impurity quenching.[115] When the $n_O \rightarrow \pi^*_{C-O}$ transition is excited in acetone vapour, the molecule dissociates mainly from vibrational levels in the $^3(n,\pi^*)$ state, and biacetyl is a product.

$$CH_3COCH_3 + h\nu \rightarrow CH_3COCH_3(T_1) \rightarrow CH_3CO\cdot + CH_3\cdot$$

$$2CH_3\dot{C}O \rightarrow (CH_3CO)_2$$

In a static system the rate of decomposition falls as the biacetyl accumulates and the vapour develops a green phosphorescence. Instead of dissociating, the triplet has photosensitized the phosphorescence of biacetyl vapour through electronic energy transfer

$$CH_3COCH_3(T_1) + (CH_3CO)_2 \rightarrow CH_3COCH_3 + (CH_3CO)_2(T_1)$$

$$(CH_3CO)_2(T_1) \rightarrow (CH_3CO)_2 + h\nu_{ph}$$

Perhaps the short lifetimes of most triplet molecules in fluid media or the vapour phase are due to bimolecular quenching processes. No doubt time will tell.

(ii) *E.S.R. of the triplet state*

The degeneracy of the sub-levels of the triplet state can be split, either by an externally applied magnetic field or by the internal electronic magnetic interactions. The latter can be separated into spin–orbit interactions (which also mix the singlet and triplet wave functions and allow inter-system crossing), and spin–spin interactions between the magnetic dipoles of the two unpaired electrons. Under the spherical symmetry of an atom the latter averages to zero, but in organic molecules there remains a zero-field splitting of the triplet sub-levels which is due almost entirely to the spin–spin interaction.* The levels are split further by an applied field, and the Hamiltonian energy operator for the interaction with a field

---

* In the aromatic hydrocarbons the spin–orbit interaction is very small; when hetero-atoms are substituted in the molecule it is much larger. The outermost electrons occupy orbitals which are partially localized on the heteroatom, and they acquire a measure of orbital angular momentum about its nucleus. There is now 'something for the spin to couple with' and this is reflected in the increased rates of intersystem crossing.

of strength H_0, takes the form

$$\hat{H}_{ext} = g\beta H_0 \cdot (S_1 + S_2)$$

The total magnetic moment can be resolved into three components $g\beta m_s = 0, \pm g\beta$ in the direction of the external field (see p. 18). Electromagnetic radiation can promote magnetic dipole transitions between them subject to the restriction $\Delta m_s = \pm 1$, and the frequency v, of an allowed transition must satisfy the equation $hv = g\beta H_0$. For a magnetic field of strength approximately 3,000 Gauss, the resonant frequency lies in the microwave region. Unfortunately, no microwave absorption from the triplet state could be detected for several years because there is a very strong anisotropy in the resonance spectrum, i.e. the energies separating the sub-levels depend very strongly on the orientation of the molecule with respect to the applied field. This arises because the electron spins are also coupled to the molecular framework, and the energies of the sub-levels are strongly dependent on the angles between H_0 and the molecular axes. The total spin can be resolved along three principal axes. If the components are S_x, S_y and S_z it can be shown that the Hamiltonian operator for the energy of the spin–spin interaction takes the form

$$\hat{H}_{ss} = -(XS_x^2 + YS_y^2 + ZS_z^2)$$

The parameters X, Y and Z can be identified with the energies of the three components of the triplet state when there is no external field: they are related as $X + Y + Z = 0$. The splittings of the triplet sub-levels in naphthalene are shown in Figure 2.53. In a randomly oriented sample, the

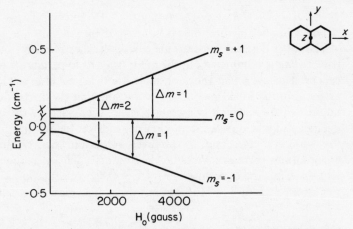

Figure 2.53 Energies of sub-levels of triplet naphthalene as a function of external magnetic field H_0 (applied with $H_0 \| y$). Transitions indicated for $v = 9650$ MHz. (After C. A. Hutchison, jun., *Rec. Chem. Progr.*, **24**, 105, (1963).)

magnetic resonance spectrum associated with '$\Delta m_s = 1$' transitions is so broad as to be undetectable; they were observed only when all the triplet molecules were regularly aligned by substitution in a crystalline 'host' matrix. The first spectrum was reported by Hutchison and Mangum in 1958, when they succeeded in detecting triplet naphthalene 'dissolved' in a single crystal of durene.[107]

Experimental life was made a great deal easier when van der Waals and de Groot[108] demonstrated that if the external field were reduced to near half its original strength, it was possible to detect transitions between the levels for which $\Delta m_s = \pm 2$. These are found to show far less anisotropy, and can readily be detected when the triplet molecules are dissolved in a glassy medium.

Since the electron spins are coupled to the molecular framework in the absence of an external field, the zero-field parameters X, Y and Z reflect the molecular symmetry. Thus if the triplet molecule has cylindrical symmetry (for example $O_2(\tilde{X}\,^3\Sigma_g^-)$) or a threefold axis (for example, triphenylene), two of the levels will be degenerate. If the axis is in the z-direction, $X = Y = \frac{1}{2}Z$. The magnitude of the parameters reflect the degree of interaction between the electrons. If they both occupy orbitals centred on the same atoms, the spin–spin interaction is strong and the parameters are large. X, Y and Z are not linearly independent and are commonly replaced by two parameters, D and E. The spin–spin Hamiltonian is rewritten

$$\hat{H}_{ss} = D(S_z^2 - \mathbf{S}^2) + E(S_x^2 - S_y^2)$$

where $X = \frac{1}{3}D - E$, $Y = \frac{1}{3}D + E$, $Z = -\frac{2}{3}D$ and $\mathbf{S}^2 = S_x^2 + S_y^2 + S_z^2$. If X and Y are degenerate, $E = 0$.

The most detailed information about the electron distribution in the triplet state is obtained from the analysis of its hyperfine structure which can be detected when the molecules are oriented in a single crystal. It arises from the weak interaction of the spinning electrons with magnetic nuclei, i.e. those which possess a nuclear spin, such as the proton. The magnetic dipole–dipole interaction depends on the relative alignment of the two spin vectors, and the sub-levels are split further. The magnitude of the splittings increases with the electron spin density at the appropriate nucleus. Thus the hyperfine structure provides information regarding the electronic structure of the triplet state, information which is highly relevant to its chemical behaviour.

(iii) *Singlet–triplet absorption*

In a molecule where the spin–orbit perturbation is very small any absorption associated with the $S_0 \rightarrow T_1$ transition is vanishingly weak. The supreme example of this is benzene, where the transition is also symmetry-forbidden: a highly purified, deoxygenated sample of the liquid showed no absorption spectrum through a 22·5 m path length. This sets an upper limit of 7×10^{-12} for the oscillator strength of the absorption band and gives the $^3B_{1u}$ state a natural radiative lifetime $\geqslant 5$ min.[116] The situation can be transformed by the introduction of a heavy atom into the molecule or into the surrounding medium, or by the addition of paramagnetic molecules such as O_2 or NO or paramagnetic transition metal ions.

Paramagnetic or O_2 effect. If the benzene is *not* deoxygenated it *is* possible to detect the $S_0 \rightarrow T_1$ absorption spectrum through long absorbing paths (though the O_2 effect was not recognized at the time). In 1957, Evans[117] discovered that the vanishingly weak absorption associated with the excitation into $^3(\pi,\pi^*)$ levels in an extensive range of olefins and aromatic hydrocarbons, could be intensified by factors $\leqslant 10^5$ under high pressures of O_2 or NO. The following year Kemula and Grabowska[118] reported a photochemical reaction between the triplet benzene and the perturbing NO.

Any perturbation which encourages the $S_0 \rightarrow T_1$ transition must also encourage the reverse process. Porter and Wright[119] found that the decay of the triplet state was increased by a range of paramagnetic species, but found that O_2 and NO were more efficient than paramagnetic ions of the first transition series and far more efficient than the paramagnetic ions of the rare earth series. Given that the quenchers were paramagnetic, their efficiencies depended on the strength of the electronic interaction between the quencher and the triplet molecule and not their magnetic moments.

Two types of perturbation mechanism can be distinguished: spin quantization can be broken down either by electron exchange within the complex with the paramagnetic species, or by electron transfer. In the one case the perturbation mixes the wave functions of the corresponding singlet and triplet states—for example,

$(^1M^* \ldots {}^3O_2)$ mixes with the triplet component of $(^3M^* \ldots {}^3O_2)$

    (triplet)   (singlet, triplet

                                                      or quintet)

and in the other both states mix with the charge-transfer state—for example, $(^1M^* \ldots {}^3O_2)$ mixes with the triplet components of

$(^2M^* \ldots {}^2O_2^-)$ (singlet or triplet)

and

$(^3M^* \ldots {}^3O_2)$ (singlet, triplet or quintet)

If the perturbing molecule were NO, electron exchange interaction mixes the doublet components of the complexes ($^3M^* \ldots {}^2NO$) and ($^1M^* \ldots {}^2NO$). The paramagnetism of the perturbing molecule or ion ensures conservation of the total spin and the $S_0 \leftrightarrow T_1$ transition is allowed to the extent that singlet character is mixed into the triplet wave function.

Heavy atom effect. The greater the spin–orbit coupling in the heavy atom introduced into the molecule or its environment, the greater the probability of intersystem crossing. The triplet yield rises at the expense of the fluorescence. Both the natural and the observed radiative lifetimes of the triplet state are reduced and the intensity of the singlet–triplet absorption is enhanced. The effect can be used to determine the quantum yield of triplet formation, ϕ_T.[121]

The effect of substituting halogen atoms into naphthalene is shown in Figure 2.54, which also shows the effect of dissolving 1-chloronaphthalene in ethyl iodide. Similar changes can be effected in solid matrices which contain, or are composed of heavy atoms.[120] For example, the lifetime of triplet naphthalene in frozen inert gas matrices falls from 1·7 sec (Ar) to 0·4 sec (Kr) to 0·07 sec (Xe). Wilkinson[121] has shown that intersystem crossings are accelerated at room temperature when the solvent is saturated with xenon.

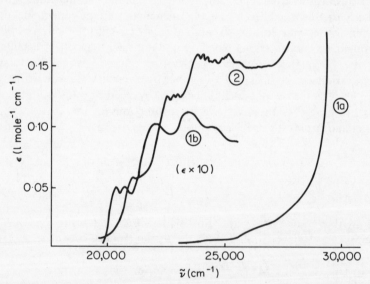

Figure 2.54 Singlet–triplet absorption in halonaphthalenes. (1) Chloronaphthalene in hexane (a) and in ethyliodide (b); (2) iodonaphthalene in hexane (after D. S. McClure, *J. Chem. Phys.*, **17**, 665 (1949).)

It is probable that both the internal and external heavy atom effects are associated with mixing of molecular electronic states with excited electronic states which are more or less localized on the heavy atom.[122] In the latter, corresponding singlet and triplet wave functions are strongly mixed by spin–orbit coupling.

Phosphorescence excitation spectra. It is much easier to measure a very low light intensity than to measure a very small change in a high light intensity due to very weak light absorption. Phosphorescent emission is far more easily observed than the reverse $S_0 \rightarrow T_1$ absorption. When the optical density of the absorbing medium is very low, the absorbed light intensity I, at a given frequency v, is given by

$$I \simeq I_0(\text{O.D.}) \simeq I_0 \varepsilon_v c l$$

where I_0 is the incident light intensity, and the other symbols have their usual significance. If v lies within the singlet–triplet absorption spectrum, I represents the rate of excitation of the absorbing molecules into the triplet state: it is directly proportional to the extinction coefficient ε_v. The rate of phosphorescence is also proportional to I, and its intensity must follow the variation in ε_v as the exciting light scans across the singlet–triplet absorption spectrum. It thus provides a very sensitive method of recording the very weak absorption. It is also possible to detect any absorption associated with excitation into higher triplet levels, provided they lie below the first excited singlet state, since all the phosphorescence is emitted by the lowest triplet regardless of which was initially populated. Kearns[123] has developed the technique and used it to determine the energies of low-lying $^3(n,\pi^*)$ and $^3(\pi,\pi^*)$ states in many aromatic ketones, (see Figure 2.55). He was able to assign the bands on the basis of heavy atom perturbation by ethyl iodide. Excitation into $^3(n,\pi^*)$ states where spin–orbit coupling is already significant was unaffected by the presence of ethyl iodide in the glassy solvent whereas the $S_0 \rightarrow T(\pi,\pi^*)$ spectra doubled their intensity.

2.5e Delayed fluorescence[124]

If the energy separating the lowest singlet and triplet levels is comparable with the energy RT, thermal activation can return a phosphorescent molecule from T_1 back to S_1, producing a delayed fluorescence

$$M(T_1) \xrightarrow{\Delta} M(S_1)$$

$$M(S_1) \rightarrow M(S_0) + h v_{fl}$$

Its spectrum is identical with the normal 'prompt' fluorescence but its lifetime is that of the phosphorescent triplet state T_1, where the molecule

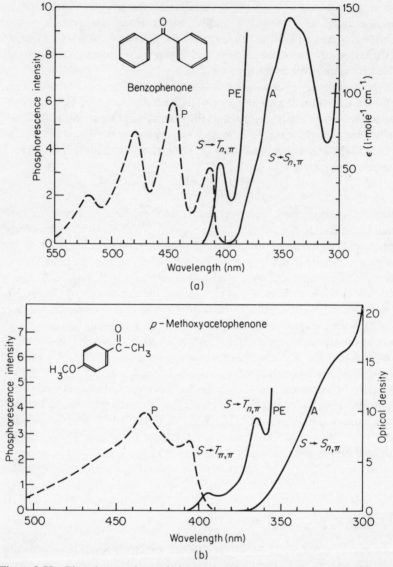

Figure 2.55 Phosphorescence emission and excitation spectra. (D. R. Kearns and W. A. Case, ref. 123.)

has temporarily resided. The quantum efficiency of the fluorescence is temperature dependent since the repopulation of S_1 requires an activation energy $\Delta E = E(S_1) - E(T_1)$. Delayed fluorescence was first observed only in solid solutions, but in 1930, Boudin[125] working in Perrin's laboratory, managed to detect the weak emission from viscous deoxy-genated solutions of eosin in glycerol at room temperature. Using only a visual phosphoroscope, she was able to measure its decay time (~ 1 m sec) and intensity with remarkable precision. In the past ten years Parker[110,124] has detected the delayed fluorescence from a variety of dyestuffs in highly purified fluid solvents, such as ethanol, using a very sensitive photo-electric spectrophosphorimeter. Since there are other processes which can generate delayed fluorescence (q.v.), he has distinguished this one as 'eosin-type' or '*E-type*'; it was originally termed '*α-phosphorescence*' and both terms remain in use.

By far the most important mechanism for the emission of delayed fluorescence, is that of '*triplet–triplet annihilation*'. The triplet state is relatively long-lived, and at a sufficient absorbed light intensity the standing concentration of triplet molecules can allow the possibility of a triplet–triplet encounter in a fluid medium or the vapour phase. Their combined energy is sufficient to excite one of them into the fluorescent state (S_1).

$$M(T_1) + M(T_1) \rightarrow M(S_1) + M(S_0)$$

$$M(S_1) \rightarrow M(S_0) + h\nu_{fl}$$

The 'triplet–triplet annihilation' results in emission of delayed fluores-cence. The process is particularly common in crystals, but here it is not the molecules which move towards each other, but the '*locale*' of the triplet excitation; triplet excitons migrate through the ordered array of molecules and when two of them combine delayed fluorescence is emitted. It can also be observed when the fluorescent molecule is dissolved in a glassy medium where translational diffusion of a large molecule cannot occur. It must result either from long range triplet–triplet energy transfer between isolated molecules or within pairs of molecules trapped in close proximity in the frozen glass.

The mechanism of the delayed emission in fluid media was established by a combination of spectroscopic and kinetic analysis. The spectrum is, of course, identical with that of the prompt fluorescence but it decays far more slowly. The intensity of the delayed fluorescence I_{df}, is proportional to the square of the standing triplet concentration $M(T_1)$. When the rate of light absorption is low, the majority of triplet molecules disappear through processes which are first order in $M(T_1)$, since $[M(T_1)]^2 \ll [M(T_1)]$. Under these conditions the decay of the triplet concentration immediately following extinction of the exciting light follows first-order kinetics

$$[M(T_1)] = [M(T_1)]^0 \exp(-t/\tau_{ph})$$

and

$$I_{df} = I_{df}^0 \{\exp(-t/\tau_{ph})\}^2$$
$$= I_{df}^0 \exp(-2t/\tau_{ph})$$

The delayed fluorescence intensity falls twice as fast as the triplet concentration and its measurement permits an estimate of the triplet lifetime, τ_{ph} in fluid media.

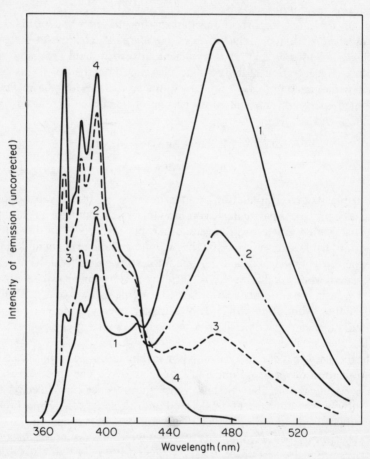

Figure 2.56 Prompt fluorescence of pyrene in ethanol. (1) 3×10^{-3} M; (2) 10^{-3} M; (3) 3×10^{-4} M; (4) 2×10^{-6} M; the instrumental sensitivity settings for curves 1 and 4 were approximately 0·6 and 3·7 times that for curves 2 and 3; the short wavelength ends of the spectra in the more concentrated solutions are distorted by self-absorption. (From C. A. Parker, ref. 124.)

   The discovery of the phenomenon dates to an original paper on the fluorescence of pyrene in ethanol by Förster and Kasper,[126] one of those rare papers that reveal an unsuspected new field of research. At low concentrations ($\sim 10^{-5}$M), the fluorescence is violet and its spectrum reflects the $S_0 \rightarrow S_1$ absorption profile. At higher concentrations ($\leqslant 10^{-3}$M) the normal fluorescence is progressively replaced by a broad featureless emission which lies at longer wavelengths and generates a brilliant sky-blue luminescence (see Figure 2.56). The concentration dependence suggests association between pairs of pyrene molecules, but since the absorption profile is independent of the concentration, the two molecules cannot both be in their ground states. Förster and Kasper attributed the new fluorescence to the decay of a stable excited dimer, or excimer as Stevens subsequently dubbed it, into a non-bonded ground state. This is consistent with its lack of structure (see Figure 2.57).

$$P(S_0) + h\nu \rightarrow P(S_1)$$

$$P(S_1) \rightarrow P(S_0) + h\nu_{fl}$$

$$P(S_1) + P(S_0) \rightarrow P_2^* \text{ (singlet)}$$

$$P_2^* \text{ (singlet)} \rightarrow 2P(S_0) + h\nu_{fl}'$$

Later studies of the decay of the monomer and dimer emission bands showed that both possessed 'prompt' and 'delayed' components. The delayed fluorescence was rationalized by Parker and Hatchard[127] in terms of a new transient excimer produced through triplet–triplet annihilation, which rapidly dissociated into normal and excited monomers.

$$P(T_1) + P(T_1) \rightarrow (P_2^{**}) \quad \text{(singlet, triplet or quintet)}$$

$$(P_2^{**}) \text{ (singlet)} \begin{cases} \nearrow 2P(S_0) + h\nu_{fl}' \\ \searrow P(S_1) + P(S_0) \end{cases}$$

$$P(S_1) \rightarrow P(S_0) + h\nu_{fl}$$

The phenomenon is a general one which has been observed in many molecules both in the liquid and vapour phases, but since the 'story' began with pyrene Parker has described this kind of delayed fluorescence as '*P-type*'.

2.5f Energy transfer and luminescence quenching

(i) *Vibrational energy transfer*
The excited molecules produced by light absorption are not only excited electronically, but their vibrational energies are also in excess of the

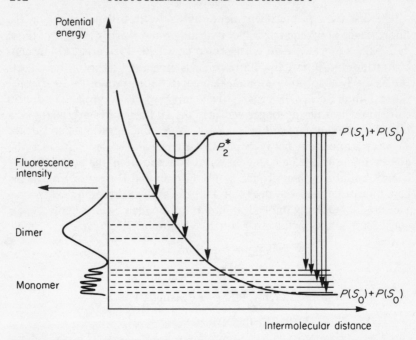

Figure 2.57 Potential curves of excited and unexcited pyrene (schematic), showing banded monomer fluorescence and the excimer fluorescent continuum. (After B. Stevens and M. I. Ban, *Trans. Faraday Soc.*, **60**, 1515 (1964).)

equilibrium distribution corresponding to thermal equilibrium with the surroundings. Intermolecular collisions may result in the transfer of both types of excitation. If the rate of conversion or transfer of electronic energy is relatively fast there may be no vibrational relaxation during the lifetime of the excited molecule at low pressures in the gas phase. For example, Calvert and his coworkers[128] find the fluorescent spectral profile of $SO_2(\tilde{A}^1B_1)$ excited by light at 287·5 nm to be independent of pressure, although the vibrational excess at this wavelength is $5{,}610\ cm^{-1}$. The rate of electronic de-excitation is too fast. However, the fluorescence efficiency is pressure dependent, increasing as the pressure falls because the excited molecule suffers self-quenching

$$SO_2(\tilde{A}^1B_1) + SO_2(\tilde{X}^1A_1) \rightarrow SO_2(\tilde{X}^1A_1) + SO_2(\tilde{X}^1A_1\ \text{or}\ \tilde{a}^3B_1)$$

Its efficiency never reaches unity, but only a limiting value of $(0·14 \pm 0·06)$, due to the competition from radiationless intersystem crossing and internal conversion. These are remarkably efficient for so simple a molecule and there is persuasive evidence that strong vibronic coupling makes up for the low density of vibrational states.

In contrast, the fluorescence efficiency of 2-naphthylamine excited at 253·5 nm in the vapour phase decreases with falling pressure.[129,130,131] At this wavelength the molecule is promoted into levels lying $\sim 10{,}200\ cm^{-1}$ above the zero vibrational level of the first excited singlet state and the fluorescence has to compete against predissociation (in fact the levels lie in the region of the second excited singlet state but internal conversion between the two is believed to be very fast). When the thermal capacity of the vapour is increased by addition of inert foreign gases (M), the fluorescence intensity is restored. Vibrational energy transfer takes place within the lifetime of the excited molecule, the rate of predissociation is retarded and the excited molecule survives long enough to fluoresce. Letting $(N^*)_{vib}$ represent a vibrationally and electronically excited molecule of naphthylamine,

$$(N^*)_{vib} \rightarrow \text{dissociation} \qquad \text{rate constant } k_d$$

$$(N^*)_{vib} \rightarrow N + h\nu_{fl} \qquad\qquad\qquad k_{fl}$$

$$M + (N^*)_{vib} \rightarrow (N^*)_{thermal} + M \qquad\qquad k_t$$

The more vibrational energy that can be transferred, the greater the chance of fluorescence since the rate of predissociation increases exponentially with the vibrational energy (see p. 166). Measurements of the dependence of the fluorescence intensity or decay time, on the exciting wavelength or the pressure of added foreign gases enables the rate of energy transfer to be estimated, using a Stern–Volmer relationship similar to that discussed in section 2.3b. The natural lifetime of the fluorescent molecule becomes $\tau_0 = 1/k_{fl} + k_t$. Schlag and his coworkers[131] have shown that as the pressure of added propylene becomes very large predissociation virtually stops, since τ_0 tends to its limiting value of $1/k_{fl}$ ($\simeq$13 nsecs). They also found that the rate constant k_t was an order of magnitude less than the specific bimolecular collision frequency $\bar{z}$. If $p = k_t/\bar{z}$ is regarded as the number of gas kinetic collisions required for complete deactivation, and the excited molecule is endowed initially with vibrational energy E, then the average energy ΔE, transferred per collision is $\Delta E = pE$. In a more rigorous treatment the kinetic description should be modified to include the possibility of stepwise deactivation down a vibrational 'ladder'.[130,132]

Boudart and Dubois,[130] reinterpreting the original data of Neporent found that ΔE increased with the number of bonds in the molecules of foreign gas: vibrational energy is more readily transferred to the larger polyatomic molecules (see Table 2.5).

In the liquid phase vibrational energy transfer is very fast. Relaxation times in the ground electronic state of molecules in the pure liquid phase

TABLE 2.5 Average amounts of vibrational energy transferred from excited β-naphthylamine per collison.[130] Total vibrational energy $= 117 \, kJ \, mole^{-1}$

Foreign gas	He	H_2	D_2	N_2	CO_2	NH_3	$CHCl_3$	$n\text{-}C_5H_{12}$
$\Delta E \, (kJ \, mole^{-1})$	0·8	0·8	0·6	2·1	6·3	12·5	14·6	19·6

have been derived from light scattering and other measurements and listed by Piercy and Hanes.[133] Typical values range from 12 nsec in CS_2 to 0·016 nsec in acetone. The measurement of vibrational relaxation times in an electronically excited molecule requires flash excitation, with pulse widths of picosecond duration (10^{-12} sec). The existence of such light sources lay in the realms of fantasy until the fantastic was actually realized by Rentzepis and others at the Bell Telephone Laboratories.[134] They demonstrated the existence of picosecond pulses of visible light at 530 nm, emitted from a Nd^{3+} doped glass in a laser cavity; their method was as simple as it is elegant.

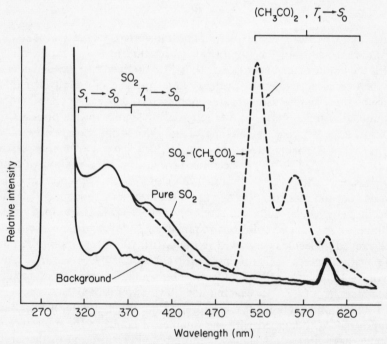

Figure 2.58 Low resolution spectrofluorimeter trace showing triplet energy transfer from SO_2 to biacetyl in the vapour phase. ——— pure SO_2 ($8·62 \times 10^{-5}$ M), − − − − − $SO_2 + (CH_3CO)_2$ ($8·62 \times 10^{-5}$ M $+ 2·18 \times 10^{-7}$ M) (note that only the phosphorescence of SO_2 is quenched by the biacetyl). (From T. N. Rao, S. S. Collier and J. G. Calvert, *J. Amer. Chem. Soc.*, **91**, 1609, (1969).)

The train of light pulses was directed through a solution of a substituted anthracene and reflected back at the far side of the cell by a plane mirror. Although the anthracene is transparent to light of normal intensity at 530 nm, in the regions where the incoming and reflected pulses are coincident, the nett intensity is sufficient to permit 'two-photon' absorption. The molecule is excited into an upper singlet state at a level corresponding to the absorption of light at $\frac{1}{2} \times 530$ nm $\equiv 265$ nm. Rapid internal conversion and relaxation into the lowest vibrational levels of the first excited singlet state is followed by fluorescence; a snapshot of the solution shows a train of fluorescent spots, corresponding to each pulse coincidence. Since the velocity of light is known, the width of the spot gives the duration of the light pulse. They were found to be $\sim 10^{-12}$ sec. By measuring the pulse shapes in experiments of this type, Rentzepis has estimated a vibrational relaxation time of $(4 \pm 1) \times 10^{-12}$ sec for the first excited singlet state of 9,10-dimethyl anthracene,[135] dissolved in a fluid medium.

The rate of vibrational relaxation in solid media at low temperatures may be considerably slower; it has been suggested that in frozen glasses radiationless intersystem crossing from the first excited singlet state could be the more rapid process.[136] It has also been possible to observe phosphorescence from vibrational levels $v' = 0 \rightarrow 6$ in $N_2(\tilde{a}^3\Sigma_u^+)$ frozen in rare gas matrices at low temperatures.[135]

(ii) *Electronic energy transfer*
Figure 2.58 shows spectrofluorimeter traces of the luminescence emitted by gaseous SO_2, excited by absorption into its first excited singlet state $(\tilde{A}^1B_1)$ at 287·5 nm. It shows very little structure since the luminescence is weak, and to collect sufficient light the spectra had to be recorded under very low resolution (10 nm). The figure also shows the dramatic result of adding only one four-hundredth the concentration of biacetyl; its intense green phosphorescence appears at the expense of that of the SO_2, while the shorter-lived fluorescence is unaffected. Biacetyl is virtually transparent to light at 287·5 nm when at such low concentration; in addition direct excitation at this wavelength causes photodissociation and not luminescence. Its phosphorescence has been sensitized by energy transfer from the triplet SO_2.

$$SO_2(S_0) + h\nu \rightarrow SO_2(S_1) \qquad \text{absorption}$$

$$SO_2(S_1) \rightarrow SO_2(T_1) \qquad \text{intersystem crossing}$$

$$SO_2(T_1) + (CH_3CO)_2(S_0) \rightarrow SO_2(S_0) + (CH_3CO)_2(T_1) \qquad \text{energy transfer}$$

$$(CH_3CO)_2(T_1) \rightarrow (CH_3CO)_2(S_0) + h\nu_{ph} \qquad \text{phosphorescence}$$

When electronic energy from a donor D is transferred to an acceptor A without the emission and subsequent re-absorption of a photon, the process is termed *radiationless resonance energy transfer*. It is a process of major importance in photochemistry, occurring in liquids and solids as well as in the gas phase. It allows the photosensitization of physical and chemical changes in the acceptor, and the competition between energy transfer and unimolecular decay processes provides a route for measuring the rates of both. If D and A are polyatomic molecules the only significant restrictions on the efficiency of radiationless energy transfer are that the electronic energy of the donor be greater than that taken up by the acceptor and that spin angular momentum be conserved.

The resonance condition which was severely restrictive in atoms (section 2.3b), is much more readily met in polyatomic molecules; when there is an electronic energy difference, vibrational (and rotational) energy can make up the balance (see Figure 2.59). Even this restriction disappears in solution since the rotational and vibronic levels are broadened by the solvent perturbation, and both absorption and emission spectra merge into continua. Where the 'virtual transition' in both the donor and the acceptor is allowed (for example where both correspond to singlet–singlet transitions) electrodynamic coupling of the electric transition dipoles leads to particularly efficient energy transfer. There is a *long-range coulombic interaction* between the coupled dipoles that remains appreciable at intermolecular distances as great as 6–7 nm, corresponding to many molecular diameters in simple molecules (cf. pp. 122–123).* When the 'virtual transition(s)' are forbidden by spin or by molecular or local symmetry, the transition dipoles are much weaker and the coulombic interaction does not penetrate very far. In this situation, resonance energy transfer can also proceed through a *short-range interaction* involving *electron exchange* within overlapping molecular orbitals of the donor and acceptor. The mechanism may operate at distances $< \sim 1 \cdot 5$ nm. Transfer at short range does not exclude a coulombic mechanism of course, since the dipole–dipole interaction also increases rapidly as the intermolecular separation is reduced. For completeness, a third, so-called 'trivial' mechanism should be included. This involves the radiative process of emission by the donor, followed by absorption by the acceptor. The process is favoured by overlapping emission and absorption spectra, and when donor and acceptor are identical, it leads to radiation imprisonment.

---

* Dipole–quadrupole of higher multipole interactions can also permit energy transfer provided the resonance condition is met, but at much shorter ranges since the interaction falls off more steeply with the distance.

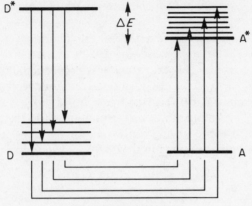

Resonant transitions

Figure 2.59   Vibronic resonance in electronic energy transfer between polyatomic molecules. (After T. Förster, ref. 137.)

Resonance energy transfer through a coulombic mechanism requires conservation of spin in both the donor and acceptor. The exchange mechanism is less restrictive since it is only necessary that the Wigner–Witmer correlation rules be satisfied to conserve a component of the nett spin (cf. p. 130). Of the steps

$$^1D^* + {}^1A \rightarrow {}^1D + {}^1A^* \tag{1}$$

$$^3D^* + {}^1A \rightarrow {}^1D + {}^3A^* \tag{2}$$

$$^3D^* + {}^3A \rightarrow {}^1D + {}^1A^* \tag{3}$$

$$^1D^* + {}^1A \rightarrow {}^1D + {}^3A^* \tag{4}$$

(1) is allowed under both coulombic and exchange mechanisms, (2) and (3) satisfy the Wigner–Witmer correlation rules (cf. p. 195) and are allowed under an exchange mechanism, and (4) is spin-forbidden. It is sometimes possible for long range energy transfer to take place even when the electronic transition in the donor is forbidden, provided the transition in the acceptor remains fully allowed. This is because the spontaneous decay of the donor is also retarded, and more time is available for the energy to be transferred. Long-range transfers of the type

$$^3D^* + {}^1A \rightarrow {}^1D + {}^1A^*$$

have been demonstrated experimentally: their efficiency is relatively low, since the probability of resonance energy transfer under a coulombic mechanism is inversely proportional to the lifetime of the excited donor (see below).

Coulombic interaction. The first theoretical treatment of dipole–dipole energy transfer in solution was provided by Förster.[137] The quantum mechanical probability of energy transfer is proportional to the square of the interaction energy (cf. p. 183): since the latter depends on the inverse cube of the distance R between the two coupled dipoles, the probability varies as R^{-6}. Assuming vibrational equilibration in the excited donor, Förster obtained the following expression for the transfer rate constant between overlapping continua in the donor and acceptor

$$k_{ET} = \frac{9,000 \ln 10}{128\, \pi^5 N} \cdot \frac{\kappa^2 \phi_D}{n^4 \tau_D R^6} \int_0^\infty f_D(\tilde{v}) \varepsilon_A(\tilde{v}) \cdot \frac{d\tilde{v}}{\tilde{v}^4}$$

where N is Avogadro's number, ϕ_D is the luminescence efficiency of the donor, n is the refractive index of the medium, and τ_D is the lifetime of the donor (reduced by the onset of electronic energy transfer); if τ_D^0 is its mean radiative lifetime, $\tau_D = \phi_D \tau_D^0$. κ is a factor determined by the relative spatial orientation of the two transition moment vectors; if molecular rotation is fast, $\kappa^2 = \frac{2}{3}$ on average. $f_D(\tilde{v})$ represents the luminescence spectral profile of the donor measured in quanta cm^{-1} and normalized to unity, (i.e. $\int_0^\infty f_D(\tilde{v}) \cdot d\tilde{v} = 1$), $\varepsilon_A(\tilde{v})$ is the molar decadic extinction coefficient of the acceptor at frequency $\tilde{v}$ (cm^{-1}). Resonance energy transfer is most efficient when the absorption spectrum of the acceptor lies slightly to the red of the donor's luminescence spectrum. The greater the overlap, measured by the integral $\int_0^\infty f_D(\tilde{v}) \varepsilon_A(\tilde{v})\, d\tilde{v}/\tilde{v}^4$, the larger the magnitude of k_{ET}. The integral is small when the two spectra are widely separated, and when the absorption spectrum of A is weak.

The equation only applies to a single pair of molecules separated by a distance R. Before it can be applied to real systems it has to be averaged over the continuous range of values which R can take, in a large assembly of molecules. It is convenient to define a mean critical separation R_0, at which the probability of transfer k_{ET} is equal to the probability of spontaneous decay $1/\tau_D$. Then

$$R^6 = \frac{1}{k_{ET}\tau_D}(R_0)^6$$

The average intermolecular separation depends on the concentration, and the critical concentration $[A]_0$ of the acceptor is related to R_0 by the expression

$$\frac{1}{[A]_0} = \frac{4}{3}\pi R_0^3$$

where $[A]_0$ is expressed in molecules cm^{-3}. R_0 can be identified as the radius of a sphere which contains one molecule of A and has a molecule D* at the centre.

The Förster treatment has been applied successfully to many systems where singlet–singlet resonance energy transfer takes place in solution. Sensitized fluorescence can be produced at concentrations $\sim 10^{-3} M$: the average intermolecular separation at this concentration is ~ 7 nm, in good agreement with Förster's prediction.* The same is also true of concentration depolarization (see p. 180) which is a special case of resonance energy transfer in which both donor and acceptor are identical. The model is inappropriate at low pressures in the gas phase since the vibronic energy levels of the luminescent molecule are likely to be discrete, and electronic energy transfer may be more rapid than the vibrational equilibration of the donor.

When long range energy transfer takes place in solution, it is not necessary for molecular diffusion to bring the donor and acceptor into close proximity before energy transfer can take place. The rate constants are found to be insensitive to variations in the viscosity of the solvent, and in the most efficient systems values of $k_{ET} \simeq 3 \times 10^{11}$ litres mole^{-1} sec^{-1} have been reported. They can be compared with the 'normal' maximum rate constant for diffusion controlled bimolecular processes of $\sim 10^{10}$ litres mole^{-1} sec^{-1} in fluid solvents. Diffusion control sets in when efficient energy transfer is possible only at short-ranges. It becomes apparent when transfer proceeds through an exchange mechanism.

Exchange interaction. Following the success of Förster's treatment of long-range electronic energy transfer, Dexter[138] extended the model to include the weaker dipole–quadrupole (energy $\propto 1/R^4$) and electron exchange interactions. The rate constant which he derived for energy transfer by the latter mechanism was

$$k_{ET} \propto \exp(-2R/L) \int_0^\infty f_D(\tilde{v}) \varepsilon_A'(\tilde{v}) . d\tilde{v} .$$

where L is a constant factor. Like the function $f_D(\tilde{v})$, the extinction coefficient $\varepsilon_A'(\tilde{v})$ is now also normalized to unity, (i.e. $\int_0^\infty f_D(\tilde{v}) . d\tilde{v} = \int_0^\infty \varepsilon_A'(\tilde{v}) . d\tilde{v} = 1$). In consequence the overlap integral is now 'unaware' of whether the electronic transition in the acceptor is allowed or not and, in contrast with the dipolar mechanism, the probability of energy transfer

---

* At a concentration of $10^{-3} M$, the molecular concentration is 6×10^{17} molecules cm^{-3}. The average volume per molecule is thus 1.4×10^{-18} cm^3 equivalent to a sphere of radius 7 nm.

does not depend on the dipole oscillator strength of the acceptor. The most important process to which it applies in photochemical systems is that of triplet–triplet energy transfer

$$^3D^* + {}^1A \rightarrow {}^1D + {}^3A^*$$

Terenin and Ermolaev were the first to demonstrate this kind of process, when in 1952 they reported the observation of sensitized phosphorescence in rigid glassy solutions at 77°K.[139,114]

To exclude any ambiguity in the mechanism by which the acceptor was excited into its triplet state, they chose donor–acceptor pairs where the energy of $^1D^* < {}^1A^*$ while $^3D^* > {}^3A^*$, (see Figure 2.60). A filter prevented

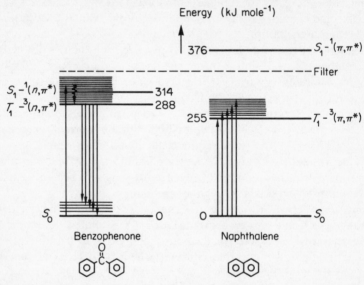

Figure 2.60 Relative energies of the lowest singlet and triplet states of benzo-phenone and naphthalene.

direct excitation of $^3A^*$ via intersystem crossing from $^1A^*$. Benzophenone and naphthalene fitted the purpose admirably, since the quantum yield of intersystem crossing in the donor is almost unity and its very small $S_1 - T_1$ separation leaves T_1 at 288 kJ mole^{-1}, some 33 kJ mole^{-1} above the lowest triplet state of naphthalene. The result of the experiment is shown in Figure 2.61. Terenin and Ermolaev estimated that efficient energy transfer occurred at an intermolecular distance of 1·3 nm, of the same order as the van der Waals separation. The transfer distance remained unchanged when halonaphthalenes were substituted for the parent molecule. Since the $S_0 \leftrightarrow T_1$ transition probabilities are increased by up to three orders of

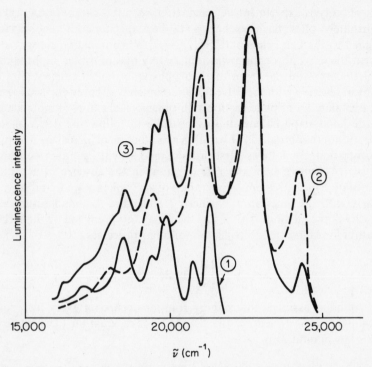

Figure 2.61 Sensitized phosphorescence of naphthalene in glassy solution at −190°C. (After A. Terenin and V. Ermolaev, *Trans. Faraday Soc.*, **52**, 1042 (1952).) Excitation of (1) Naphthalene alone at short wavelengths (2) benzophenone alone at 366 nm (3) benzophenone in the presence of naphthalene, at 366 nm.

magnitude in the substituted naphthalenes (see Figure 2.54), this clearly demonstrated the lack of any dependence on the oscillator strength of the transition in the acceptor.

A few years later, Bäckström and Sandros[140] discovered that the same type of process can take place in fluid solution. In their first experiments they studied the quenching of the phosphorescence of biacetyl and concluded that quenching could occur either chemically, through hydrogen atom abstraction from the solvent

$$CH_3COCOCH_3(T_1) + HR \rightarrow CH_3\dot{C}(OH)COCH_3 + R\cdot$$

or physically through triplet energy transfer. Subsequently they reversed the role of the biacetyl when they used triplet benzophenone to photosensitize its phosphorescence. Porter and Wilkinson[141] returned to benzophenone and naphthalene to demonstrate triplet energy transfer in a flash photolysis experiment. A transient absorption band photographed immediately after flash excitation of benzophenone in alcohol was assigned to the ketyl radical $(C_6H_5)_2\dot{C}OH$. When naphthalene was also present the ketyl absorption was replaced by that of triplet naphthalene. Triplet energy transfer from the benzophenone was so efficient that its chemical reaction with the solvent was completely suppressed, (see Figure 2.62 (see plate IX of plate section)). In other systems (for example, naphthalene-1-iodonaphthalene) it was possible to follow the decay of the triplet donor (naphthalene) absorption as a function of increasing acceptor concentration. It follows first-order kinetics with a rate 'constant' $k_0 + k_{ET}[A]$, where k_0 is the rate constant in the absence of acceptor. Quantitative measurements gave values of k_{ET}, which were found to vary linearly with the function T/η, where T and η are the temperature and viscosity of the medium. This is just the behaviour predicted by the Debye equation for diffusion-controlled bimolecular processes

$$k_{ET} = \frac{8RT}{3,000\eta} \, 1 \, mole^{-1} \, sec^{-1}$$

The equation assumes that energy transfer occurs on every molecular encounter and is based on the Smoluchowski equation (derived from Fick's law of diffusion),

$$k_{ET} = \frac{8\pi\sigma DN}{1,000} \, 1 \, mole^{-1} \, sec^{-1}$$

where σ is the encounter radius and D is the diffusion coefficient. D is assumed to be given by the Einstein–Stokes equation

$$D = \frac{RT}{6\pi\eta N} \left\{ \frac{1}{r_M} + \frac{1}{r_S} \right\}$$

where r_M and r_S are the radii of the diffusing molecules (assumed equal) and the solvent molecules. If they are comparable in size, $r_M \simeq r_S$ and

$$D \simeq \frac{RT}{3\pi\eta r_M}$$

whence

$$k_{ET} \simeq \frac{8RT}{3,000\eta} \cdot \frac{\sigma}{r_M}$$

The Debye equation assumes that the encounter and diffusion radii σ and r_M are equal. This is quite reasonable when energy transfer follows an exchange mechanism and there is close agreement between the measured and calculated values of k_{ET}. For example in the case of naphthalene and 1-iodonaphthalene in glycerol ($\eta = 512 \, cp$), $k_{ET} = 2.1 \times 10^7 \, l \, mole^{-1} \, sec^{-1}$ (experimental) and $1.3 \times 10^7 \, l \, mole^{-1} \, sec^{-1}$ (calculated).[142] Better agreement is obtained if it is assumed that the diffusing molecules are 'slippery' and slide smoothly past each other.[142] This yields the modified expression $k_{ET} = 8RT/2{,}000\eta$.

In general triplet energy transfer seems to be diffusion-controlled with a unit collision efficiency so long as the transfer is exothermic,* i.e. when the triplet energy of the donor is greater than that of the acceptor. It has been exploited by many photochemists, particularly Hammond and co-workers,[58,143] as a method of photosensitizing chemical change in the lowest triplet state of the acceptor while eliminating the possibility of interference from the excited singlet state. Since it occurs with unit efficiency, measurement of the quantum efficiency of photosensitization can also be used to measure the triplet yield ϕ_T, of the donor. The efficiency of photosensitization decreases when the triplet energy of the donor approaches that of the acceptor, and falls to zero when triplet energy transfer is endothermic, providing a useful means of locating the energy of the lowest triplet state in the acceptor. Unfortunately the energies have not always agreed with those estimated from the singlet–triplet absorption or emission spectra. The estimates assumed that the spectral origins can be assigned to the (0,0) transition, but may be in error if there is a large change in the equilibrium geometry.

For example, energy transfer to cis-stilbene remains efficient even when the energy of the triplet donor is less than the 'spectroscopic' triplet energy of the cis-stilbene. This has led to considerable discussion as to the possible consequences of a change in the equilibrium geometry of the acceptor during excitation. Hammond[143] has argued that the 'spectroscopic' measurements may be too high, if stilbene adopts a twisted conformation in its triplet state, because of the operation of the Franck–Condon Principle. If energy transfer occurs in times $\sim 10^{-10}$ sec there is no reason why the acceptor should not relax into its equilibrium conformation during that time and make up the apparent energy deficit. However, there may be an alternative explanation for some of the apparent anomalies, if the photosensitizing species is an *excited* triplet state of the donor. For example, Liu and Edman[144] have presented convincing evidence for energy transfer from the second triplet state in anthracene and some of its derivatives. Since the level T_2 lies only just below S_1 in these molecules (see pp. 184 and 189 and Figure 2.50), intersystem crossing

---

* Though its efficiency may also be sensitive to the relative orientations of the donor and acceptor.

favours the initial population of T_2 rather than T_1. If the rate of internal conversion from $T_2 \rightarrow T_1$ is relatively slow, resonance energy transfer may be able to compete against it. In the case of anthracene the radiationless transition is electronically forbidden by the u $\leftrightarrow$ g selection rule.

(iii) *Luminescence quenching*

A quenching process can be defined as one which shortens the natural lifetime of the excited molecule. The process may be collisional or 'dynamic', perhaps involving the formation of an intermediate association complex in the excited state, 'long-range' as in the case of coulombic resonance energy transfer, or 'static' involving the formation of an encounter complex or some donor–acceptor complex prior to the act of light absorption. Intermolecular processes in the excited state may involve atom, electron or proton transfer, or association to form transient and perhaps permanent adducts, or they may be purely physical processes such as energy transfer or collisional perturbations which allow the molecule to decay into a lower electronic state (not necessarily the ground state). In many cases the luminescence is not completely extinguished but transformed into one associated with some other species present in the system. A few examples will illustrate some of the possibilities.

(a) Aromatic ketones are phosphorescent when dissolved in fully fluorinated solvents such as perfluoromethyl cyclohexane at room temperature. Parker[145] has reported a phosphorescent lifetime of 0·7 msec for benzophenone. In alcohol solution, no phosphorescence can be detected and the lifetime of the triplet state estimated from competitive energy transfer experiments is ~ 10 nsec. The phosphorescence is quenched through rapid intermolecular hydrogen atom transfer.

$$\phi_2 CO(T_1) + HR \rightarrow \phi_2 \dot{C}OH + R\cdot$$

(b) Anthracene fluoresces strongly in hexane solution, but as its concentration increases the fluorescence is quenched and white crystals of a 'sandwich' dimer precipitate out of solution. There is a linear relation between the reciprocal of the fluorescence intensity and the anthracene concentration, which satisfies the Stern–Volmer relation (p. 117). Excited singlet anthracene suffers self-quenching to form a non-fluorescent dimer linked in the 9,10 positions, which decays into a stable ground state.[146]

$$A(S_1) + A(S_0) \rightleftharpoons A_2^* \text{ (singlet)} \rightarrow A_2(S_0)$$

This represents a variation on the theme discussed for pyrene, where self-quenching produced a fluorescent excimer which was non-bonded in the ground state (section 2.5e).

(c) Anthracene does not fluoresce when dissolved in carbon tetrachloride, because of a chemical reaction in the excited singlet state. Direct abstraction of a chlorine atom has been proposed[147]

$$A(S_1) + CCl_4 \rightarrow \cdot ACl + \cdot CCl_3$$

to produce the 9-chloroanthracyl radical, but the nett reaction could plausibly proceed through an electron transfer mechanism (see below)

$$A(S_1) + CCl_4(S_0) \rightarrow \{A^+ \cdot CCl_4^-\} \rightarrow \{A^+ \cdot Cl^- + \cdot CCl_3\}$$
$$\rightarrow \cdot ACl + \cdot CCl_3$$

(d) When anthracene or biphenyl is dissolved in a non-polar solvent together with an easily oxidized amine such as NN'-diethylaniline, its fluorescence is replaced by a broad, structureless emission spectrum lying at longer wavelengths. It is associated with the formation of an excited, fluorescent charge-transfer complex.[60b,148] For example

$$\phi_2(S_1) + \phi N(C_2H_5)_2(S_0) \rightleftharpoons \{\phi N(C_2H_5)_2^+ \cdot \phi_2^-\} \quad \text{(singlet)}$$
$$\phi_2(S_0) + \phi N(C_2H_5)_2(S_0) + h\nu_{fl}$$

There is no evidence of complexing in the ground state, since the absorption spectra of the two components are additive. The excited complex can be classified as an '*exciplex*' to distinguish it from the special case of an excimer where both components of the excited complex happen to be identical.

As the polarity of the solvent is increased the charge-transfer fluorescence fades (and shifts to the red because of solvation). In a very polar solvent such as acetonitrile, it is possible to detect the transient absorption spectra of the solvated molecular ions immediately following flash excitation, and in addition that of the aromatic hydrocarbon in its lowest triplet state. The following scheme, presented by Weller, accommodates all his observations (A represents the electron acceptor and D the donor),[148]

$$A(S_1) + D(S_0) \quad\begin{array}{l} (A^-D^+)^* \text{ (singlet)} \\ \downarrow \\ (A^-D^+)^*_{solv} \text{ (singlet)} \\ \downarrow \\ (A^-)_{solv} \cdots (D^+)_{solv} \end{array}$$

completely suppressed in a highly polar solvent

solvated radical–ion pair: two doublets

$$A(S_0) + D(S_0) \qquad A(T_1) + D(S_0) \quad \text{(or } A(S_0) + D(T_1))$$

In a strongly polar solvent, ion-pair formation is very fast, and the rate of fluorescence quenching is diffusion controlled. Production of the triplet acceptor via electron transfer has been confirmed by the observation of a phosphorescent chemiluminescence on mixing solutions of stable molecular radical–ions.

(e) In aqueous solution phenols, naphthols and aromatic amines are very weak acids. For example, β-naphthol has a pK_a value of 9·5* and in neutral aqueous solution it is virtually undissociated. However, absorption in the near ultraviolet excites two fluorescence bands, one in the near-ultraviolet associated with the neutral molecule and the other in the blue emitted by the naphthoate ion. In the first excited singlet state the acidity constant increases by nearly seven orders of magnitude. The new equilibrium

$$ROH(S_1) + H_2O \rightleftharpoons (RO^- \ldots H_3O^+)^* \rightarrow (RO^-)^* + H_3O^+$$

is established rapidly and many of the excited molecules ionize before returning to the ground state.[149] As the pH is increased, making the solution more alkaline, the blue component increases in intensity at the expense of the ultraviolet one and eventually replaces it entirely.

The large increase in acidity can be understood if there is appreciable charge transfer from the heteroatom into the ring in the first excited singlet state. A valence-bond description must include structures such as

and   or in general

In contrast the electronic configuration in the lowest triplet state must be quite different, since its acidity is only a little greater than that of the ground state. An experimental study of the pH dependence of the triplet–triplet absorption spectrum of β-naphthol led to a pK_a value of 8·1, compared with 9·5 in the ground state.[150] β-Naphthylamine shows similar behaviour.

If an electron withdrawing group is substituted into the ring as in β-naphthoic acid, the pK_a values change in the opposite sense on excitation.

---

* pK_a values are given by log K_a, where K_a is the equilibrium constant of the acid dissociation $HA + H_2O \rightleftharpoons H_3O^+ + A^-$.

In the excited singlet state the acid is readily protonated

$$RC\overset{\displaystyle O}{\underset{\displaystyle OH}{\diagdown}} (S_1) + H_3O^+ \rightleftharpoons RC\overset{\displaystyle \overset{+}{O}H}{\underset{\displaystyle OH}{\diagdown}} + H_2O \qquad K_{eq} \simeq 1$$

while the triplet state remains a weak acid.

(f) Dissolved oxygen often quenches fluorescence as well as phosphorescence, and in many cases the quenching leads to photooxidation. For example, an oxygenated solution of anthracene readily forms the transannular peroxide

when irradiated in the near-ultraviolet. Its quantum yield varies with the anthracene concentration [A] as

$$\phi = \phi' \cdot \frac{k_2[A]}{k_1 + k_2[A]}$$

The kinetics require the participation of an unexcited anthracene molecule in the nett reaction. The ratio k_2/k_1 represents the relative probability of reaction and decay of the intermediate oxidizing species, and the proportionality constant ϕ' is the quantum efficiency of its formation.[146]

Anthracene can also photosensitize the oxidation of a substrate and the reaction has been developed as a general method for the production of peroxides, especially by Schenck and coworkers.[151] For example

A wide range of photosensitizing dyes have been employed. Two general types of mechanism have been proposed, both of which are kinetically indistinguishable. In the first the oxygen transferring species is assumed to be an intermediate complex

$$A(T_1) + O_2(\tilde{X}^3\Sigma_g^-) \rightarrow (A\cdots O_2)^*$$

$$(A\cdots O_2)^* \xrightarrow{k_1} A(S_0) + O_2(\tilde{X}^3\Sigma_g^-)$$

$$(A\cdots O_2)^* + A(S_0) \text{ or } M(S_0) \xrightarrow{k_2} A(S_0) + AO_2(S_0) \text{ or } MO_2(S_0)$$

The alternative mechanism assumes that 'singlet' oxygen is produced through resonance energy transfer from the triplet sensitizer via the exchange interaction.

$$A(T_1) + O_2(\tilde{X}^3\Sigma_g^-) \rightarrow (A\cdots O_2)^* \text{ (singlet, triplet or quintet)}$$

$$(A\cdots O_2)^* \text{ (singlet)} \rightarrow A(S_0) + O_2(\tilde{a}^1\Delta_g \text{ or } \tilde{b}^1\Sigma_g^+)$$

$$O_2(\tilde{a}^1\Delta_g \text{ or } \tilde{b}^1\Sigma_g^+) \xrightarrow{k_1} O_2(\tilde{X}^3\Sigma_g^-)$$

$$O_2(\tilde{a}^1\Delta_g \text{ or } \tilde{b}^1\Sigma_g^+) + A(S_0) \text{ or } M(S_0) \xrightarrow{k_2} AO_2(S_0) \text{ or } MO_2(S_0)$$

The singlet oxygen molecules are metastable since their radiative decay into the ground state is doubly forbidden. They lie respectively, 96 kJ mole^{-1} and 163 kJ mole^{-1} above the ground state (see p. 30).

For many years the 'complex mechanism' was favoured as the more likely pathway. However in 1964, Foote and Wexler were able to assign the red chemiluminescence excited by the action of hydrogen peroxide on hypochlorite ions to emission from $O_2(\tilde{a}^1\Delta_g)$ molecules.[152]

$$ClO^- + H_2O_2 \rightarrow O_2(\tilde{a}^1\Delta_g) + H_2O + Cl^-$$

The singlet oxygen was found to be a reactive species readily forming peroxides identical with those produced through photosensitized oxidation, and at the same relative rates. These experiments have been followed by a flood of others and the evidence for a common intermediate in the two types of reaction is now overwhelming (though as a space scientist was reported as saying on learning that the evidence for moon rock having a volcanic origin was overwhelming, 'It depends on who is being overwhelmed'). There seems little doubt that most, and perhaps all of the sensitized photooxidations follow an energy transfer mechanism. Very recently, Snelling has been able to detect the phosphorescence of $O_2(\tilde{a}^1\Delta_g)$ molecules, excited by energy transfer from triplet benzene in the vapour phase.[153]

REFERENCES

1. E. U. Condon and G. H. Shortley, *Theory of Atomic Spectra*, Cambridge University Press, London, 1951
2. H. F. Hameka, The Triplet State (Ed. by A. B. Zahlan), Cambridge University Press, London, 1967
3. S. P. McGlynn, T. Azumi and M. Kinoshita, *Molecular Spectroscopy of the Triplet State*, Prentice-Hall, New Jersey, 1969
4. L. Pauling and E. B. Wilson, Jr., *Introduction to Quantum Mechanics*, McGraw-Hill, New York, 1935
5. H. G. Kuhn, *Atomic Spectra*, Longmans, London, 1962
6. G. Porter and F. Wright, *Discussions Faraday Soc.*, **14**, 23 (1953)
7. R. A. Durie and D. A. Ramsay, *Canadian J. Phys.*, **36**, 35 (1958)

8. G. Herzberg, Spectra of Diatomic Molecules, 2nd Ed., Van Nostrand, Princeton, New Jersey, 1950
9. N. Bayliss, Proc. Roy. Soc., A, **158**, 551 (1937)
10. V. Henri, Compt. Rend., **177**, 1037 (1923)
11. R. W. Fair and B. A. Thrush, Trans. Faraday Soc., **65**, 1208 (1969)
12. G. Herzberg and L. Mundie, J. Chem. Phys., **8**, 263 (1940)
13. L. A. Turner and E. W. Samson, Phys. Rev., **37**, 1684 (1931)
14. E. Wasserman, W. E. Falconer and W. A. Yager, Ber., **72**, 248 (1968)
15. (a) L. A. Turner, Phys. Revs., **41**, 627 (1933)
    (b) E. O. Degenkolb, J. I. Steinfeld, E. Wasserman and W. Klemperer, J. Chem. Phys., **51**, 615 (1969)
    (c) A. Chutjian and T. C. James, J. Chem. Phys., **51**, 1242 (1969)
16. J. Franck and R. W. Wood, Phil. Mag., **21**, 314 (1911)
17. J. I. Steinfield and W. Klemperer, J. Chem. Phys., **42**, 3475 (1965)
18. L. Brewer, R. A. Berg and G. M. Rosenblatt, J. Chem. Phys., **38**, 1381 (1963)
19. H. P. Broida and T. Carrington, J. Chem. Phys., **38**, 136 (1963)
20. E. Durand, J. Chem. Phys., **8**, 46 (1940)
21. H. E. Gunning and O. P. Strausz, Advances in Photochemistry, Vol. 1, Interscience, New York, p. 209
22. R. J. Cvetanovic, Progress in Reaction Kinetics (Ed. G. Porter), Vol. 2, Pergamon, New York, 1964, p. 39
23. C. G. Matland, Phys. Revs., **92**, 637 (1953)
24. (a) A. J. Yarwood, O. P. Strausz and H. E. Gunning, J. Chem. Phys., **41**, 1705 (1964)
    (b) K. Yang, J. Amer. Chem. Soc., **88**, 4575 (1966)
25. R. J. Cvetanovic, J. Chem. Phys., **23**, 1203 (1955)
26. M. D. Scheer and J. Fine, J. Chem. Phys., **36**, 1264 (1962)
27. A. B. Callear and R. G. W. Norrish, Proc. Roy. Soc., A, **266**, 299 (1962)
28. (a) A. Jablonski, Z. Physik., **70**, 723 (1931)
    (b) H. E. Gunning, S. Penzes, H. S. Sandhu and O. P. Strausz, J. Amer. Chem. Soc., **91**, 7684 (1969)
29. A. C. G. Mitchell and M. W. Zemansky, Resonance Radiation and Excited Atoms, Cambridge University Press, London, 1934: Reprinted 1961
30. G. Cario and J. Franck, Z. Physik., **17**, 202 (1923)
31. D. R. Bates, Discussions Faraday Soc., **33**, 7 (1962)
32. H. Beutler and B. Josephi, Z. Physik., **53**, 747 (1929)
33. G. Karl, P. Kruse and J. C. Polyani, J. Chem. Phys., **46**, 224 (1967)
34. T. G. Slanger, J. Chem. Phys., **48**, 586 (1968)
35. A. B. Callear and G. J. Williams, Trans. Faraday Soc., **60**, 2158 (1964)
36. G. Karl, P. Kruse, J. C. Polyani and I. W. M. Smith, J. Chem. Phys., **46**, 244 (1967)
37. K. Laidler, The Chemical Kinetics of Excited States, Clarendon Press, Oxford, 1955
38. E. W. R. Steacie, Atomic and Free Radical Reactions, 2nd Ed., Vols. 1 and 2, Reinhold, New York, 1954
39. L. Landau and E. Teller, Phys. Z. Sowjetunion, **10**, 34 (1936)
40. T. L. Cottrell and J. C. McCoubrey, Molecular Energy Transfer in Gases, Butterworths, London, 1961
41. B. Stevens, Chemical Activation in Gases, Pergamon, Oxford, 1967

42. A. G. Gaydon, *Dissociation Energies and Spectra of Diatomic Molecules*, 3rd Ed., Chapman and Hall, London, 1968
43. R. S. Mulliken, *Phys. Revs.*, **51**, 310 (1937)
44. R. J. Donovan and D. Husain, *Trans. Faraday Soc.*, **62**, 1050 (1966)
45. J. Solomon, *J. Chem. Phys.*, **47**, 889 (1967)
46. G. E. Busch, R. T. Mahoney and K. R. Wilson, *IEEE J. Quantum Electronics*, **QE-6**, 171 (1970)
47. J. N. Murrell, *The Theory of the Electronic Spectra of Organic Molecules*, Methuen, London, 1963
48. M. Kasha, *Discussions Faraday Soc.*, **9**, 14 (1950)
49. E. Clar, *Aromatische Kohlenwasserstoffe*, Springer Verlag, Berlin, 1952
50. J. R. Platt, *J. Chem. Phys.*, **22**, 1448 (1954)
51. M. Kasha and H. R. Rawls, *Photochemistry and Photobiology*, **7**, 561 (1968)
52. R. S. Mulliken and C. A. Rieke, *Reports on Progress in Physics*, **8**, 231 (1941)
53. A. J. Merer, *Discussions Faraday Soc.*, **35**, 127 (1963)
54. A. E. Douglas and E. R. V. Milton, *J. Chem. Phys.*, **41**, 357 (1964)
55. A. N. Terenin, *Acta Physicochim., U.R.S.S.*, **18**, 210 (1943) in English
56. G. N. Lewis and M. Kasha, *J. Amer. Chem. Soc.*, **66**, 2100 (1944)
57. G. Porter and M. Windsor, *Discussions Faraday Soc.*, **17**, 178 (1954); *Proc. Roy. Soc. A*, **245**, 238 (1958)
58. P. J. Wagner and G. S. Hammond, *Advances in Photochemistry*, Vol. 5, Interscience, New York, 1968, p. 21
59. N. S. Bayliss and E. G. McRae, *J. Phys. Chem.*, **58**, 1002 (1954)
60. (a) E. Lippert, *Z. Elektrochem.*, **61**, 962 (1957)
    (b) H. Beens, H. Knibbe and A. Weller, *J. Chem. Phys.*, **47**, 1183 (1967)
61. P. Suppan, *J. Molecular Spec.*, **30**, 17 (1969)
62. R. S. Mulliken, *J. Amer. Chem. Soc.*, **72**, 600 (1950); **74**, 811 (1952); *J. Phys. Chem.*, **56**, 801 (1952)
63. R. S. Mulliken, *Reactivity of the Photoexcited Molecule*, Interscience, London, 1967, p. 289
64. G. Briegleb and J. Czekalla, *Angew. Chem.*, **72**, 401 (1960)
65. E. M. Kosower, *J. Amer. Chem. Soc.*, **80**, 3253 (1958)
66. E. M. Kosower, J. A. Skorcz, W. M. Schwarz, Jr., and J. W. Patton, *J. Amer. Chem. Soc.*, **82**, 2188 (1960)
67. E. M. Kosower and L. Lindquist, *Tetrahedron Letters*, 4481 (1965)
68. M. J. Blandamer, T. R. Griffiths, L. Shields and M. C. R. Symons, *Trans. Faraday Soc.*, **60**, 1524 (1964)
69. (a) E. J. Hart and J. W. Boag, *J. Amer. Chem. Soc.*, **84**, 4090 (1962)
    (b) M. S. Matheson, W. A. Mulac and J. Rabani, *J. Phys. Chem.*, **67**, 2613 (1963)
    (c) H-I. Joschek and L. I. Grossweiner, *J. Amer. Chem. Soc.*, **88**, 3261 (1966)
70. (a) L. I. Grossweiner and M. S. Matheson, *J. Phys. Chem.*, **61**, 1089 (1957)
    (b) F. H. C. Edgecombe and R. G. W. Norrish, *Proc. Roy. Soc., A*, **253**, 154 (1959)
    (c) J. Jortner, A. Ottolenghi and G. Stein, *J. Phys. Chem.*, **67**, 1271 (1963)
71. F. S. Dainton, *J. Chem. Soc.*, **1533** (1952)
72. J. B. Coon, R. E. DeWames and C. M. Loyd, *J. Molecular Spec.*, **8**, 285 (1962)
73. D. P. Craig, *J. Chem. Soc.*, 2146 (1950)
74. G. Herzberg, *Electronic Spectra and Electronic Structure of Polyatomic Molecules*, Van Nostrand, Princeton, New Jersey, 1966

75. F. A. Johnson and C. B. Colburn, *J. Amer. Chem. Soc.*, **83**, 3043 (1961)
76. J. P. Simons, *J. Chem. Soc.*, 1005 (1965)
77. M. Ito, P. C. Huang and E. M. Kosower, *Trans. Faraday Soc.*, **57**, 1662 (1961)
78. B. R. Henry and M. Kasha, *Ann. Rev. Phys. Chem.*, **19**, 161 (1968)
79. (a) G. W. Robinson and R. P. Frosch, *J. Chem. Phys.*, **37**, 1962 (1962)
    (b) C-H. Ting, *Photochem. and Photobiol.*, **9**, 17 (1969)
80. A. Jablonski, *Zeit. f. Physik*, **94**, 38 (1935)
81. D. Phillips, J. Lemaire, C. S. Burton and W. A. Noyes, Jr., *Advances in Photochemistry*, Vol. 5, Interscience, New York, 1968, p. 329
82. R. H. S. Liu and J. R. Edman, *J. Amer. Chem. Soc.*, **91**, 1492 (1969)
83. C. N. Hinshelwood, *Proc. Roy. Soc.*, A, **113**, 230 (1927)
84. L. S. Kassel, *Kinetics of Homogeneous Gas Reactions*, Reinhold, New York, 1932, Chap. 5.
85. O. K. Rice and H. C. Ramsperger, *J. Amer. Chem. Soc.*, **49**, 1617 (1927); **50**, 617 (1928)
86. R. A. Marcus and O. K. Rice, *J. Phys. Chem.*, **55**, 894 (1951); G. M. Wieder and R. A. Marcus, *J. Chem. Phys.*, **37**, 1835 (1962)
87. N. B. Slater, *Theory of Unimolecular Reactions*, Cornell University Press (1959)
88. H. S. Johnston, *Gas Phase Reaction Rate Theory*, Ronald Press, New York, 1966, Chap. 15
89. R. C. Mitchell and J. P. Simons, *Discussions Faraday Soc.*, **44**, 208 (1967)
90. A. J. Yarwood and J. P. Simons, *Trans. Faraday Soc.*, **59**, 90 (1963)
91. R. J. Campbell and E. W. Schlag, *J. Amer. Chem. Soc.*, **89**, 5098; 5103 (1967)
92. M. A. Pollack, *Appl. Phys. Letters*, **9**, 94 (1966)
93. A. N. Terenin, *Usp. Fiz. Nauk.*, **36**, 292 (1948)
94. I. Tanaka, T. Carrington and H. P. Broida, *J. Chem. Phys.*, **35**, 750 (1961)
95. R. J. Donovan and D. Husain, *Trans. Faraday Soc.*, **62**, 11 (1966)
96. J. V. V. Kasper and G. C. Pimentel, *Appl. Phys. Letters*, **5**, 252 (1964); J. V. V. Kasper, J. H. Parker and G. C. Pimentel, *J. Chem. Phys.*, **43**, 1827 (1965)
97. R. G. W. Norrish, *J. Chem. Soc.*, 1611 (1929)
98. D. Neuberger and A. B. F. Duncan, *J. Chem. Phys.*, **22**, 1693 (1954)
99. (a) E. M. Anderson and G. B. Kistiskowsky, *J. Chem. Phys.*, **48**, 4787 (1968)
    (b) A. E. Douglas and C. W. Mathews, *J. Chem. Phys.*, **48**, 4788 (1968)
100. L. Kaplan and K. E. Wilzbach, *J. Amer. Chem. Soc.*, **90**, 3291 (1968)
101. F. Perrin, *Ann. Phys.*, (Paris), **12**, 169 (1929)
102. M. Bixon and J. Jortner, *J. Chem. Phys.*, **48**, 715 (1968)
103. P. M. Rentzepis, *Chem. Phys. Letters*, **3**, 717 (1969)
104. P. A. Geldof, R. P. H. Rettchnick and G. J. Hoytink, *Chem. Phys. Letters*, **4**, 59 (1969)
105. (a) G. N. Lewis, D. Lipkin and T. T. Magel, *J. Amer. Chem. Soc.*, **63**, 3005 (1941)
    (b) G. N. Lewis and M. Calvin, *J. Amer. Chem. Soc.*, **67**, 1232 (1945)
    (c) G. N. Lewis, M. Calvin and M. Kasha, *J. Chem. Phys.*, **17**, 1804 (1949)
106. D. F. Evans, *Nature*, **176**, 777 (1955)
107. C. A. Hutchison, Jr. and B. W. Mangum, *J. Chem. Phys.*, **29**, 952 (1958); **34**, 908 (1961)
108. J. H. van der Waals and M. S. DeGroot, *Mol. Phys.*, **2**, 33 (1959)
Refs. 107 and 108: see also review articles in *The Triplet State* (Ed. A. B. Zahlan), Cambridge University Press, London, 1967
109. H. Linschitz, C. Steel and J. A. Bell, *J. Phys. Chem.*, **66**, 2574 (1962)

110. C. A. Parker, *Chem. Brit.*, **2**, 160 (1966)
111. (a) G. Porter and F. Wright, *Trans. Faraday Soc.*, **51**, 1205 (1955)
     (b) P. G. Bowers and G. Porter, *Proc. Roy. Soc.*, *A*, **296**, 435 (1967)
112. S. K. Lower and M. El-Sayed, *Chem. Revs.*, **66**, 199 (1966)
113. G. Porter and P. Suppan, *Trans. Faraday Soc.*, **61**, 1664 (1965)
114. V. L. Ermolaev, *Soviet Physics* (*Uspekhi*), **6**, 333 (1963)
115. B. Stevens, M. S. Walker and E. Hutton, *The Triplet State* (Ed. A. B. Zahlan), Cambridge University Press, London, 1967, p. 239
116. D. P. Craig, J. M. Hollas and G. W. King, *J. Chem. Phys.*, **29**, 974 (1958)
117. D. F. Evans, *J. Chem. Soc.*, 1351, 3885 (1957)
118. W. Kemula and A. Grabowska, *Bull. L'Acad. Pol. Sci.*, *Ser. sci.*, *chim.*, *geol.*, *geogr.*, **6**, 12 (1958)
119. G. Porter and M. Wright, *Discussions Faraday Soc.*, **27**, 18 (1959)
120. G. W. Robinson, *J. Molecular Spec.*, **6**, 58 (1961)
121. A. R. Horrocks, A. Kearvell, K. Tickle and F. Wilkinson, *Trans. Faraday Soc.*, **62**, 3393 (1966)
122. G. W. Robinson, *J. Chem. Phys.*, **46**, 572 (1967)
123. D. R. Kearns and W. A. Case, *J. Amer. Chem. Soc.*, **88**, 5087 (1966)
124. C. A. Parker, *Photoluminescence of Solutions*, Elsevier, Amsterdam, 1968
125. S. Boudin, *J. Chim. Phys.*, **27**, 285 (1930)
126. Th. Förster and K. Kasper, *Z. Elektrochem.*, **59**, 976 (1955)
127. C. A. Parker and C. G. Hatchard, *Proc. Roy. Soc. A*, **269**, 574 (1962)
128. (a) T. N. Rao, S. S. Collier and J. G. Calvert, *J. Amer. Chem. Soc.*, **91**, 1609 (1969)
     (b) H. D. Mettee, *J. Chem. Phys.*, **49**, 1784 (1968)
129. (a) B. S. Neporent, *Zhur. fiz. Khim.*, **24**, 1219 (1950)
     (b) B. S. Neporent and S. O. Mirumyants, *Opt. and Spec.*, **8**, 635 (1960)
130. (a) M. Boudart and J. T. Dubois, *J. Chem. Phys.*, **23**, 223 (1955)
     (b) B. Stevens, *Chem. Revs.* **57**, 439 (1957)
     (c) B. Stevens, *Canadian J. Chem.*, **36**, 96 (1958)
131. E. W. Schlag, H. von Weyssenhof and M. E. Starzak, *J. Chem. Phys.*, **47**, 1860 (1967)
132. G. H. Kohlmaier and B. S. Rabinovitch, *J. Chem. Phys.*, **38**, 1692, 1709 (1963)
133. J. E. Piercy and G. R. Hanes, *J. Chem. Phys.*, **43**, 3400 (1965)
134. J. A. Giordmaine, P. M. Rentzepis, S. L. Shapiro and K. W. Wecht, *Appl. Phys. Letters*, **11**, 216 (1967); P. M. Rentzepis and M. A. Duguay, *Appl. Phys. Letters*, **11**, 218 (1967)
135. Quoted by R. G. Gordon, W. Klemperer and J. I. Steinfeld, *Ann. Revs. Phys. Chem.*, **19**, 215 (1968)
136. H. Y. Sun, J. Jortner and S. A. Rice, *J. Chem. Phys.*, **44**, 2539 (1966)
137. Th. Förster, *Discussions Faraday Soc.*, **27**, 7 (1959); *Naturwiss.*, **33**, 166 (1946)
138. D. L. Dexter, *J. Chem. Phys.*, **21**, 836 (1953)
139. V. Ermolaev and A. Terenin, *Doklady Acad. Sci. U.R.S.S.*, **85**, 547 (1952)
140. H. L. J. Bäckström and K. Sandros, *Acta. Chem. Scand.*, **12**, 823 (1958); **14**, 48 (1960)
141. G. Porter and F. Wilkinson, *Trans. Faraday. Soc.*, **57**, 1686 (1961)
142. A. D. Osborne and G. Porter, *Proc. Roy. Soc.*, *A*, **284**, 9 (1965)
143. G. S. Hammond, *Reactivity of the Photoexcited Molecule*, Interscience, London, 1967, p. 119
144. R. H. S. Liu and J. R. Edman, *J. Amer. Chem. Soc.*, **91**, 1492 (1969)

145. C. A. Parker, Chemical Society Anniversary Meeting, Nottingham, April, 1969
146. E. J. Bowen and D. W. Tanner, *Trans. Faraday Soc.*, **51,** 475 (1955)
147. E. J. Bowen and K. K. Rohatgi, *Discussions Faraday Soc.*, **14,** 146 (1953)
148. (a) H. Leonhardt and A. Weller, *Z. Physik. Chem. (Frankfurt)*, **29,** 277 (1961)
     (b) H. Leonhardt and A. Weller, *Ber Bunsenges. Physik. Chem.*, **63,** 791 (1963)
     (c) H. Knibbe, K. Röllig, F. P. Schäfer and A. Weller, *J. Chem. Phys.*, **47,** 1184 (1964)
149. A. Weller, *Progress in Reaction Kinetics*, Vol. 1, Pergamon Press, Oxford, 1961, p. 187
149. (a) H. Beens, K.-H. Grellmann, M. Gurr and A. Weller, *Discussions Faraday Soc.*, **39,** 183 (1965)
150. G. Jackson and G. Porter, *Proc. Roy. Soc.*, *A*, **260,** 13 (1961); T. S. Godfrey, G. Porter and P. Suppan, *Discussions Faraday Soc.*, **39,** 194 (1965)
151. K. Gollnick and G. O. Schenk, *Pure and Appl. Chem.*, **9,** 507 (1964)
152. C. S. Foote and S. Wexler, *J. Amer. Chem. Soc.*, **86,** 3897 (1964)
153. D. R. Snelling, *Chem. Phys. Letters*, **2,** 271 (1968)

PROBLEMS

1. Which of the following excited atoms are metastable, and why, $Br(4^2P_{\frac{1}{2}})$, $O(2^1D_2)$, $S(3^1S_0)$, $Na(3^2P_{\frac{1}{2}})$, $C(2^3P_2)$, $Hg(6^1P_1)$, $N(2^2D_{\frac{3}{2}})$?

2. Calculate the Doppler width at $300°K$, of the near infrared emission line at $7{,}603\ cm^{-1}$, associated with $5^2P_{\frac{1}{2}} \rightarrow 5^2P_{\frac{3}{2}}$ transition in atomic iodine.

3. What conclusions could be drawn from the fact that the $\tilde{B} \leftarrow \tilde{X}$ electronic absorption spectra of S_2 and SO both consist of very long vibrational progressions? Do you expect Q-branches to form in their rotational fine structure?

4. What is the configuration of the ground electronic state of NH? On the basis of the spectrum shown in Figure 2.8, suggest a possible assignment for the transition centred at $336{\cdot}0$ nm.

5. The near-ultraviolet absorption spectrum of ICN is a broad continuum with a profile very like that of ICH_3. What conclusions can be drawn about the probable electronic assignment?

6. Although the $\tilde{B} \leftarrow \tilde{X}$, Schumann–Runge band system in oxygen extends right out to the dissociation limit in absorption, no emission can be detected beyond vibrational levels with $v' > 3$ in the upper state. Excitation in the banded part of the spectrum can lead to the formation of ozone. Comments?

7. Which of the following processes satisfy the demands of symmetry and/or spin conservation

$$N_2O(^1\Sigma^+) \rightarrow N(^4S) + NO(^2\pi)$$
$$\rightarrow N_2(^1\Sigma_g^+) + O(^3P)$$
$$\rightarrow N_2(^1\Sigma_g^+) + O(^1D)$$
$$N_2(^1\Sigma_g^+) \rightarrow N(^4S) + N(^2D)$$
$$HI(^1\Sigma^+) \rightarrow H(^2S) + I(^2P)$$
$$COS(^1\Sigma^+) \rightarrow CO(^1\Sigma^+) + S(^1S)$$
$$NO(^2\Pi) \rightarrow N(^4S) + O(^3P)$$
$$\rightarrow N(^4S) + O(^1D)$$
$$H_2O(^1A_1) \rightarrow H_2(^1\Sigma_g^+) + O(^3P)$$

8. Although the two dissociations

$$H_2O(\tilde{X}^1A_1) \begin{cases} \nearrow H_2(\tilde{X}^1\Sigma_g^+) + O(2^1D) \\ \searrow H_2(\tilde{X}^1\Sigma_g^+) + O(2^1S) \end{cases}$$

both satisfy the symmetry correlation rules the second cannot occur. Why?

9. The nitroso chromophore $(-N{=}O)$ is associated with a weak absorption band in the red $(\lambda_{max} \simeq 665\ nm,\ \varepsilon_{max} \simeq 151\ mole^{-1}\ cm^{-1})$, in addition to a relatively stronger band in the near ultraviolet $(\lambda_{max} \simeq 280\ nm,\ \varepsilon_{max} \simeq 80\ 1\ mole^{-1}\ cm^{-1})$. In dimeric nitroso compounds $(RNO)_2$, the visible absorption band disappears

and the near-ultraviolet absorption intensifies. What can be inferred from these results in regard to the possible structure of the dimer?

10. Calculate the value of the bimolecular rate constant for diffusion controlled processes in the following solvents: water at 20°C ($\eta = 10^{-2}$ poise), ethanol at -98°C ($\eta = 0.44$ poise), glycerol at 20°C, 0°C and -42°C ($\eta = 15$, 121 and 6.7×10^4 poise).

11. A triplet molecule has a natural phosphorescent lifetime of 1 sec, and is quenched by oxygen at a diffusion controlled rate. Estimate the mean lifetime of the triplet molecule in each of the solvents listed above when they contain dissolved oxygen at a concentration of 10^{-5}M.

12. Bromobenzene is capable of quenching the fluorescence of anthracene and some of its derivatives. The relative fluorescence intensity of a de-gassed solution of 9-phenyl anthracene in liquid paraffin ($\eta = 2 \times 10^4$ poise), varied with bromobenzene concentration in the following way*

$\dfrac{\text{(Intensity)}_0}{\text{(Intensity)}_Q}$	Concentration [Q] (moles litre^{-1})
1.00	0.00
1.12	0.48
1.26	0.96

Estimate the mean lifetime of the excited singlet 9-phenyl anthracene on the assumption of diffusion controlled quenching. Do you think the assumption was justified? Suggest a mechanism for the quenching process.

* T. Medinger and F. Wilkinson, *Trans. Faraday Soc.*, **61** 620, (1965).

CHAPTER THREE

THE CHEMICAL CONSEQUENCES OF LIGHT ABSORPTION: A SELECTION OF CASE HISTORIES

3.1   LIGHT ABSORPTION BY ATOMS

Left to itself an electronically excited atom would have no choice other than to re-emit the absorbed photon as fluorescence or phosphorescence (or ionize if its energy lies above an ionization limit). If, however, it should encounter a molecule during its brief life, there is a possibility of its promoting chemical change in that molecule. The likelihood of such a sensitized chemical change depends on the product of the encounter rate and the probability of energy transfer or chemical interaction during the encounter as compared with the natural fluorescence rate. If the two are not comparable then the more rapid process will predominate.

If the atom has been excited by the absorption of resonance radiation, its natural lifetime is determined by the transition probability for decay back to the ground state and is typically $\sim 10^{-7}$ sec. The encounter rate in the gas phase at S.T.P. is commonly $\sim 10^{10}$ sec^{-1}; the probability of deactivation during the encounter is expressed by the quenching cross-section (see section 2.3b). The presence of foreign gases at pressures of a few centimetres of mercury is often sufficient to quench atomic resonance radiation, and it is clear that the quenching cross-sections are commonly of a magnitude which permits deactivation to become significant at convenient pressures. This greatly facilitates the study of atom photosensitized reactions. The photon energy of the resonance radiation absorbed by the most common atomic photosensitizers is indicated in Table 3.1. Mercury is far and away the most popular because of the ready availability of low pressure mercury vapour lamps, its significant vapour pressure at room temperature, which is quite sufficient to allow complete light absorption in path lengths of a millimetre or so, and its chemically useful energy content in the 6^3P_1 state. Cadmium and zinc have much lower vapour pressures and cannot be used at room temperature, and the rare gases do not absorb until well down in the vacuum ultraviolet, though the advent of microwave-powered rare gas discharge lamps which emit high intensities of resonance radiation has removed the experimental difficulties once associated with photochemical studies in this region.

TABLE 3.1 Resonance wavelengths and equivalent energies of atomic photo-sensitizers

Atom	Resonance wavelength (nm)	Equivalent energy[a] (kJ mole^{-1})
Hg (6^3P_1)	253·7	471·1
(6^1S_1)	184·9	646·2
Cd (5^3P_1)	326·1	366·6
(5^1P_1)	228·8	522·5
Zn 4^3P_1	307·6	388·3
4^1P_1	213·9	558·9
Xe[b] $(5p)^5(6s)^1$	147·0	813·0
	129·6	922·5
Kr[b] $(4p)^5(5s)^1$	123·6	967·3
	116·5	1025·8
Ar[b] $(3p)^5(4s)^1$	106·7	1120·2
	104·8	1140·3

[a] To be strict, kJ einstein^{-1}.   [b] See discussion on p. 245.

Despite the wide use of excited atoms as photosensitizers in chemical kinetic studies, the details of the primary photochemical step are often hazy, even though the outlines are fairly clear. Possibilities may include one or more of the following:

(i) Energy transfer into a repulsive electronic state, or a continuous part of an attractive state of the acceptor

$$A* + QX \rightarrow A + Q + X$$

(ii) Energy transfer into an attractive electronic state to produce an electronically and/or vibrationally excited acceptor

$$A* + QX \rightarrow A + (QX)*$$

(iii) Atom transfer $A* + QX \rightarrow AX + Q$

(iv) Association $A* + QX \rightarrow (AQX)*$

(A* represents the excited atom and QX the acceptor, usually polyatomic in nature.)

Steps (i) and (iii) obviously lead to chemical change; step (iv) may lead to energy transfer if the association complex dissociates

$$(AQX)* \rightarrow A + (QX)*$$

to give the unexcited atom, but chemical change may follow if the excited acceptor subsequently suffers isomerization or predissociation or takes part in some bimolecular reaction. The same may also be true for an acceptor excited through direct energy transfer. If the atom transfer step (iii) is exothermic, it is possible that the product AX may carry sufficient vibrational energy to dissociate; the nett change is then indistinguishable from step (i). The nett heat of reaction in step (iii) is boosted by the strength of the new bond formed. Primary steps in which total spin is conserved will be more probable and hence more rapid than those in which it is not. Thus if A* occupies a triplet state and QX a ground singlet state, electronic energy transfer should produce (QX)* and A in triplet and singlet states, respectively. When A* and QX form an excited complex (AQX)*, intersystem crossing may occur within the lifetime of the complex releasing both A and QX in singlet electronic states (cf. $Hg(6^3P_1)$ and $CO(\tilde{X}^1\Sigma^+)$ discussed in section 2.3b).

The identification of the primary products is a problem which has taxed the ingenuity of many experimental photochemists, since they are seldom stable molecular species. Their identity can sometimes be established by direct methods such as the mass spectrometric technique developed by Lossing (see section 3.1a), or else inferred by working backwards from the final product analysis, the reaction kinetics and the energy requirements of individual steps. The occurrence of atom transfer in the primary step can be revealed by an ingenious isotopic labelling technique developed by Gunning and his coworkers, provided that abstraction leads to the formation of an involatile mercury compound in the final products. These methods are discussed in the next section.

3.1a Experimental: the study of primary processes

Mass spectrometry
Conventional analytical mass spectrometry is directed towards the identification of stable molecules and is widely used in determining the final products of chemical reaction. In order to detect the short-lived intermediates which may be produced in the course of a chemical change it is necessary to leak the components of a reacting system directly into the ionization chamber of the mass spectrometer, and take precautions to avoid removal of free radicals through surface reactions or the production of 'spurious' radicals through pyrolysis on the hot filament which provides the source of ionizing electrons. Eltenton[1] and later Lossing and Tickner[2] developed inlet systems which achieved this end, and Lossing and his coworkers[3] have used it to identify the free radicals produced through mercury photosensitized reactions. The reactant gas is carried at a partial

pressure of $\sim 10^{-3}$ mm mercury on a fast current of helium, and flows over a pool of mercury where hopefully it is saturated with mercury vapour. It then passes along the axis of an annular mercury resonance lamp which provides a concentrated source of radiation at 253·7 nm. The entire flow system and lamp are held at a common temperature of 55°C. After an exposure time in the 'reaction zone' of ~ 1 msec, the flowing gas can leak into the mass spectrometer.

Under these conditions atomic photosensitization is an ideal means of producing free radicals, since atomic absorption lines are so intense that high absorbed light intensities can still be maintained at very low pressures. Even if the reactant molecules can absorb the resonance radiation, direct photodissociation can be neglected since molecular absorption will always be virtually zero at such low pressures. The carrier gas increases the flow rate and reduces diffusion to the walls, while the low total pressure reduces the chance of collisional stabilization of an excited molecule. The low molecular concentration also favours radical–radical rather than radical–molecule reactions, and in this respect the technique is akin to that of flash photolysis (see section 3.2a). This has been used to advantage to 'label' primary free radical products with methyl radicals derived from the addition of mercury dimethyl to the gas stream. The additive provides a clean source of the radicals, which combine with others produced in the primary process(es). If the mass of the methyl group is 'subtracted' from that of the final molecular product, the primary radicals can be identified.

Isotopic labelling[4,5]
When the resonance absorption or emission of mercury at 253·7 nm is examined under very high spectral resolution, the 'single' line is found to possess hyperfine structure. The component lines are associated with the excitation or decay of the individual mercury isotopes present in a normal sample of mercury. If the mercury vapour is exposed to resonance radiation from an isotopically pure mercury lamp, it is possible to excite just one of the isotopes in the normal mercury sample. Should the excited mercury atoms abstract an atom from some substrate, producing a final solid such as HgO, the product will be enriched in the particular mercury isotope that was excited. If the substrate is not decomposed by atom transfer, no enrichment will result. For example, with Hg^i representing the isotopic mercury and Hg^n the normal sample having a spread of isotopic abundances, a non-enriched product could be formed through the sequence

$$(Hg^i)^* + QX \longrightarrow Hg^i + Q + X$$
$$Hg^n + X \xrightarrow{(+M)} Hg^nX$$

energy transfer association

Atoms produced in the primary step react with each of the isotopes present in the normal sample.

In order for the experiment to be successful, several precautions have to be taken. Since the hyperfine splitting is only of the order 10^{-3} nm, the spectral line widths both in the lamp and the absorbing vapour cannot be allowed to broaden. This requires precise temperature control of the lamp, which runs near 30°C, and a low total pressure in the absorption cell. If the pressure rises much above 25 mm mercury, the Lorentz broadening begins to exceed the hyperfine splitting. The isotopic content of the solid product is found spectroscopically by resonance radiation absorptiometry. However, the degree of enrichment actually measured can be reduced because of adsorption of normal mercury on the enriched product, or by isotope exchange.

Optical spectroscopy

One of the most successful techniques for identifying the primary products of a photochemical reaction is the flash photolysis method. Absorption of a brief, very intense light flash produces a very high, transient concentration of free radicals, atoms or molecules, possibly carrying electronic or vibrational excitation, which can be detected through their characteristic absorption in the ultraviolet or visible regions of the spectrum (or very occasionally, in the infrared). The technique is discussed in section 3.2a. The exciting flash is usually a polychromatic source, and if the light is focussed into an atomic vapour where all the absorption is concentrated into a few narrow lines, most of the wavelengths in the flash will pass straight through the absorption cell. The absorbed light intensity will be relatively low. However, if a large pressure of some inert gas is added to the atomic vapour, the absorption is greatly increased by the Lorentz broadening and enough light can be absorbed to produce measurable concentrations of excited atoms or transient products formed through photosensitization. For example, the transient ultraviolet absorption of the HgCl radical has been detected following the flash excitation of mercury in the presence of CF_3Cl and Ar at a total pressure just less than atmospheric.[6] By kinetic measurements it was shown that the HgCl was produced in the sequence

$$Hg(6^3P_1) + CF_3Cl \longrightarrow Hg(6^1S_0) + CF_3 \cdot + Cl \cdot$$

$$Hg(6^1S_0) + Cl \cdot \xrightarrow{+Ar} \cdot HgCl$$

Callear[7] has shown that the presence of mercury vapour in the flash lamp greatly increases its emission around 253·7 nm, when the flash discharge is excited by a brief microwave pulse, rather than a capacitor discharge. He was able to detect the HgH radical in absorption following the flash photolysis of a Hg + H_2 mixture; the radical had been detected in emission many years earlier.[8] The excited HgH could only be formed by association of a hydrogen atom and an excited mercury atom, but unexcited HgH could arise either through its subsequent fluorescent decay or it could be produced directly through a primary atom transfer

$$Hg(6^3P_1) + H_2 \rightarrow \cdot HgH + H\cdot$$

At the present time, the consensus of opinion seems to favour the latter alternative.

3.1b Results and discussion

(i) *Photosensitization by metal vapours*

Many molecules whose vapours quench the fluorescence emitted by excited atoms also suffer photosensitized decomposition as a consequence. We begin our discussion with the simplest molecules and gradually work towards the more complex ones.

Diatomic molecules. The simplest stable molecule is hydrogen, and the manner in which it quenches the emission from $Hg(6^3P_1)$ and $Cd(5^3P_1)$ has long been debated. In both cases quenching leads to the primary production of hydrogen atoms, and in cadmium certainly this must occur through an atom transfer mechanism, since its 5^3P_1 level lies 71 kJ mole^{-1} below the dissociation energy of H_2.

$$Cd(5^3P_1) + H_2(\tilde{X}^1\Sigma_g^+) \rightarrow \cdot CdH(\tilde{X}^2\Sigma^+) + H\cdot(1^2S)\Delta H = +4.2 \text{ kJ mole}^{-1}$$

However, $Hg(6^3P_1)$ has an excess of 33 kJ mole^{-1}: if the primary step involves atom transfer, the excess increases to 69 kJ mole^{-1} which is almost twice the dissociation energy of the weakly bound HgH radical. The radical might well be endowed with sufficient of the excess to dissociate within its first vibration, and the nett process would be indistinguishable from photosensitization through energy transfer into the repulsive $\tilde{a}^3\Sigma_u^+$ electronic state of H_2, (see section 1.3a).

$$Hg(6^3P_1) + H_2(\tilde{X}^1\Sigma_g^+) \rightarrow Hg(6^1S_0) + [H_2(\tilde{a}^3\Sigma_u^+) \rightarrow 2H(1^2S)]$$

It is difficult to distinguish between the two possibilities, though the recent detection of HgH in absorption following flash photosensitization favours atom transfer.

The mercury-photosensitized decomposition of O_2 also presents some unresolved problems. The final products of the reaction are O_3 and HgO. Volman[9] proposed that the HgO was produced in secondary reactions, rather than through atom transfer in a primary step, and there is no doubt that much of it can be formed through thermal reaction with ozone ('tailing of mercury'). However, Gunning and Strausz[10] have found considerable isotopic enrichment in the HgO produced in a mono-isotopic sensitization study (which corrected an earlier false lead). This might be interpreted in terms of a primary atom transfer step

$$Hg(6^3P_1) + O_2 \rightarrow HgO + O(2^3P)$$

were it not for the fact that the relative ozone yields are reduced by the addition of inert foreign gases such as Ar or CO_2. This observation points to the intermediacy of an excited molecule which can be deactivated through energy transfer; for example, the reaction sequence

$$Hg(6^3P_1) + O_2 \rightarrow Hg(6^1S_0) + O_2^*$$

$$O_2^* + O_2 \begin{array}{l} \nearrow O_3 + O(2^3P) \rightarrow O_2 + O_2 \\ \searrow O_2 + O_2 \end{array}$$

$$O_2^* + M \rightarrow O_2 + M$$

$$O + O_2 + M \rightarrow O_3 + M$$

includes the possibility of the deactivation of an excited oxygen molecule produced in the primary step. The O_2^* could be electronically and/or vibrationally excited. Unfortunately such a mechanism will not lead to isotopic enrichment.

A way out of the impasse may be provided if the primary step involves association to form an excited complex $(Hg \cdot \cdot O_2)^*$ which can either react with a second molecule of oxygen, or dissociate following collisional relaxation into its non-bonding ground state.

$$Hg(6^3P_1) + O_2 \rightarrow (Hg \cdot \cdot O_2)^*$$

$$(Hg \cdot \cdot O_2)^* + O_2 \rightarrow HgO + O_3$$

$$(Hg \cdot \cdot O_2)^* + M \rightarrow Hg(6^1S_0) + O_2 + M$$

This allows the possibility of isotopic enrichment and collisional deactivation.

When NO quenches the 6^3P_1 state of mercury, the photosensitized decomposition is very inefficient and the quantum yield of NO consumption is only $\sim 10^{-3}$. The volatile products include N_2 and N_2O, together with higher oxides of nitrogen.[11] Despite the low quantum yields, the quenching cross-section is high, $\sim 2.5 \times 10^{-15}$ cm^2, and infrared emission has been detected from NO molecules excited into vibrational levels as high as $v'' = 16$ in the ground electronic state (see section 2.3b). Whether the levels are populated in a single step or through the decay of the metastable $\tilde{a}^4\Pi$ state of NO produced by electronic energy transfer is uncertain

$$Hg(6^3P_1) + NO(\tilde{X}^2\Pi) \rightarrow Hg(6^1S_0) + NO(\tilde{a}^4\Pi)$$

Strausz and Gunning[11] assume that the latter mechanism does operate and in order to accommodate the observed kinetics and final products, propose the secondary production of an excited dimer

$$NO(\tilde{X}^2\Pi) + NO(\tilde{a}^4\Pi) \rightarrow (NO)_2^*$$

Simple m.o. arguments predict some stability for such a species. Once formed, the excimer either dissociates, disproportionates or reacts with a third, unexcited NO molecule

$$(NO)_2^* \nearrow \begin{array}{l} 2NO(\tilde{X}^2\Pi)_{v'' \geqslant 0} \\ \rightarrow N_2 + O_2 \\ (+NO) \searrow N_2O + NO_2 \end{array}$$

Since the nett reaction is inefficient, the first, self-quenching step must be the most probable.

If a stream of HCl is presaturated with mercury vapour and passed along a quartz tube through a source of mercury resonance radiation at 253·7 nm, it decomposes to produce hydrogen and a deposit of calomel, which appears on the walls of the tube immediately following the illuminated zone.[5] The primary process is thought to include both atom and energy transfer steps, since mono-isotopic photosensitization with ^{202}Hg increases the isotope abundance in the calomel from 29·8 per cent to ~ 40 per cent. Neglecting any isotope exchange during recovery of the combined mercury, and assuming a steady state in the flow system, the relative quantum efficiencies of decomposition by atom and energy transfer have been estimated as $\sim 0\cdot4$.

Triatomic molecules. The $Hg(6^3P_1)$-sensitized decomposition of water vapour follows a somewhat similar course to that of HCl, but although the mercury is oxidized to a solid product, little of it is formed in the primary process.[5,12] Mono-isotopic sensitization only leads to a marginal increase in the relevant isotope abundance, and the principal primary process would seem to be

$$Hg(6^3P_1) + H_2O \rightarrow Hg(6^1S_0) + H\cdot + \cdot OH$$

although it is 29 kJ mole^{-1} endothermic.

The HgH spectrum has been detected in emission in the Hg/H_2O system but not in absorption, and there is no evidence for a primary hydrogen atom transfer.

When nitrous oxide quenches the emission from $Hg(6^3P_1)$ atoms, it dissociates and the final products are almost exclusively nitrogen and oxygen produced in a ratio of $2:1$ or slightly greater. Some HgO is also produced as the decomposition progresses.[13,14] The closeness of the $N_2:O_2$ yield to the stoichiometric ratio is consistent with the primary process

$$Hg(6^3P_1) + N_2O(\tilde{X}^1\Sigma) \rightarrow Hg(6^1S_0) + N_2(\tilde{X}^1\Sigma_g) + O(2^3P)$$

in which the nett spin is conserved. The HgO is not produced in a primary step, since there is no enrichment when ^{202}Hg is used as the sensitizer, and it presumably arises through a reaction with atomic oxygen. Despite the low N_2–O dissociation energy of 150 kJ mole^{-1}, abstraction of its oxygen atom usually requires a considerable activation energy, so the result is not too surprising. In addition the reaction

$$Hg(6^3P_1) + N_2O(\tilde{X}^1\Sigma) \rightarrow HgO(\tilde{X}^1\Sigma) + N_2(\tilde{X}^1\Sigma_g)$$

is spin-forbidden.

Cvetanovic[15] has very usefully exploited the $Hg(6^3P_1)$ sensitized decomposition as a relatively 'clean' source of $O(2^3P)$ atoms. In a series of kinetic studies the reaction of oxygen atoms with a range of paraffins and olefins has been investigated. Attention has been focussed on the fate of energy-rich oxygen atom–olefin adducts, and the competition between collisional transfer of their excess vibrational energy and unimolecular isomerization or decomposition.

In 1939 Cline and Forbes reported that the decomposition of CO_2 could be sensitized by $Hg(6^1P_1)$ atoms but not by $Hg(6^3P_1)$ atoms.[16] Since the OC–O bond dissociation energy is 531 kJ mole^{-1} and the energies of the singlet and triplet mercury atoms lie respectively well above and well below it, the conclusion was perfectly acceptable. However, in 1961 Strausz and Gunning[17] established that decomposition *did* occur on quenching $Hg(6^3P_1)$ atoms, albeit very inefficiently, and CO was produced with a quantum yield $\leqslant 10^{-2}$. Some HgO was also formed, but it could not be isotopically enriched; O_2 was not detected in the final products. The apparent energy deficit was accommodated when it was discovered that the rate of decomposition increased roughly as the *square* of the absorbed light intensity, and decreased with increasing pressure. The observations indicate the operation of a two photon process involving an intermediate excited molecule. Thus the sequence

$$Hg(6^3P_1) + CO_2 \rightarrow Hg(6^1S_0) + CO_2^*$$

$$Hg(6^3P_1) + CO_2^* \rightarrow Hg(6^1S_0) + CO + O$$

would provide sufficient energy for the decomposition and be consistent with the observations. The identity of the CO_2^* is a matter for conjecture, though its assignment to a low-lying triplet state is attractive.

Addition of nitrogen was found to increase the rate of decomposition. N_2 is known to be effective in promoting the relaxation of $Hg(6^3P_1)$ into the metastable 6^3P_0 level where decay into the ground state is temporarily arrested. If it were assumed that the $Hg(6^3P_0)$ atoms were equally capable of initiating reaction, the increased steady state population atoms should have increased the rate far more than was actually observed. The metastable atoms must be relatively inefficient photosensitizers.

In an elegant experiment, the $Hg/CO_2/N_2$ system was cross-irradiated with light at 253·7 nm and 404·7 nm. The latter is associated with the transition $7^3S_1 \leftarrow 6^3P_0$ in mercury; with a sufficient steady state concentration in the lower level, stepwise excitation into the 7^3S_1 level is possible, producing mercury atoms with 765 kJ mole^{-1}. These have sufficient energy to decompose CO_2 in a single primary step, and an increase in the rate of decomposition was, in fact, recorded.

Polyatomic molecules. Though the cross-sections for the quenching of $Hg(6^3P_1)$ atoms by paraffin molecules are generally low, they are decomposed with near unit quantum efficiency. They all suffer C–H bond scission in the primary chemical process, but there is no evidence for rupture of a C–C bond, although it is energetically feasible. The nett quenching cross-sections increase as the C–H bond dissociation energies fall (see Table 2.1(a)), and can be resolved into 'group contributions' associated with each group of primary, secondary and tertiary C–H bonds present in the molecule. The relative efficiencies of C–H scission within a given molecule also vary with the nature of the C–H bond and show strong deuterium isotope effects. For example, a study of the primary dissociation of propane using Lossing's mass spectrometric method, produced the following figures for the relative yields of isopropyl and n-propyl radicals in the primary step. The ratios are a measure of the relative rates of dissociation at secondary and primary C–H bonds.[18]

$$\frac{\text{Sec. C–H}}{\text{Prim. C–H}} \qquad 9\cdot1 \qquad\qquad 7\cdot0 \qquad\qquad 30 \qquad\qquad 0\cdot6$$

$$(CH_3CH_2CH_3)(CD_3CD_2CD_3)(CD_3CH_2CD_3)(CH_3CD_2CH_3)$$

The primary photochemical process is

$$Hg(6^3P_1) + HR \rightarrow Hg(6^1S_0) + H\cdot + \cdot R \text{ or } HgH\cdot + R\cdot$$

There is no experimental evidence for the production of ·HgH nor is there any evidence of any 'pressure dependent' process involving an excited intermediate. However, a weak fluorescent emission which can be detected in the range 240–280 nm in the presence of high pressures of paraffins[19] has been ascribed to the decay of an association complex $(Hg(6^3P_1)\cdot\cdot HR)^*$. Yang[20] has proposed that the excited complex leads either to radiative or chemical quenching.

$$(Hg \cdots HR)^* \Big\langle {\begin{array}{l} \nearrow Hg(6^1S_0) + HR + h\nu \\ \\ \searrow \cdot HgH + R\cdot \ (or\ Hg(6^1S_0) + H\cdot + R\cdot) \end{array}}$$

Since the quantum yields of chemical quenching commonly lie in the range 0·8–1·0, the physical route can only be of very minor importance. Experimental observation shows that the more readily the paraffin dissociates, the less the contribution from radiative quenching. Although ·HgH has not been detected, it is possible that the radical does have a fleeting existence in the reaction process. $Cd(5^3P_1)$ atoms also photosensitize the decomposition of paraffins, and in their case the primary production of CdH is essential if the energy deficit is to be overcome (cf. H_2).

With the exception of cyclopropane, the primary steps in the photosensitized decomposition of the cyclic alkanes seem to be identical to those in the paraffins, and the same is probably true of the saturated hydrides of atoms other than carbon. For example, the $Hg(6^3P_1)$-sensitized photolysis of NH_3 leads to the formation of H_2 and N_2H_4. In a fast flow reactor (which avoids any accumulation of reaction products in the illuminated zone), the N_2H_4 accounts for 95 per cent of the NH_3 consumed[21] and the primary chemical step must be

$$Hg(6^3P_1) + NH_3 \rightarrow Hg(6^1S_0) + \cdot NH_2 + H\cdot\ or\ \cdot HgH + \cdot NH_2$$

There is some uncertainty about the magnitude of its quantum efficiency,[22] and it is possible that quenching again proceeds through an intermediate association complex which has alternative physical and chemical decay routes. The behaviour of cyclopropane is much more complex: it closely resembles that of ethylene and is discussed below.

The olefins are very efficient quenchers of both $Hg(6^3P_1)$ and $Cd(5^3P_1)$ atoms, with cross-sections at least ten times those of the paraffins. The chemical consequences of quenching depend on the nature of the olefin and the energy content of the excited atom. For example, $Hg(6^3P_1)$ atoms sensitize the molecular elimination of H_2 from ethylene[23] and promote interchange among its deuterio-isomers;[24] $Cd(5^3P_1)$ atoms behave similarly, but the elimination of H_2 proceeds with a very much reduced quantum yield.[24] There is no evidence of hydrogen atom abstraction by the excited cadmium and, in the case of $Hg(6^3P_1)$, the production of hydrogen atoms accounts for no more than one per cent of the dissociation products.[25] In the higher olefins, quenching of $Hg(6^3P_1)$ leads to both C–H and C–C bond cleavage and to geometrical and valence isomerization. In all cases the quantum yields decrease with increasing pressure (either of the olefin itself or of added, chemically inert gases), and there is competition between collisional deactivation and chemical change in some excited intermediate(s).[26]

The most likely intermediates would seem to be the lowest triplet states of the olefins, since they all lie at energies below those of mercury or cadmium, and can be populated through electronic energy transfer in the primary step. That is not the whole story because the triplet olefins adopt a 90° twist about their C=C bonds in the equilibrium conformation (see p. 51). If excitation occurred through direct electronic energy transfer, the relevant energies would be those of the planar triplet molecules, (i.e. the 'vertical' singlet–triplet excitation energies of the olefins measured at the absorption maximum). In the case of $C_2H_4(\tilde{a}^3B_{1u})$, this lies $\sim 443\,\mathrm{kJ\,mole}^{-1}$ above the ground state, still below that of $Hg(6^3P_1)$ but well above the $368\,\mathrm{kJ\,mole}^{-1}$ of $Cd(5^3P_1)$; yet both atoms are strongly quenched. Electronic energy transfer must occur through an association complex to give the olefin time to twist into its more stable configuration.[27] The primary products are the atom in its ground state, and the triplet olefin (Ol) endowed with excess vibrational energy.[26] For example

$$Ol(S_0) + Hg(6^3P_1) \rightarrow Ol(T_1)_{vib} + Hg(6^1S_0)$$

(the alternative of energy transfer into high vibrational levels of the ground singlet state is not supported by the kinetic evidence).

The triplet olefin may decay by several alternate paths. The most direct is collisional relaxation back into isomeric conformations of the ground electronic state. The molecular elimination of H_2 from ethylene is thought to occur in a second excited state since kinetic measurements of its suppression by increasing pressures show that two steps are involved in the deactivation process.[28] In 1956, Callear and Cvetanovic reported[29] a study of the kinetics of the $Hg(6^3P_1)$-sensitized isomerization and decomposition of cis-ethylene-d_2 and found the following mechanism gave the best agreement with observation:

$$C_2HD + HD \text{ or } C_2H_2 + D_2 \text{ or } C_2D_2 + H_2 \text{ dissociation}$$

The intermediate production of a triplet ethylidene would account for the 1.2 H atom shift, revealed by the production of the *unsym.* isomer.

'Activated' ethylene molecules have since been produced by alternative routes, including $Cd(5^3P_1)$ sensitization, photosensitization by triplet organic molecules such as benzene, and in direct photolytic reactions. In all cases there is competition between collisional energy transfer and the molecular elimination of H_2. Their relative probabilities can be calculated on the basis of unimolecular reaction rate theories, given the initial excess vibrational energy content of the ethylene. The calculations agree with experiment only when the dissociating molecules are assumed to occupy high vibrational levels in the ground electronic state. Hunziker[27] has accommodated the later results by a slight modification of the original mechanistic scheme.

$$(Ol)^*_{vib} \rightleftharpoons Ol(T_1)_{vib} \xrightarrow{(+Ol)} \text{isomerization, } sym \text{ and } unsym$$

$$Ol(S_0)_{vib} \rightarrow \text{molecular elimination of } H_2$$

$$\downarrow$$

collisional relaxation

Primary dissociation products in the $Hg(6^3P_1)$-sensitized decomposition of higher olefins have been identified by Lossing and his coworkers,[3] using the low pressure, mass spectrometric technique (p. 228). In general the molecules prefer to dissociate at β_{C-C} and β_{C-H} bonds, since this generates a resonance-stabilized allylic radical; ethylene of course, has no β-bonds and suffers molecular dissociation. For example, propylene gives principally allyl-1,5-hexadiene (diallyl) and hydrogen[3]

Butene-2 produces a radical of mass 55 and its dimer, 2,6-octadiene, while isobutene produces an isomeric radical of the same mass and *its* dimer 2,5-dimethylhexadiene-1,5.

In butene-1 the most abundant dissociation products are methyl and allyl radicals:[30] the methylallyl radical was detected in rather lower concentration. The results indicate two competing dissociation modes involving C–C or C–H bond scission.

In contrast to the other butenes the former is dominant.

The results of kinetic studies made under conventional conditions at higher pressures are in essential agreement with those of Lossing. However, at pressures of millimetres rather than micrometres the competition between collisional relaxation and dissociation becomes apparent. In addition isomeric olefins are found among the final products. For example, butene-2 undergoes *cis–trans* isomerization while butene-1 isomerizes to butene-2 and in much greater yield, to methyl cyclopropane.[31]

Pentene-1 undergoes a similar 'back-biting' isomerization to give ethyl cyclopropane. It seems likely that the vibrationally excited triplet olefins decay into at least one other excited intermediate, and that the primary reaction pathways resemble those followed by triplet ethylene.

The same is probably true of the cyclopropanes, which behave chemically in much the same way as ethylene, despite their relatively low quenching cross-sections toward $Hg(6^3P_1)$. Both cyclopropane-d_2 and 1,2-dimethyl cyclopropane suffer *cis–trans* isomerization and also form polymers produced through free radical reactions.[32] Scavenging experiments with added ethylene ($^{14}C_2H_4$) indicate C–H scission in a primary process,[33] though the majority of radicals were trapped as allyl rather than cyclopropyl. The isomerization is a 'molecular' rather than a radical-catalysed process, since dilution with D_2 causes little isotope exchange in the isomer and does not alter its quantum yield. The possibility that isomerization follows energy transfer into high vibrational levels of the ground state has been rejected by Setser, Rabinovitch and Spittler.[32] Calculations based on kinetic data for the rate of thermal unimolecular isomerization in 1,2-dimethyl cyclopropane show that deactivation ought to have been important at the pressures employed. In fact, the quantum yield of iso-merization was independent of the pressure.

Primary steps in the $Hg(6^3P_1)$]-photosensitized decomposition of 1,2 and 1,3-butadiene have been identified by the mass spectrometric technique.[34] The primary products are remarkably similar, encouraging the conclusion that the nuclear configuration in the excited molecule is common to both isomers. Both produce CH_3 (and C_2H_6), C_3H_3 (and C_6H_6), and C_4H_4 and H_2: in particular some of the 1,2 isomer is obtained from 1,3-diene while the vinyl radical is not observed. Primary steps which account for the observed mass peaks are included in the following com-petitive scheme:

$$C_4H_4 \text{ (vinyl acetylene?)} + H_2$$
molecular dissociation

$$\cdot CH_3(\rightarrow C_2H_6) + \cdot C_3H_3(\rightarrow C_6H_6)$$
radical dissociation

collisional relaxation
(and isomerization)

Boue and Srinivasan[35] have found cyclopropane derivatives in the final products from 1,3-butadiene, and suggest that the radical C_3H_3 may be cyclopropenyl.

As with ethylene and cyclopropane the spectrum of products probably reflects the various decay paths of vibrationally excited triplet molecules produced through electronic energy transfer. The lowest triplet state of 1,3-butadiene lies some 209 kJ mole^{-1} below that of mercury. In complex molecules the kinetics of the photosensitized reactions become increasingly involved, since there are many alternative decay processes and they occur at rates which depend on the magnitude of the excess energy and its distribution within the excited molecule.

The diversity of products obtained from more complex molecules is well illustrated by the series of 1,4, 1,5 and 1,6-dienes studied by Srinivasan and Colclough.[36] The general mechanistic scheme which they proposed on the basis of product analyses and reaction kinetics included the following steps:

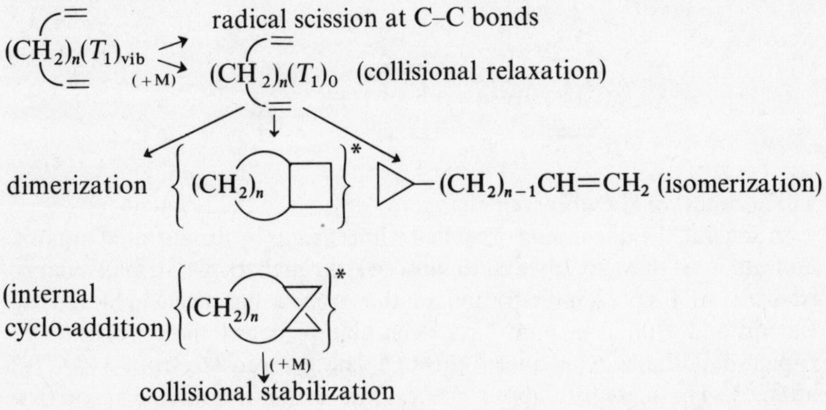

$$(CH_2)_n(T_1)_{vib} \underset{(+M)}{\overset{\text{radical scission at C–C bonds}}{\rightleftarrows}} (CH_2)_n(T_1)_0 \quad \text{(collisional relaxation)}$$

dimerization $\left\{ (CH_2)_n \bigcirc\!\square \right\}^*$ $\triangleright\!\!-(CH_2)_{n-1}CH{=}CH_2$ (isomerization)

(internal cyclo-addition) $\left\{ (CH_2)_n \bigcirc\!\boxtimes \right\}^*$

$\downarrow (+M)$

collisional stabilization

In the 1,4 and 1,6-dienes, the monocyclobutane adduct was formed preferentially, but in the 1,5-dienes the situation was reversed: in 1,5-cyclooctadiene the crossed, bicyclobutane adduct was the only one recovered. If the initial step in the cycloaddition always involves the formation of a C_5 ring, the behaviour can be rationalized

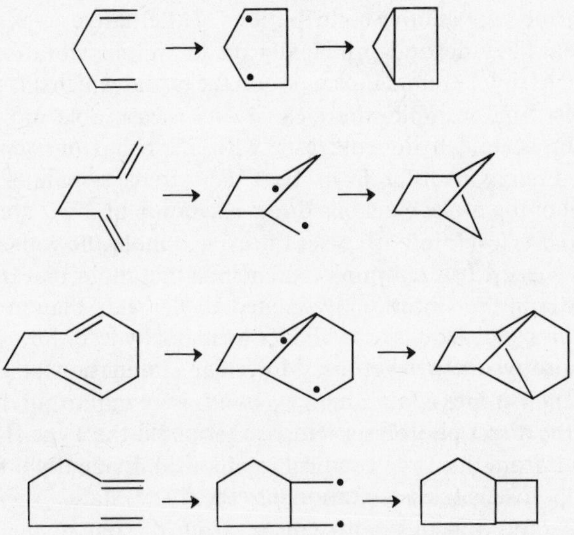

The isomerization of 1,4-dienes to vinyl cyclopropanes is well known in organic photochemistry. It is more complex than might be imagined at first, since the 'ordering' of the carbon atom chain is altered in the isomer.

The bond switching process

will account for the observed change.

In general, the quenching of excited triplet atoms by unsaturated organic molecules is thought always to proceed through triplet–triplet energy transfer. A direct demonstration of the process has been achieved by Burton and Hunziker, who have been able to detect the absorption of triplet naphthalene produced through energy transfer from $Hg(6^3P_0)$ atoms.[37] The mercury vapour was excited with a modulated resonance lamp and the absorption spectrometer was tuned to the modulating frequency in order to amplify the signal:noise ratio and permit the very low concentration of excited molecules to be detected.

Primary steps in the $Hg(6^3P_1)$-sensitized decompositions of carbonyl compounds have been identified, mainly by Lossing and coworkers again using the mass spectrometric method.[3,38–40] Some results are collected in Table 3.2. In view of the kinetic complexities mentioned above, generalizations concerning the relative contributions of alternative steps are fraught with danger. They depend principally on the initial vibrational energy content of the triplet excited molecule and the rate at which it is transferred in collisions. For example, the lack of any measurable molecular dissociation in acetaldehyde contrasts with its behaviour under direct excitation. Energy transfer from $Hg(6^3P_1)$ atoms populates the $T_1 - {}^3(n,\pi^*)$ level of the molecule while direct excitation at 253·7 nm populates the $S_1 - {}^1(n,\pi^*)$ level: in both cases the excited molecule will carry excess vibrational energy. It is tempting to conclude that molecular dissociation is much faster in the vibrationally excited singlet state than in the triplet. However, this conclusion is not valid for formaldehyde, or for n-butyraldehyde or n-propyl methyl ketone. Molecular elimination of C_2H_4 from aldehydes and ketones containing γ_{C-H} bonds is an important dissociation process in the direct photolysis: Norrish termed it the Type II process to distinguish it from the Type I radical mechanism. Evidently it also occurs following photosensitized excitation into the $^3(n,\pi^*)$ state.

There is some difficulty with conservation of spin in the molecular elimination since the dissociation products all have singlet ground states. It is possible that one or other of them is produced initially in a triplet state, but in some molecules the residual energy may not be sufficient to achieve this. The production of $CO(\tilde{a}^3\Pi)$ from formaldehyde demands an

TABLE 3.2 Primary dissociation products in the $Hg(6^3P_1)$-sensitized decomposition of some aldehydes, ketones, carboxylic acids and esters

	Molecular	Radical	Comments
HCHO	$H_2 + CO$ (40%)	$H\cdot + HCO$ (60%); $2H\cdot + CO$	The molecular products must be formed in their ground singlet states
CH_3CHO	nil	$CH_3\cdot + \cdot CHO$ (95%)	Molecular dissociation has a quantum efficiency of $\sim 0\cdot 66$ in the direct photolysis at 253·7 nm
$n\text{-}C_3H_7CHO$	$C_2H_4 + CH_3CHO$	$C_3H_7\cdot + \cdot CHO$	Majority of final products polymeric
⟨C₆H₅⟩-CHO	⟨C₆H₅⟩ + CHO	⟨C₆H₅⟩$\cdot + \cdot CHO$	
CH_3COCH_3	nil	$CH_3\cdot + \cdot(CH_3CO)_{vib}$; $CH_3\cdot + CO$	$\sim 25\%$ of $(CH_3CO)_{vib}$ survive
$n\text{-}C_3H_7COCH_3$	$C_2H_4 + CH_3COCH_3$	$n\text{-}C_3H_7\cdot + \cdot COCH_3$	Molecular split dominant
$CH_2=CHCHO$ $\left.\begin{array}{}\\\end{array}\right\}$ $CH_3CH=CHCHO$	$RH + CO$	$R\cdot + \cdot CHO$; $(R\dot{C}O + H?)$	
$CH_2=CO$	nil	$CH_2(\tilde{X}^3\Sigma_g^-?) + CO \rightarrow \cdot CH_2 + CO$	No C—H scission
$CH_3COCOCH_3$	nil	$2(CH_3\dot{C}O)_{vib} \rightarrow CH_3COCH_2\cdot + \cdot COCH_3$	$\sim 25\%$ of $(CH_3\dot{C}O)_{vib}$ survive: some sensitization by $Hg(7^3S_1)$?
$CH_3COCH_2COCH_3$	nil	$CH_3COCH_2\cdot + \cdot COCH_3$	—
$CH_3CO(CH_2)_2COCH_3$	nil	$CH_3CO(CH_2)_2\cdot + \cdot COCH_3$	—
		$CH_3CO\dot{C}HCH_3$	
HCOOH	$H_2O + CO$; $H_2 + CO_2$	nil	—
$HCOOCH_3$	$CH_3OH + CO$ (12%)	$H\dot{C}O + \cdot OCH_3$ (60%); $H\cdot + (\cdot COOCH_3 \rightarrow CO_2 + \cdot CH_3)$	no HCOO· radical?
CH_3COOH	$CH_2CO + H_2O$ (24%); $CH_4 + CO_2$ (19%)	$H\cdot + (\cdot COOH \rightarrow CO_2 + \cdot H)$; $CH_3\cdot + (\cdot COOH \rightarrow CO_2 + \cdot H)$	no $CH_3COO\cdot$ radical?

extra $577 \, \text{kJ mole}^{-1}$.

$$H_2CO \left\langle \begin{array}{ll} H_2(\tilde{X}^1\Sigma_g^+) + CO(\tilde{X}^1\Sigma^+) & \Delta H = +4 \, \text{kJ mole}^{-1} \\ H_2(\tilde{X}^1\Sigma_g^+) + CO(\tilde{a}^3\Pi) & \Delta H = +581 \, \text{kJ mole}^{-1} \end{array} \right.$$

The triplet formaldehyde must suffer intersystem crossing into high vibrational levels of its ground singlet state before the molecular fragments separate.

This difficulty does not arise in the dissociation of ketene since the CH_2 fragment has a triplet ground state ($\tilde{X}^3\Sigma_g^-$): the lowest singlet state ($\tilde{a}^1A_1$) lies a few kJ mole^{-1} above it. In the direct photodissociation at 253·7 nm almost all the CH_2 radicals are produced in the singlet state. Dissociation of the molecule is too fast to permit much competition from intersystem crossing. Unfortunately, the relative proportions of singlet and triplet CH_2 radicals produced through $Hg(6^3P_1)$-photosensitization remains uncertain at present. Cvetanovic has suggested that they all appear in the triplet state since their addition to *cis* or *trans*-butene-2 is non-stereospecific.[40a]

$$CH_2(\tilde{X}^3\Sigma_g^-) + \quad \text{(structures)}$$

Addition of singlet CH_2 is known to preserve the stereoisomerism. On the other hand Frey has produced strong evidence for the presence of singlet CH_2 in the $Hg(6^3P_1)$-sensitized system. It could be formed directly in the primary step, or by intersystem crossing from vibrationally excited levels in its triplet state, i.e.

$$Hg(6^3P_1) + CH_2CO \rightarrow Hg(6^1S_0) + CH_2(^3\Sigma_g^-)_{vib} + CO$$

$$CH_2(^3\Sigma_g^-)_{vib} + M \rightarrow CH_2(^1A_1) + M$$

The extent to which each may contribute is not known.

So far the discussion has been concerned almost entirely with the photochemical behaviour of $Hg(6^3P_1)$ atoms. Reports of photosensitization by singlet excited metal atoms are very scarce, although the resonance radiation from the lowest $^1P_1$ states of mercury, cadmium and zinc can be quenched by foreign gases. There are two reasons for this. The radiative lifetime of the singlet states are considerably shorter so that energy transfer or reaction has to be much more efficient to compete, and there may be experimental difficulties in preventing any simultaneous excitation of the $^3P_1$ levels, since the resonance lamps emit both resonance lines and the one of longer wavelength has to be filtered out.

The $Hg(6^1P_1)$-sensitized decomposition of *cis* and *trans*-deuterated ethylenes has been reported, as well as mixtures of C_2H_4 and C_2D_4.[41] Very little HD was produced from the latter in comparison with the H_2 and D_2, indicating molecular elimination of hydrogen as the dominant dissociation process. The efficiency decreased with increasing pressure, as with $Hg(6^3P_1)$-sensitization. If the molecular elimination occurs in high vibrational levels of the ground state, the similar behaviour with both triplet and singlet sensitizers would be plausible, but the detailed nature of the primary process is not known. $Hg(6^1P_1)$ atoms sensitize the decomposition of CO_2 and form a solid product, identified as mercuric oxalate, $Hg(C_2O_4)$.[42] An excited association complex $Hg(CO_2)^*$ has been suggested as a likely precursor.

(ii) *Photosensitization by rare gas atoms*
The resonance absorption lines of rare gases lie in the vacuum ultraviolet region and are all associated with electronic transitions of the type

$$\ldots (np)^5 \overline{(n + 1s)^1} \leftarrow \ldots (np)^6, n^1S_0$$

where n is the principal quantum number of the outermost occupied shell. The upper term symbols have not been indicated because the spin–orbit coupling scheme is complex. While the 'core electrons' follow Russell–Saunders coupling, their interaction with the 'excited electron' follows the (j,j) scheme. Instead of ^1P and ^3P terms, the excited configuration generates two doublets associated with different values of the total orbital angular momentum vector J. Their energies are given in Table 3.1.

Direct photoexcitation populates the upper components of each doublet since the lower transitions are forbidden. However, the doublet splittings are only $\sim 10\,kJ\,mole^{-1}$, and as the gas pressure is increased collisional relaxation into the lower components becomes increasingly likely. Further increase leads to the formation of excimers, and when xenon absorbs resonance radiation at 147 nm its resonance emission is replaced by a broad continuum, with a maximum intensity $\sim 172\,nm$.[43] These phenomena can complicate the study of photosensitization since there may be three distinct photosensitizing species in the irradiated gas. There is also the possibility of 'trivial' photosensitization involving the direct photolysis of a molecule by absorption of 'imprisoned' resonance radiation or the excimer radiation. Since the emission frequencies lie in the vacuum ultraviolet, where all molecules have absorption bands, this possibility must be carefully excluded before any decomposition can be assigned to a true photosensitization.

Quantitative experimental studies remain relatively rare, and those that have been made owe much to the development of high intensity, microwave-powered resonance discharge lamps. However, the earliest studies were made by Groth and coworkers, before the lamps became part of the photochemists' 'artillery'. One of the first was krypton-photosensitized dissociation of N_2. A mixture of Kr, $^{14}N_2$ and $^{15}N_2$ was found to contain some $^{14}N^{15}N$ after exposure to the resonance radiation at 123·6 and 116·5 nm,[44] and in the presence of H_2, ammonia and hydrazine were also produced.[45] In contrast, the less energetic excited xenon atoms endowed with 813 or 923 kJ mole^{-1} as against 967 or 1026 kJ mole^{-1} (see Table 3.1), were found to be ineffective. The null result provided the first strong evidence against acceptance of the 'low' value of 711 kJ mole^{-1}, which had been proposed for the dissociation energy of N_2. Groth and Oldenburg[46] suggested that the N_2 molecules were excited by resonance energy transfer into the repulsive $\tilde{a}^3\Sigma_u$ electronic state. The xenon atoms carry insufficient energy for this, though energy transfer into triplet levels of CO (isoelectronic with N_2) has since been observed (see p. 123). More recently it has been shown that excited krypton and xenon atoms are both capable of mixing the isotopes in a mixture of H_2 and D_2.[43]

A careful quantitative study of the xenon-photosensitized decomposition of CH_4 at 147 nm has been reported by Sieck.[43] The molecule is virtually transparent at this wavelength and direct photolysis can be neglected. The primary products included H_2 and CH_2 (whether in the singlet or triplet state is not known); when the pressures were sufficient to relax the excited krypton atoms into the lower, metastable state, $\cdot CH_3$ radicals were also released and the kinetic measurements pointed to hydrogen atom abstraction in the primary process

$$Xe^* + CH_4 \rightarrow Xe + CH_2 + H_2$$
$$Xe^{**} + CH_4 \rightarrow \cdot XeH + CH_3\cdot$$

Miller and Dacey[47] have studied the xenon-photosensitized decomposition of fluoroalkanes at 147 nm, and find the infrared absorption bands of XeF_2 among the final products. They were able to rationalize the pattern of fluorocarbon products if the primary photochemical process included both energy transfer and F_2 transfer. Not so long ago of course, they might have been 'laughed off the stage' for daring to suggest the formation of a stable rare gas compound. Like the aristocracy the noble gases are now rather more rare than inert.

In some cases the energy transferred from the excited atom may exceed the ionization potential of the acceptor: it is bound to be the case for helium. In these circumstances, sensitized photoionization results. For example krypton and argon have been used to sensitize the ionization of

NO.[48] Subsequent electron–ion recombination leads to the dissociation of the molecule.

3.2 Some diatomic and simple polyatomic molecules

When photochemists visit other photochemists' laboratories they are often heard to remark that 'It looks familiar'. Not only does this give them a sense of security but it also reveals the existence of an immediately recognizable set of 'tools of the trade'. They can be divided into several varieties of which the most common are

(i) sources of light for photoexcitation in the ultraviolet, vacuum ultraviolet or visible, either continuous or pulsed,

(ii) optical filters and monochromators for isolating given wavelengths from their emission spectra,

(iii) photometers or chemical actinometers for measuring their intensities at the given wavelength(s),

(iv) lenses, mirrors, optical benches, etc. for concentrating the light into the absorption cell, with fused silica the most common material since its transmission extends to wavelengths beyond the oxygen 'cut-off' (~ 190 nm), and

(v) spectrometers and spectrographs for recording ultraviolet and visible absorption and emission spectra.

We shall not concern ourselves here with a detailed account of the types of equipment presently available since it lies outside our terms of reference and a very helpful and informative survey is given in Calvert and Pitts' book.[49] Our intention is to discuss some of the more general techniques available for the experimental study of the chemical consequences of light absorption, and in particular those which help to identify primary photochemical processes.

For a proper understanding of the complete mechanism of a photochemical process, we should want to know the initial effect of the absorption of photons of given wavelength on the absorbing species; their subsequent fates; the rates and quantum efficiencies of each alternative;* the identity of unstable photoproducts which take part in the photochemical change; the sequence of chemical steps initiated by the primary photoproducts and the identities and quantum yields* of the final

---

* Note the fine distinction between quantum efficiency and quantum yield. The first relates to the relative probabilities of alternative decay paths in the photoexcited molecule, while the second denotes the number of product molecules formed per photon absorbed. If ϕ_i is the quantum efficiency of the ith decay path $\sum_i \phi_i = 1$ and $\phi_i \leqslant 1$; this is an expression of Einstein's Law of Photochemical Equivalence. On the other hand a quantum yield may take any value; that of HCl for example, when formed through the photochemical reaction of Cl_2 and H_2, may rise to $\sim 10^6$, since it is formed in a chain reaction initiated by the photodissociation of a Cl_2 molecule.

products. As far as diatomic and simple polyatomic molecules are concerned, the identification of final products presents no serious difficulty. The advent of gas chromatographic separation techniques and the development of sensitive analytical spectrometers, particularly mass spectrometers has made this almost a routine procedure, which we shall not discuss. Much more difficult are the problems of unravelling the sequence of chemical changes which produce them, detecting the reactive intermediates and identifying the competing primary photochemical processes which follow light absorption. The philosophy of many experimental studies involves a gradual regression from the 'long-lived' parts of the photochemical reaction system (i.e. the final products) back to the most transient (i.e. the primary steps). Much of the renaissance in photochemistry which followed the 1940's was due to the development of experimental techniques designed for direct study of the transient stages. Amongst these, the techniques of flash photolysis is the most powerful and versatile.

3.2a Experimental: The identification of primary products

Flash photolysis

The basis of the technique, developed originally by Norrish and Porter[50] at Cambridge, is quite simple. A source which emits a single high intensity pulse of light is used to promote a large amount of photochemical excitation or decomposition in a time which is of the same order or less than the lifetime of the intermediates. If these are atoms or free radicals produced following photodissociation, their lifetimes are typically $\sim 10^{-2}$ to 10^{-5} sec. If they are molecular (or atomic) excited states their lifetimes are generally much shorter, unless they occupy some metastable electronic state (for example, a triplet state). Until the development of giant-pulsed lasers (see p. 80), all the flash sources of sufficient intensity to produce measurable concentrations of intermediates detectable in absorption, had lifetimes $\geqslant 10^{-6}$ sec: most flash photolysis studies have been concerned with photodissociation, or the behaviour of the triplet state in polyatomic molecules. However, the past year or so has seen a jump in time resolution from the microsecond into the nanosecond region, a jump which carries us much closer to the time scale of the primary photochemical process itself.

By far the best way of following the transient stages of the photochemical change is to record the changing absorption spectrum of the reaction system during, and immediately following the photolysis flash, i.e. by kinetic spectroscopy. Ideally this would involve continuous measurement of the entire absorption spectrum, but since this is difficult to do without sacrificing resolution, either in the time or the absorption wavelength measurements, the two are usually separated. In the *photographic method* the photolysis flash is used to trigger a second 'spectroscopic' flash at a known delay (typically $\sim 10^{-6}$ to 10^{-1} sec). This provides a 'white-light' continuum which acts as a background throughout the visible and ultraviolet regions, against which transient absorption spectra can be photographed on a spectrograph. A series of separate experiments produces a series of spectral 'snapshots' of the photochemical system. The *photoelectric method* involves continuous monitoring of the changes in transmission at a selected wavelength, and requires a steady background light source. The changes are displayed on an oscilloscope trace. Figure 3.1 (see plates X and XI of plate section) shows the two alternative configurations and typical photographic records.

In the 'conventional' apparatus, the light pulses are provided by gas-filled silica flash-discharge tubes similar in design to the Edgerton flash tubes used in photography. Electrical energy (commonly ~ 1 to 4 kJ), is stored in a bank of high voltage capacitors and discharged through the gas to generate a light flash with a peak intensity as much as a million times greater than a steady light source, and extending throughout the visible and ultraviolet regions. Its duration is governed principally by the capacity and the self and mutual inductance in the discharge circuit.* Since the stored energy is given by $\frac{1}{2}CV^2$, where C and V are the capacity and voltage respectively, the briefest and brightest flashes are obtained with high voltage and low capacitances. The inductance can be minimized by careful attention to the geometry of the discharge system; a coaxial configuration is the most efficient. It is possible to 'overdo' the improvements and reduce the electrical damping too far. In this situation a 'ringing' or oscillatory discharge results, which can be a nuisance if kinetic measurements are to be made.

Absorption of the flash can concentrate very large energies into the system in a very short time. If the absorbing molecules are in solution, the energy is rapidly transferred to the solvent which acts as a heat sink and maintains isothermal conditions. However, when they are in the vapour phase the system has a low thermal capacity and the temperature may increase by several thousand degrees under effectively adiabatic conditions. Pyrolysis takes the place of photolysis! To maintain near isothermal conditions it is essential to dilute the vapour with a large excess of some transparent, inert foreign gas (such as N_2 or Ar). A limitation in the technique which can sometimes be more serious, is the polychromatic nature of the flash. If the photochemical primary process is wavelength

* The flash duration is a function of the product $(LC)^{\frac{1}{2}}$ where L is the circuit inductance.

dependent (as is usually the case), several alternatives may occur at once. If a narrow wavelength band is isolated from it the absorbed light intensity is usually so reduced that there is insufficient decomposition per flash to detect any transient intermediates. This problem was partially overcome by Claesson and Lindqvist[51] at Uppsala by developing a 'monster' of a flash, which dissipated ~ 100 kJ across a spark-gap in air. The resulting discharge occupied a sphere of ~ 3 cm diameter and was so intense that a high proportion of the light could be filtered out and still leave sufficient intensity both to produce measurable decomposition in a single flash* and to allow quantum yields to be measured at different wavelengths. However, no attempt was made to record any transient spectral changes; only the final products were analysed.

In a photodissociating system this is no great disadvantage, since high absorbed light intensities can produce very high transient free radical concentrations. Radical–radical reactions are favoured at the expense of radical–molecule reactions and the result is a considerable reduction in complexity of the final product distribution. For example, in the flash photolysis of acetone, the principle final products are C_2H_6 and CO; CH_4 is only a minor product. The majority of $CH_3\cdot$ radicals produced in the primary steps associate, rather than abstracting hydrogen atoms from the parent molecule. The final products clearly indicate the identities of the radical fragments formed in the primary process, even though the fragments were not detected as transients.

Since the initial development of the flash photolysis technique the range over which transient absorption spectra may be photographed, has been extended into the vacuum ultraviolet and rapid scanning spectrometers have been developed for recording transient absorption in both the ultraviolet and visible, and in the infrared.[53] Unfortunately there can be serious experimental drawbacks. In the vacuum ultraviolet all molecules have intense Rydberg transitions and it is not always possible to detect transient absorption spectra which lie in their shadow. For example, the long search made by Herzberg and his collaborators for the transient vacuum ultraviolet absorption spectrum of CH_2 (in its triplet ground state, $\tilde{X}^3\Sigma_g^-$) was fruitless until CH_2N_2 was substituted for CH_2CO as the parent molecule.[52] Although flash photolysis will produce CH_2 from either molecule, its absorption is obscured by that of CH_2CO; it was originally preferred to CH_2N_2, since the latter is inclined to explode with very little provocation! In the infrared the problem is inverted, since vibrational transitions have intensities typically two or three orders of magnitude less than electronic transitions. The only free radicals for which infrared

---

* Its intensity was sufficient to illuminate an office separated from the flash by a corridor with a right angle bend and two closed doors—but with open key holes!

absorption spectra have so far been recorded in the vapour phase, are CF_2 and CF_3. The high polarity of C–F bonds leads to particularly intense absorption.

The flash photolytic system has been extended into the vacuum ultraviolet by setting LiF windows into the walls of the reaction vessel, and combining the flash discharge tube and reaction cell into a single concentric unit.[54] A simplified design is shown in Figure 3.2. The latest (and most exciting) development is the 'giant-pulse laser' flash photolysis system which combines the virtues of monochromatic output, very high intensity and short duration. For example, a 'giant-pulse' ruby laser can readily deliver 1–10 J of *light* at 694 nm in ~10 nanosec. However, the laser becomes a far more useful photochemical source when its light is passed through a suitably cut doubly refracting crystal, since such a crystal has the property of generating optical harmonic frequencies. In this way some 10–15 per cent of the red light emitted by a ruby laser at 694 nm, can be converted into 'frequency-doubled' ultraviolet light at 347 nm. Using this type of flash, Windsor, Porter, Lindqvist and their coworkers[55] have been

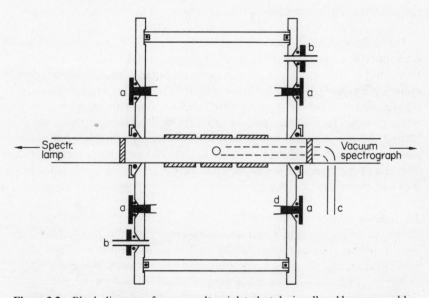

Figure 3.2 Block diagram of vacuum ultraviolet photolysis cell and lamp assembly: (a) copper electrodes; (b) gas inlet and pumping port for lamp; (c) gas inlet for photolysis cell; (d) alumina antisputter shields. LiF windows indicated by cross hatching. (From R. J. Donovan, ref. 65.)

* But a very large number have been detected in solid rare gas matrices (see p. 253).

able to detect transient absorption from the first excited singlet states of several aromatic hydrocarbons in solution (see p. 298) and follow the kinetics of intersystem crossing into the triplet state.

Modulated photolysis

The problem of spectroscopically detecting very low concentrations of transient photoproducts is basically one of signal-to-noise ratio. Flash photolysis 'cuts the Gordian knot' by producing a high concentration of products in a very short time and greatly increasing the signal. However, there are other methods of amplification. The simplest is to increase the length of the optical path through the irradiated sample, either by having a much longer cell or more elegantly, by use of multiple-reflecting mirror systems. Ramsay[56] has been able to detect the absorption of CN and NH_2 radicals by this technique, during continuous photolysis of $(CN)_2$ and NH_3 vapours. The alternative route is to integrate the results of many repeated experiments. By repetitive pulsing of the photolysis lamp and repetitively recording any transient absorption, the integrated measurement will eventually become significantly greater than the noise level, provided the noise is random. Under this condition the signal-to-noise ratio increases in proportion to the square root of the number of measurements.

This approach has been developed by Johnston.[57] The photolysis light is chopped at a known (but variable) frequency, and the light transmission through the cell at any given wavelength is recorded with a phase-sensitive system. Any change in the monitoring light intensity due to absorption by transient photoproducts must be correlated with the chopping frequency, while changes in the background intensity will be random and uncorrelated. After a sufficient integration time the signal will climb out of the noise level. By varying the sector frequency it is also possible to measure the lifetimes of the transient products, make reaction rate measurements and pick out transients of different lifetimes when more than one is present.

Photofluorescence

When photodissociation produces electronically excited products in the vapour phase, they can often be detected in emission (see p. 170), either photographically or more commonly, photoelectrically using a photomultiplier. In either case the cell is 'viewed' along a direction perpendicular to the exciting beam, in order to reduce the background light intensity. If the decay is associated with a strong electronic transition the radiative lifetime is likely to be shorter than the interval between collisions. The emission spectra of the excited products are easily recorded, and if they are molecular products the vibrational and rotational structure faithfully reflects their initial internal energy distribution. When the decay is forbidden, collisional energy transfer will tend to relax the original distribution and also quench the emission of both molecular and atomic products. Reducing the pressure of the vapour does not necessarily increase the emission intensity since the rate of light absorption is reduced as well as the collision frequency. Despite these difficulties it has been possible to detect emission from, for example, the metastable 2^1D and 2^1S states of atomic oxygen produced in the vacuum ultraviolet photolysis of molecules such as O_2, N_2O and CO_2 (see section 3.2b).

Photolysis in a rigid matrix
The study of photodissociation is greatly simplified if the ramifications of the secondary reactions which may follow can be prevented. Irradiation of the photosensitive substance dispersed at low temperatures in a transparent rigid matrix provides a simple means to this end. The rigid matrix impedes diffusion of primary products and the low temperature can reduce the rates of any secondary chemical reactions, (though not if they involve 'hot' primary photoproducts). Ideally it should be possible for the primary products to diffuse apart, but not together again. Transfer of their initial excess energy to the surrounding matrix allows some diffusion to occur, but as the heat is conducted away to the refrigerant, molecular diffusion rapidly ceases and any radical or molecular products are left isolated in the matrix. Unless it is allowed to soften and anneal, they may be stored for hours or longer, and their absorption spectra can be recorded at leisure when a sufficient concentration has accumulated. When the trapped products are paramagnetic it may be possible to detect their e.s.r. spectra, though the signals may not be well-resolved in the rigid environment.

The technique was initially developed in the 1940's by G. N. Lewis and coworkers[58] though they were primarily interested in detecting ultraviolet and visible absorption spectra of phosphorescent polyatomic molecules. Their matrices were alcoholic or paraffinic solvents, frozen into a rigid glass at 77°K and incorporating low concentrations ($\sim 10^{-3}$ M) of the solute. The frozen solvents remain transparent throughout the quartz ultraviolet and visible regions, and absorption (or emission) spectra could be measured both during and after irradiation. Figure 3.3 shows the ultraviolet absorption of CS produced through photodissociation of CS_2 under these conditions. However, the photodissociation of simple molecules is much better studied by photolysis in a frozen rare gas matrix, since the rare gases remain transparent over an immense spectral range stretching from the infrared through to the vacuum ultraviolet and are also chemically inert to the photolysis products. The technique of 'rare gas matrix isolation' has been developed by Pimentel[59] and subsequently Jacox and

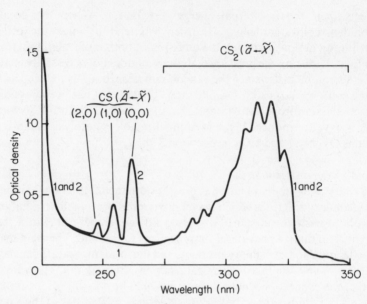

Figure 3.3 Ultraviolet absorption of CS trapped in a 3-methyl pentane glass at 77°K. (1) Solution before irradiation; (2) after 20 mins exposure to high pressure Hg arc lamp radiation (cf. Figure 2.6).

Milligan[60] to a fine art, and infrared or ultraviolet absorption spectra of at least forty diatomic and polyatomic free radicals have so far been detected in this way.

The matrices are condensed from the gas phase onto a transparent refrigerated window held at temperatures in the range 4–20°K. They form a microcrystalline film which tends to scatter light of short wavelengths but remains clear in the infrared. Figure 3.4 shows the infrared absorption of the $\cdot CH_3$ radical produced through photolysis of CH_4 in a N_2 matrix at 14°K.[60] The technique can be extended by substituting a reactive matrix for the inert gas, and trapping the primary product in chemical combination with the substrate. For example, the infrared absorption of the HCO radical was detected after photolysis of HI and HBr in a frozen CO matrix.[61]

These techniques are very powerful, since they give direct information relating to the primary photochemical process, but it must always be kept in mind that the process is one occurring in the solid phase at low temperatures and extrapolation to the vapour phase at normal temperatures may not always be valid. In particular, any weak intermolecular attraction between the solute molecule due, for example, to dipole–dipole interactions

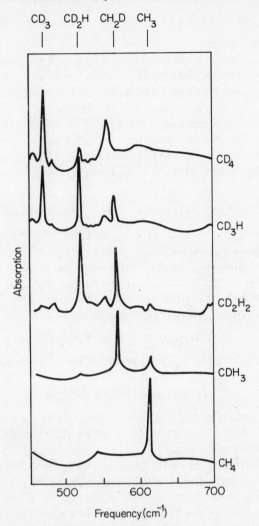

Figure 3.4 Infrared absorption of methyl radicals isolated in a N_2 matrix at 14°K, (produced by vacuum-ultraviolet photolysis of methane). (Taken from D. E. Milligan and M. E. Jacox, *J. Chem. Phys.*, **47**, 5146 (1967).)

or hydrogen bonding, may lead to association complexes or clusters in the frozen matrices. In this situation they are no longer isolated from each other, and their photochemical behaviour may be quite distinct from that in the vapour phase.

3.2b Results and discussion

(i) Diatomic molecules

The simplest diatomic molecules are hydrogen and the hydrogen halides. H_2 remains transparent until far down in the vacuum ultraviolet and need not concern us, but HBr and HI both present broad absorption continua in the near-ultraviolet, beginning at ~ 285 nm and ~ 360 nm respectively, and peaking at ~ 180 nm and ~ 218 nm. HCl remains transparent down to ~ 230 nm and its maximum lies well into the vacuum ultraviolet. Their absorption intensities are rather low; for HI, $\varepsilon_{max} \simeq 180 \, \mathrm{l \, mole^{-1}}$ $\mathrm{cm^{-1}}$ and the bands have been assigned to transitions of the type $n_X \rightarrow \sigma^*_{H-X}$,[62] i.e.

$$\ldots (z\sigma)^2(y\sigma)^2(x\sigma)^1(u\pi)^3, \, {}^1\Pi_1 \text{ or } {}^3\Pi_{2,1,0} \longleftarrow \ldots (z\sigma)^2(y\sigma)^2(x\sigma)^0(u\pi)^4, \, \tilde{X}^1\Sigma^+$$

(see p. 30). Excitation into the ${}^1\Pi_1$, ${}^3\Pi_1$ and ${}^3\Pi_0$ components contributes most of the intensity in HI and HBr. Since the nett absorption bands are completely continuous, the molecules must be excited into strongly repulsive parts of the upper electronic states, and photodissociation probably proceeds with unit quantum efficiency. The ${}^3\Pi_1$ and ${}^1\Pi_1$ components correlate with ground state products

$$HX({}^3\Pi_1) \text{ or } ({}^1\Pi_1) \rightarrow H({}^2S) + X({}^2P_{\frac{3}{2}}) \tag{1a}$$

but dissociation from the ${}^3\Pi_0$ state generates halogen atoms in the excited $({}^2P_{\frac{1}{2}})$ state

$$HX({}^3\Pi_0) \rightarrow H({}^2S) + X({}^2P_{\frac{1}{2}}) \tag{1b}$$

Mulliken suggested that the ${}^3\Pi_0$ state was populated most efficiently near the maximum of the absorption band, while the other two were more readily populated in the wings.[62]

Flash photolysis of HI and HBr at wavelengths $\geqslant 165$ nm, has confirmed that both normal and excited halogen atoms are primary products, since they have both been detected in absorption during the flash.[63] The Lyman α-band of the hydrogen atom has also been photographed. Spectra recorded at longer delay times showed that normal and excited halogen atoms are also produced in secondary reactions, since their concentrations continue to increase after the extinction of the flash. The final products are H_2 and I_2 or Br_2 produced in the reaction sequence

$$H\cdot + HX \rightarrow H_2 + X\cdot \tag{2}$$

$$X\cdot + X\cdot + M \rightarrow X_2 + M \tag{3}$$

where M is some third body, and it must be possible for steps of type (2) to generate the X atom in either the ${}^2P_{\frac{3}{2}}$ or ${}^2P_{\frac{1}{2}}$ states. The production of $I(5^2P_{\frac{1}{2}})$ atoms during continuous photolysis of HI at 254 nm has been demonstrated by recording the infrared emission at 1.315 μm, associated with radiative decay into the ground state.[64]

Because of the disparity in the masses of hydrogen and iodine or bromine atoms, the hydrogen atoms are initially endowed with high translational energies (see p. 132). Their excess energy can be rapidly redistributed in collisions with neighbouring molecules until their energy distribution matches that of the surrounding heat bath. On the other hand, if translational excitation helps to surmount the activation energy of a secondary reaction such as (2), a high proportion of the 'hot' hydrogen atoms may be removed in chemically reactive collisions. The specific reaction rate will be quite different from the rate constant averaged over the thermal distribution.* It was in a study of the photolysis of HI, that the enhanced reactivity of 'hot' atoms was first recognized.[66] It was discovered in the following way.

In the reaction sequence (1–3), two molecules of HX are consumed for each photon absorbed, but as the photolysis products accumulate the quantum yield begins to fall. The reaction

$$\text{H}\cdot + \text{X}_2 \rightarrow \text{HX} + \text{X}\cdot \qquad (4)$$

competes with (2) for the removal of hydrogen atoms. A steady state kinetic treatment based on steps (1–4) and assuming a unit quantum yield for the production of hydrogen atoms gives the expression

$$\frac{1}{\phi_{\text{H}_2}} = 1 + \frac{k_4[\text{X}_2]}{k_2[\text{HX}]}$$

for the quantum yield of H_2. From the measured value, k_4/k_2 was determined to be 3·5 for HI photolysed at 254 nm. The ratio was independent of temperature in the range 102–189°C. However, the addition of inert diluents increases the ratio to a limiting value of 12[68] (see Figure 3.5), in close agreement with the figure obtained for the relative rate constants measured for hydrogen atoms in thermal equilibrium (the agreement favours the assumed unit quantum efficiency of the primary photodissociation). The most effective diluent was helium, a result readily understood if it absorbs the excess translational energy of the hydrogen atoms.[67,68] The closer the masses of the colliding particles, the more efficient is the translational energy transfer (conservation of momentum). By varying the wavelength of the absorbed light, the initial energies of the

---

* When it is possible to inject hydrogen atoms (or any other atoms, radicals or molecules) into a system all at the same initial energy E, a rate 'constant" has no real significance, since its value varies with the energy distribution.[65] It is better to discuss reaction probabilities in terms of a reaction cross-section $\sigma(E)$: it can be defined in an analogous way to the extinction coefficient $\varepsilon(v)$ which represents a photon capture cross-section for photons of frequency v.

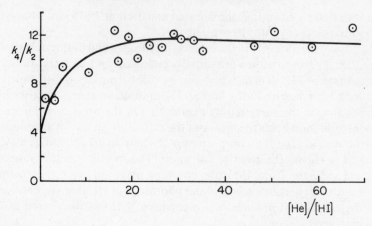

Figure 3.5 Moderation of 'hot' hydrogen atoms in the photolysis of HI (λ = 254 nm; T = 102°C). (After Holmes and Rodgers, ref. 68.)

hydrogen (or deuterium) atoms produced from HI (or DI) can be varied, and their relative reactivities measured as a function of their energy content.[65] Experiments of this type provide a simple means of exploring the finer details of the reactive collisions and they are becoming increasingly popular. The only snag is the uncertainty about the proportion of $^2P_{\frac{1}{2}}$ and $^2P_{\frac{3}{2}}$ halogen atoms produced in the primary step, since this determines the initial translational energy distribution in the hydrogen atoms.

All the halogens absorb light in the visible and near-ultraviolet regions, associated with electronic transitions of the type

$$\ldots (x\sigma_g)^2(w\pi_u)^4(v\pi_g)^3(u\sigma_u)^1, \, ^1\Pi_{1u}, \, ^3\Pi_{2u}, \, ^3\Pi_{1u}, \, ^3\Pi_{0u}^+ \leftarrow$$

$$\ldots (x\sigma_g)^2(w\pi_u)^4(v\pi_g)^4, \tilde{X}^1\Sigma_g^+ \qquad \text{(largely } n_X \rightarrow \sigma_{X-X}^*)$$

Relevant details are summarized in Table 3.3, and Figure 3.6 shows the absorption spectra and potential energy curves of the states involved (taken from diagrams prepared by Wilson[69]). In Cl_2, Br_2 and I_2 the upper states all lie close together, and their absorption spectra are composed of three overlapping transitions, (cf. the hydrogen halides). Wilson and coworkers have been able to identify each one unequivocally by observing the energy distribution in the photodissociation products at a series of wavelengths across the absorption spectrum.[69] (See p. 133 for a brief outline of the experimental method.)

The first transition $^1\Pi_{1u} \leftarrow \, ^1\Sigma_g^+$, is continuous at all wavelengths and results in dissociation into two unexcited atoms, $X(^3P_{\frac{3}{2}})$. The banded parts of the absorption spectra, associated with the $^3\Pi_{0u}^+ \leftarrow \, ^1\Sigma_g^+$ transition,

TABLE 3.3 Visible and near-ultraviolet absorption in the halogens

Halogen	Transitions	Comments
F_2	$^1\Pi_{1u} \leftarrow {}^1\Sigma_g^+$	Continuum, 220–360 nm; λ_{max}, 290 nm; ε_{max}, 6 l mole^{-1} cm^{-1}
Cl_2	$^1\Pi_{1u} \leftarrow {}^1\Sigma_g^+$	Continuum, 250–400 nm; λ_{max}, 330 nm; ε_{max}, 64 l mole^{-1} cm^{-1}
	$^3\Pi_{0u}^+$	Weak bands ~500 nm, converging at 478·5 nm
Br	$^1\Pi_{1u}$	Continuum, 510–300 nm; λ_{max}, 418 nm; ε_{max}, 162 l mole^{-1} cm^{-1}
	$^3\Pi_{0u}^+ \leftarrow {}^1\Sigma_g^+$	Bands beginning ~600 nm, converging at 514 nm
	$^3\Pi_{1u}$	Weak continuum, partially overlapping the banded absorption, but persisting further into the red
I_2	$^1\Pi_{1u}$	Continuum, lying beneath banded system
	$^3\Pi_{0u}^+ \leftarrow {}^1\Sigma_g^+$	Prominent band system, beginning ~650 nm, converging to a continuum at 499·5 nm; λ_{max}, 500 nm; ε_{max}, 560 l mole^{-1} cm^{-1}
	$^3\Pi_{1u}$	Continuum, partly overlapping banded system, terminating in the near-infrared

converge to a second continuum where one of the atoms is produced in the excited $^2P_{\frac{1}{2}}$ state. Br_2 and I_2 also dissociate at wavelengths to the red of the convergence limit, again producing two unexcited atoms. For example, Kistiakowsky and Sternberg[70] studied the photobromination of cinnamic acid at a series of wavelengths lying between 482 nm and 680 nm.* The efficiency remained effectively constant over the entire range, although the convergence limit in Br_2 lies at 510 nm. There are two mechanisms for photodissociation of Br_2 and I_2 in the banded region. One involves collisional perturbation of bound levels in the $^3\Pi_{0u}^+$ state, which induces predissociation (see p. 109); the other involves direct excitation into the continuum of the $^3\Pi_{1u}$ state, which overlaps the $^3\Pi_{0u}^+$ band system, but can be revealed by careful study of the absorption spectrum. The latter mechanism will obviously occur in the absence of collisions. Tiffany[71] has used monochromatic ruby laser light at wavelengths between 693 and 694 nm to demonstrate that $Br(^2P_{\frac{3}{2}})$ atoms are also produced when Br_2 is excited into bound levels of the $^3\Pi_{1u}$ state, provided they lie within

* At 680 nm the dissociation is actually ~15 kJ mole^{-1} endothermic and results from the photoselection of vibrationally excited molecules from the ground electronic state.

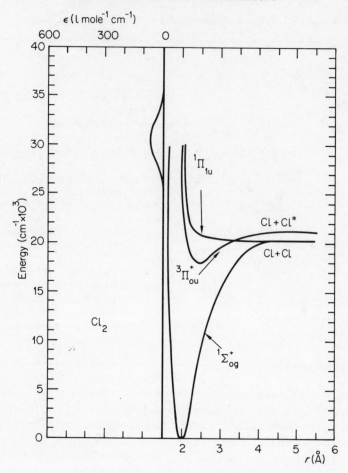

Figure 3.6(a)   Potential curves for Cl_2. (After K. R. Wilson, ref. 69.)

$\sim 10\,kJ\,mole^{-1}$ of the dissociation limit. Collisional energy transfer from neighbouring molecules makes up the deficit.

If there is no substrate with which the halogen atoms can react, then they must recombine, either in the gas phase or at a surface. Recombination in the gas phase requires collision with a third atom or molecule to act as a 'chaperone', accepting some of the vibrational energy of the newly formed halogen molecule and preventing it from re-dissociating. Under steady illumination, a photostationary state results. Rabinowitch and Wood showed that when the light was sufficiently intense, it was possible to estimate the rate of recombination from the slight reduction in optical

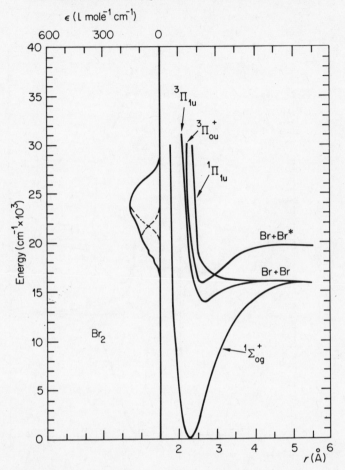

Figure 3.6(b) Potential curves for Br_2. (After K. R. Wilson, ref. 69.)

density of the molecular absorption.[72] The advent of flash photolysis provided a more accurate method of following the kinetics of recombination, since the changes in optical density were much larger and the rates of recombination could be measured directly.

The results of many studies have shown that for the great majority of chaperones, the rate of recombination is controlled principally by the formation of an intermediate 'halogen atom–chaperone complex' rather than collisional energy transfer.[73] Bromine and iodine atoms are 'sticky' reagents, and the more polarizable the molecule with which they collide the more stable is the complex they form with it. In some cases, the binding

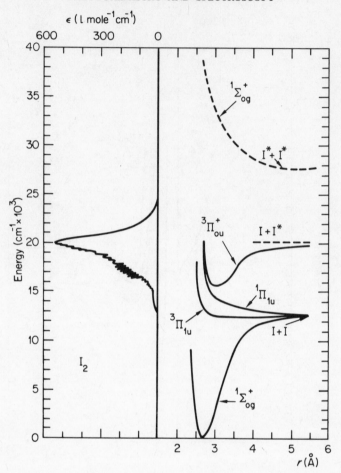

Figure 3.6(c) Potential curves for I_2. (After K. R. Wilson, ref. 69.)

is particularly strong and cannot be explained in terms of van der Waals interactions alone. I_2 is unique among the halogens in being specifically effective in catalysing the recombination of its atoms and the radical I_3 must be particularly stable (infrared absorption spectra of other trihalogen 'radicals' such as Cl_3 have been recorded in frozen rare gas matrices, but they do not remain bound at more elevated temperatures[74]). Benzene and its methyl derivatives form charge transfer complexes with bromine and iodine atoms which can actually be detected in absorption following flash photolysis in solution.[75] In the vapour phase transient absorption spectra attributed to INO and I·CH$_3$I have been detected in the visible.[76] The reaction with NO represents a special case where an unstable molecular species is formed rather than a transient complex. Molecular iodine is regenerated through the disproportionation

$$2INO \rightarrow 2NO + I_2$$

as well as the atom transfer

$$I + INO \rightarrow I_2 + NO$$

In discussing the recombination we have tacitly forgotten that excitation in the continua of the $^3\Pi_{0u}^+$ states produces excited $(^2P_{\frac{1}{2}})$, as well as normal halogen atoms. If their recombination is to produce ground state molecules relaxation must occur at some stage, either in the excited atoms before recombination or in the excited molecules afterwards. Relaxation of the metastable $^2P_{\frac{1}{2}}$ states in I or Br atoms is slow (see Table 2.1b), and it has been possible to observe fluorescence[77] from I_2 and transient absorption from Br_2 and Cl_2 $(^3\Pi_{0u}^+)$[78] molecules, all produced in the termolecular association

$$X(^2P_{\frac{1}{2}}) + X(^2P_{\frac{1}{2}}) + M \rightarrow X_2(^3\Pi_{0u}^+) + M$$

At low pressures emission can also be detected during the recombination of unexcited atoms, but this is due to the bimolecular radiative recombination

$$X(^2P_{\frac{3}{2}}) + X(^2P_{\frac{3}{2}}) \rightarrow X_2(^1\Sigma_g^+) + h\nu$$

The emission is continuous and the process is the reverse of photo-dissociation, (see Figure 3.7).

The recombination of halogen atoms in the presence of oxygen follows a somewhat complex path and its exploration has prompted several research papers. The first appeared in 1953, when Porter and Wright[79] reported that although flash photolysis of Cl_2–O_2 mixtures in the near-ultraviolet produces no permanent chemical change, the ClO· radical can be detected as a transient in absorption. Its production in the single step

$$Cl\cdot + O_2 \rightarrow ClO\cdot + O(2^3P)$$

can be discounted, since the reaction would be 226 kJ mole^{-1} endothermic. The intermediate formation of a chloro-peroxy radical ClOO· would surmount this problem by providing a lower energy pathway which is actually 38 kJ mole^{-1} exothermic

$$Cl\cdot + O_2 + M \rightarrow ClOO\cdot + M$$

$$ClOO\cdot + Cl\cdot \rightarrow ClO\cdot + ClO\cdot$$

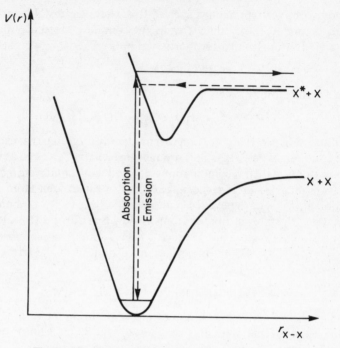

Figure 3.7 Radiative recombination of halogen atoms.
$\longrightarrow$ Dissociation $X_2 + h\nu \rightarrow X + X^*$.
$--\leftarrow--$ Association $X + X^* \rightarrow X_2 + h\nu$.

Unfortunately no evidence of a transient absorption attributable to ClOO· could be detected, but this has recently been remedied by Johnston[80] using his extremely sensitive modulated photolysis technique (see p. 252). The radical shows broad continuous absorption, partially overlapping the bands of ClO·, with a maximum ~ 250 nm. Decay of the ClO· takes several milliseconds and the rate follows second-order kinetics. After some controversy it has now been established[81] that the decay effectively reverses the reaction sequence which forms ClO·, i.e.

$$ClO\cdot + ClO\cdot \rightarrow ClOO\cdot + Cl\cdot$$

$$ClOO\cdot + Cl\cdot \rightarrow Cl_2 + O_2$$

$$ClOO\cdot + M \rightarrow Cl\cdot + O_2 + M$$

The photochemical behaviour of Br_2–O_2 and I_2–O_2 mixtures appears to be similar at first sight. The transient spectra of BrO· and IO· radicals can be seen during flash photolysis, but now the nett reactions

$$2X\cdot + O_2 \rightarrow 2XO\cdot$$

are endothermic, and the mechanism(s) by which the XO· radicals are formed has not been established.[82] It is possible that excited $^2P_{\frac{1}{2}}$ atoms are involved; $I(5^2P_{\frac{1}{2}})$ atoms lie $85.7 \text{ kJ mole}^{-1}$ above the ground state and although the energy can be transferred to other molecules on collision, it can also be used to promote chemical reaction. For example, the hydrogen atom abstractions

$$I(5^2P_{\frac{1}{2}}) + CH_3CH_2CH_3 \longrightarrow HI + \cdot CH_2CH_2CH_3$$
$$\searrow HI + CH_3\dot{C}HCH_3$$

are both strongly endothermic, by 92 and 105 kJ mole^{-1}, respectively. Add the $85.7 \text{ kJ mole}^{-1}$ of the $I(5^2P_{\frac{1}{2}})$ atom, and abstraction from the secondary carbon atom is almost thermoneutral. Unless the reaction has a large potential energy barrier, its rate should be accelerated by many orders of magnitude. Filseth and Willard[83] discovered that isopropyl iodide was indeed produced when I_2 was exposed to visible light in the presence of propane.* Subsequently, Callear showed that addition of gases known to deactivate excited iodine atoms reduced its quantum yield, and the kinetics of the inhibition correlated with the rates of deactivation measured by flash absorption spectroscopy.[84] The observations confirmed the participation of $I(5^2P_{\frac{1}{2}})$ atoms.

The photodissociation of oxygen by sunlight is the major photochemical process occurring in the earth's upper atmosphere. It results in the photosynthesis of ozone since oxygen atoms associate not only with each other

$$O + O + M \longrightarrow O_2 + M$$

but also with molecular oxygen

$$O + O_2 + M \longrightarrow O_3 + M$$

The concentration of ozone in the atmosphere is very small but it absorbs so strongly in the near-ultraviolet that it screens the earth's surface from light of wavelengths $\geqslant 290$ nm. Without this shield life on the earth would rapidly be destroyed and for this reason, if for no other, the photochemistry of the oxygen–ozone system has been studied for many years. It has been discovered that the atmospheric ozone concentration is regulated through a complex interplay of competing elementary processes involving atomic and molecular oxygen both in their ground and low-lying metastable, electronic states (see pp. 29–30 and Table 1.1). Because these can act as

---

* n-Propyl iodide was also formed but this was later shown to arise through a heterogenous reaction on the vessel wall.

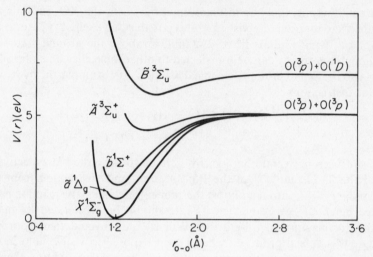

Figure 3.8 Potential curves of a few of the low-lying electronic states of O_2. (After F. R. Gilmore, Memorandum R–4034–PR, The RAND Corp., June (1964).)

photosensitizers as well as chemically reactive intermediates, it is not easy to reconstruct the web of competing steps.

The potential energy curves of some of the electronic states in O_2 are shown in Figure 3.8. All those which correlate with two unexcited atoms $(O(2^3P))$ are metastable; since transitions to or from the ground state are forbidden they are not readily accessible in absorption. For example, the transition into the highest metastable state

$$\ldots (w\pi_u)^3(v\pi_g)^3, \tilde{A}^3\Sigma_u^+ \longleftarrow \ldots (w\pi_u)^4(v\pi_g)^2, \tilde{X}^3\Sigma_g^-$$

can only be detected through the very long absorbing path provided by atmospheric oxygen. The $\pi \rightarrow \pi^*$ electron promotion produces a large increase in the equilibrium bond length, and the transition gives rise to a long progression of absorption bands which converge to a continuum beyond 242 nm, ($\equiv$ 493 kJ mole^{-1}, the dissociation energy of $O_2(\tilde{X}^3\Sigma_g^-)$).

The first intense transition, $\tilde{B}^3\Sigma_u^- \longleftarrow \tilde{X}^3\Sigma_g^-$, has its origin at 202·6 nm and is associated with the same $\pi \rightarrow \pi^*$ promotion. It produces the Schumann–Runge band system (shown in Figure 1.32) which merges into a continuum beyond 175·9 nm, where the molecule dissociates, and reaches a maximum intensity at $\sim$ 145 nm.

$$O_2(\tilde{X}^3\Sigma_g^-) \xrightarrow{h\nu \ < \ 175\cdot9 \ nm} O(2^3P) + O(2^1D)$$

Excited $O(2^1D)$ atoms are metastable; the radiative lifetime is so long that unless the pressure of oxygen is very low, the atoms are deactivated by the energy transfer

$$O(2^1D) + O_2(\tilde{X}^3\Sigma_g^-) \rightarrow O(2^3P) + O_2(\tilde{b}^1\Sigma_g^+)$$

(it is very nearly resonant if the $O_2(\tilde{b}^1\Sigma_g^+)$ is excited into its second vibrational level). The final product is ozone, produced with a quantum yield of 2·0. A transient broadening of its absorption spectrum detectable during the flash photolysis of oxygen at wavelengths > 160 nm, suggests that the ozone is formed initially in high vibrational energy levels, from which it subsequently relaxes through collisional energy transfer.[85]

$$O(2^3P) + O_2 \rightarrow (O_3)_{vib}$$

$$(O_3)_{vib} + M \rightarrow \rightarrow \rightarrow O_3 + M \text{ (many steps)}$$

At wavelengths < 134 nm, a third electronic transition causes the photodissociation

$$O_2(\tilde{X}^3\Sigma_g^-) \xrightarrow{h\nu\ <\ 134\ nm} O(2^3P) + O(2^1S)$$

Deactivation of the metastable $O(2^1S)$ atoms by O_2 is $\sim 10^2$ slower than of $O(2^1D)$, and they have been detected in the emission following flash photolysis in the vacuum ultraviolet[86] (but even then, only under collisional perturbation from added inert gases). In contrast, emission from electronically excited oxygen, both atomic and molecular, is well known in the spectrum of the night airglow. This originates in the upper atmosphere where the pressure is so low that radiative decay can compete favourably with collisional deactivation.

It is also possible to dissociate oxygen by absorption in the banded region of the spectrum. The strong smell of ozone produced by mercury vapour resonance lamps is familiar to all photochemists, but the lamps do not emit light at wavelengths < 184·9 nm. The $\tilde{A} \leftarrow \tilde{X}$ continuum is far too weak to contribute to the absorption at this wavelength, and the decomposition is due to a predissociation of the $\tilde{B}^3\Sigma_u^-$ electronic state. There is a measurable broadening of the rotational fine structure in the (4,0) and (12,0) bands of the Schumann–Runge system and fluorescent emission has only been observed from levels with $v' \leqslant 3$ in the upper state. Murrell and Taylor[87] have discussed the mechanism of the predissociation and conclude that a repulsive $^3\Pi_u$ state crosses the $\tilde{B}$ state near the level $v' = 4$. The second predissociation at levels near $v' = 12$ is due to overlap of subsidiary maxima in the vibrational wave functions, rather than a second intersection. The products are two atoms in the ground electronic state.

(ii) *Triatomic molecules*

H_2O remains transparent in the vapour phase until wavelengths < 185 nm, when the photon energy is sufficient to excite the $n_O \rightarrow \sigma^*_{O-H_2}$ transition (see p. 37)

$$\ldots (1b_1)^1 (4a_1)^1, \tilde{A}^1 B_1 \longleftarrow \ldots (1b_1)^2, \tilde{X}^1 A_1$$

The upper orbital is strongly antibonding, and it is not surprising to find that the first absorption band is entirely continuous, though there is a suggestion of very broad shallow structure on the absorption profile. The continuum reaches a maximum intensity at 165 nm.

At wavelengths < 177 nm the energy of the absorbed photon (677 kJ mole^{-1}) would be sufficient to promote any of the following primary steps

$$H_2O \begin{array}{l} \nearrow \\ \rightarrow \\ \searrow \end{array} \begin{array}{ll} H(1^2S) + \cdot OH(\tilde{X}^2 \Pi) & \Delta H = +493 \text{ kJ mole}^{-1} \quad (1) \\ H_2(\tilde{X}^1 \Sigma_g^+) + O(2^3 P) & \Delta H = +489 \text{ kJ mole}^{-1} \quad (2) \\ H_2(\tilde{X}^1 \Sigma_g^+) + O(2^1 D) & \Delta H = +677 \text{ kJ mole}^{-1} \quad (3) \end{array}$$

though (2) would be spin-forbidden of course. In fact, the radical split (1) seems to be the sole important primary process at wavelengths > 165 nm. The transient absorption of the hydroxyl radical has been detected during flash photolysis, and interestingly it carries no excess vibrational or rotational energy.[88] The majority of the excess energy (~ 170–210 kJ mole^{-1}) is carried off by the hydrogen atom is translation in much the same way as it was in the photolysis of HI. A second region of continuous absorption beginning at 141 nm, has been associated with the excitation of an electron out of the inner 'lone-pair orbital', $3a_1 \rightarrow 4a_1$. The continuum has a long progression of diffuse bands superimposed upon it, corresponding to the bending frequency in the upper state. The excited molecule is either linear or nearly so and this is thought to account for the high rotational excitation of the fluorescent $\cdot OH$ radicals produced by absorption of krypton resonance radiation at 123·6 nm (see p. 170)*

$$H_2O \rightarrow H(1^2S) + \cdot OH(\tilde{A}^2 \Sigma^+)_{rot}$$

At the shorter wavelengths, the molecular dissociation (3) also contributes since H_2 is still found among the final products when the molecule is photolysed in the presence of the radical scavenger C_2D_4[89] (similarly D_2 is found when D_2O is photolysed in the presence of C_2H_4). Formation of oxygen atoms has been inferred from the increased yield of OH during

---

* At wavelengths $< 125·6$ nm, a series of narrow intense, Rydberg bands lie superimposed on the second continuum. Although the Rydberg transitions populate bent upper states, the sharp resonance line at 123·6 nm just misses the first member of the series, which is centred at 124·0 nm (see ref. 90, p. 265).

the flash photolysis of H_2O in the presence of a large excess of H_2.[88b] Since the atom transfer step

$$O(2^3P) + H_2 \longrightarrow \cdot OH + H\cdot$$

is $4\,kJ\,mole^{-1}$ endothermic it was concluded that the oxygen atoms occupied the 2^1D state. In the absence of H_2, some $\cdot OH$ radicals are also produced in the exothermic reaction

$$O(2^1D) + H_2O \longrightarrow \cdot OH + \cdot OH$$

The molecular dissociation becomes more probable as the absorbed wavelength is reduced, a trend which is common to many simple polyatomic molecules.

Walsh's orbital correlation diagram for molecules of the type AB_2 (or BAC) which was discussed in section 1.3c and represented in Figure 1.21, is very helpful in any discussion of their electronic spectra. The m.o.'s of stable triatomic molecules (as opposed to radicals or ions), are occupied by between 16 and 20 valency-shell electrons, and their number and disposition determine the equilibrium geometry of the molecule. Some examples are collected in Table 3.4.

The AB_2 molecules with 16 valence electrons are in some ways the most complex; the first electronic transition $1\pi_g \rightarrow 2\pi_u$, promotes an electron from a non-bonding orbital centred on the terminal atom into a π-antibonding orbital. In the linear molecule, the degenerate $(1\pi_g)^3(2\pi_u)^1$ electronic configuration generates Δ_u, Σ_u^- and Σ_u^+ electronic states (see

TABLE 3.4 Electronic configurations of the ground electronic state in some triatomic molecules

Shape	Number of valency electrons	Electronic configuration	Examples
Linear	16	$\ldots (1\pi_u)^4(1\pi_g)^4, {}^1\Sigma_g^+$	$CO_2, CS_2, COS^a, N_2O^a,$ $BrCN^a$
Bent	17	$\ldots (1a_2)^2(4b_2)^2(6a_1)^1, {}^2A_1$	NO_2
	18	$\ldots (1a_2)^2(4b_2)^2(6a_1)^2, {}^1A_1$	$O_3, SO_2, ClNO^b$
	19	$\ldots (4b_2)^2(6a_1)^2(2b_1)^1, {}^2B_1$	ClO_2
	20	$\ldots (4b_2)^2(6a_1)^2(2b_1)^2, {}^1A_1$	Cl_2O, F_2O

[a] The subscripts u and g do not apply.

[b] Bent molecules of the type BAC possess only a single plane of symmetry, to which all the orbitals are either symmetric (a') or antisymmetric (a''). The electronic configuration in the ground state of ClNO, becomes $\ldots (2a'')^2(6a')^2(7a')^2, {}^1A'$.

p. 39), but the Walsh diagram predicts that the molecule should bend to lower the energy of the antibonding orbital. The bending splits the degeneracies of the $1\pi_g$ and $1\pi_u$ orbitals to produce four possible electronic states (see Figure 1.2). The correlations are shown in order of increasing energy below.

$$
\begin{array}{l}
\Sigma_u^+ - B_2 - (1a_2 \rightarrow 2b_1) \\
\quad\nearrow \\
\uparrow \quad (1\pi_g \rightarrow 2\pi_u) - \Sigma_u^- - A_2 - (4b_2 \rightarrow 2b_1) \\
\text{Energy} \quad\quad\quad\quad\quad\searrow \\
\quad\quad\quad\quad\quad\quad\quad\quad \nearrow A_2 - (1a_2 \rightarrow 6a_1) \\
\quad\quad\quad\quad\quad\quad \Delta_u \\
\quad\quad\quad\quad\quad\quad\quad\quad \searrow B_2 - (4b_2 \rightarrow 6a_1)
\end{array}
$$

Lines $\longleftrightarrow$ Bent

Table 3.5 summarizes the absorption wavelengths of some transitions which have been identified at the present time. In themselves they should lead only to a reduction in the bond angle and a moderate increase in the equilibrium bond lengths in the upper states, but in fact, apart from the singlet–triplet transition in CS_2, each one results in photodissociation. The reason is not hard to find since the absorption bands lie in the vacuum ultraviolet, where the absorbed energy lies well above the minimum dissociation threshold. Rotational fine structure can only be resolved in CS_2 : in all the other molecules the upper states are predissociated by repulsive electronic states derived from more highly excited electronic configurations. It is these which determine the nature of the primary photoproducts. Absorption spectroscopy tells us very little about them and if we want to know more we are forced to do some photochemistry!

One of the first questions that might be asked is 'How important is the conservation of spin in the photodissociation?' since for each molecule the formation of ground state products is spin-forbidden.

$$CO_2(\tilde{X}^1\Sigma_g^+) \rightarrow CO(\tilde{X}^1\Sigma^+) + O(2^3P) \quad \Delta H = +531 \text{ kJ mole}^{-1}$$

$$N_2O(\tilde{X}^1\Sigma^+) \rightarrow N_2(\tilde{X}^1\Sigma_g^+) + O(2^3P) \quad \Delta H = +167 \text{ kJ mole}^{-1}$$

$$\rightarrow NO(\tilde{X}^2\Pi) + N(2^4S) \quad \Delta H = +481 \text{ kJ mole}^{-1}$$

$$CS_2(\tilde{X}^1\Sigma_g^+) \rightarrow CS(\tilde{X}^1\Sigma^+) + S(3^3P) \quad \Delta H = +460 \text{ kJ mole}^{-1}$$

$$COS(\tilde{X}^1\Sigma^+) \rightarrow CO(\tilde{X}^1\Sigma^+) + S(3^3P) \quad \Delta H = +301 \text{ kJ mole}^{-1}$$

$$\rightarrow CS(\tilde{X}^1\Sigma^+) + O(2^3P) \quad \Delta H = +686 \text{ kJ mole}^{-1}$$

None of them can correlate with a singlet electronic state, and excitation into any of the singlet states associated with the lowest $\pi \rightarrow \pi^*$ transitions

TABLE 3.5 Probable assignments of some of the ultraviolet absorption bands associated with the lowest $\pi \rightarrow \pi^*$ transitions in 16-electron AB_2 and BAC molecules.

Upper state	CO_2	CS_2	COS	N_2O	Comments
$^1B_2(^1\Delta_u)$	Diffuse bands (overlapping weak continuum) 140–175 nm, λ_{max}, 148 nm	—	Broad continuous absorption, λ_{max}, 225 nm and superimposed vibrational band structure	Weak continuum 160–260 nm, λ_{max}, 182 nm	Electronically forbidden in linear AB_2
$^1A_2(^1\Delta_u)$ $^1A_2(^1\Sigma_u^-)$	—	—	—	—	Electronicall forbidden in both linear and bent AB_2
$^1B_2(^1\Sigma_u^+)$	^{a}Strong diffuse bands 124–139 nm, λ_{max}, 133·5 nm	Sharp bands 180–220 nm, λ_{max}, 197 nm	Strong diffuse bands 144–157 nm λ_{max}, 151 nm	Diffuse bands 140–152 nm, λ_{max}, 145·5 nm	Electronically allowed in both linear and bent AB_2

Note Rotational analysis has confirmed the assignment in CS_2, and also used to identify the $\bar{a}^3A_2(^3\Delta_u$ or $^3\Sigma_u^-) \leftarrow \bar{X}^1\Sigma_g^+$ transition, which generates a weak system of sharp absorption bands between 330 and 430 nm. The u and g subscripts do not apply to the non-centrosymmetric molecules.
^{a}This assignment seems preferable to the $2\pi_u \leftarrow 1\pi_u$ transition suggested by Herzberg,[90] populating bent upper states derived from the linear configuration $(1\pi_u)^3(1\pi_g)^4(2\pi_u)^1$, Σ_g^+, Σ_u^-, Δ_g. Transitions from the $\bar{X}^1\Sigma_g^+$ ground state are strongly forbidden in the linear molecule, by the g $\leftrightarrow$ g selection rule, yet the bands at 124–139 nm are twice as strong as those at longer wavelength.

is bound to produce electronically excited products, (for example, O or $S(^1D$ or $^1S)$, $N(^2D)$), unless intersystem crossing can occur within the dissociative lifetime. The molecule for which this seems most likely is CS_2; the resolved rotational structure of its $\tilde{A}^1B_2 \leftarrow \tilde{X}^1\Sigma_g^+$ transition indicates a long lifetime in the upper state, and the sulphur atoms should encourage spin–orbit coupling. Experiments confirm the prediction. The ultraviolet absorption of both $S(3^3P)$ atoms and $CS(\tilde{X}^1\Sigma^+)$, (in vibrational levels $v'' \leqslant 7$), can be seen during flash photolysis of CS_2 at wavelengths $\geqslant 190$ nm, but there is no trace of absorption from sulphur atoms in the excited 3^1D state. In the presence of a large excess of H_2, no ·SH radicals are formed, although the reaction

$$S + H_2 \rightarrow \cdot SH + H\cdot$$

is 25 kJ mole^{-1} exothermic for $S(3^1D)$; for ^3P sulphur atoms it is 88 kJ mole^{-1} endothermic. When COS is substituted for CS_2, there is no doubt that $S(3^1D)$ atoms are produced at wavelengths > 190 nm, since ·SH radicals are produced both from H_2 and CH_4. As their pressures are increased, the yield of ·SH increases at the expense of S_2, reflecting competition against sulphur atom abstraction from the parent molecule.[91] In order to conserve spin the S_2 is produced initially in its metastable $\tilde{a}^1\Delta_g$ state.

$$S(3^1D) + SCO \rightarrow S_2(\tilde{a}^1\Delta_g) + CO$$

$$S_2(\tilde{a}^1\Delta_g) + M \rightarrow S_2(\tilde{X}^3\Sigma_g^-)$$

Flash excitation through lithium fluoride windows promotes COS into the second $^1B_2(^1\Sigma_u^+)$ electronic state, where sufficient energy is absorbed to produce sulphur atoms in the 3^1S level. The excited atoms can be detected in absorption immediately following the flash, together with CO, (endowed with $\leqslant 2$ vibrational quanta).[92]

$$COS(\tilde{X}^1\Sigma^+) \rightarrow CO(\tilde{X}^1\Sigma^+) + S(3^1S) \quad \Delta H = +568 \text{ kJ mole}^{-1}$$

Although they are metastable and carry 155 kJ mole^{-1} more than atoms in the 3^1D state, they are far less reactive toward hydrogen atom abstraction. The same is also true of the 2^1S states of oxygen and carbon atoms. Donovan and Husain[93] attribute the low reactivity to the requirement of orbital symmetry correlation. In an adiabatic reaction, i.e. one occurring on a single potential energy surface without any intervening radiationless transition, the total symmetry must be conserved.

As a H_2 molecule approaches the excited atom the lowered symmetry splits the five-fold degeneracy of the ^1D state, producing two states which are antisymmetric to reflexion in the plane of the complex (A'') and three which are symmetric (A'). The reaction

$$S(3^1D) + H_2(\tilde{X}^1\Sigma_g^+) \rightarrow SH(\tilde{X}^2\Pi_r) + H(1^2S)$$

leading to electronically unexcited products can take place adiabatically over one surface of each type. The other three lead to excited products. Atoms in the ^1S state have no degeneracy and in their case reaction with H_2 could begin only on a single, symmetric potential surface $(^1A')$. Unfortunately this one cannot correlate with unexcited products because of the two other surfaces of the same symmetry which lie at lower energies (non-crossing rule, see p. 26). The adiabatic reaction would lead to excited products and is probably endothermic. Thus the ^1S atoms are unreactive.

Excitation of CO_2 and N_2O into their first excited $^1B_2$ levels produces $O(2^1D)$ atoms,

$$CO_2 \rightarrow CO + O(2^1D)$$

$$N_2O \rightarrow N_2 + O(2^1D)$$

The nett quantum efficiency of photodissociation is probably unity at all absorbed wavelengths though some atoms may be formed in the 2^3P ground state, following intersystem crossing, while photons absorbed by N_2O at wavelengths <210 nm carry sufficient energy to produce $O(2^1S)$ atoms. For CO_2 the threshold lies at 127·3 nm, within the band system associated with the second $^1B_2 \leftarrow \tilde{X}^1\Sigma_g^+$ transition. $O(2^1S)$ atoms have been detected in emission during continuous photolysis.[94] In N_2O, the equivalent transition produces $O(2^1S)$ atoms with a quantum efficiency of $\sim$0·6.[94] The remaining molecules dissociate along the following paths

$$N_2O \nearrow \begin{matrix} N_2(\tilde{X}^1\Sigma_g^+) + O(2^1D) \\ \rightarrow N_2(\tilde{A}^3\Sigma_u^+) + O(2^3P) \\ \searrow N(2^2D) + NO(\tilde{X}^2\Pi) \end{matrix}$$

with the first probability being the most favoured. The final products include N_2, O_2 and NO produced in reactions such as

$$O(2^1D \text{ or } 2^1S) + N_2O \rightarrow N_2 + O_2(\tilde{a}^1\Delta_g \text{ or } \tilde{b}^1\Sigma_g^+ ?) \text{ or } 2NO$$
$$N(2^2D) + N_2O \rightarrow N_2 + NO$$

The major products from the photolysis of CO_2 are CO and O_2, and ideally their relative quantum yields should be in the ratio 2:1. In practice the ratio is often very much larger, depending on the experimental conditions under which the photolysis is effected. The small amount of ozone that is also produced does not quantitatively account for the oxygen deficiency, but it has been shown that $O(2^1D)$ atoms can add to CO_2 to form metastable CO_3.[95] The molecule has not positively been identified in the vapour phase, but has been isolated in a rigid matrix at low temperatures and characterized through its infrared absorption. The oxygen deficiency has been attributed to reaction of the CO_3 at the surfaces of the containing vessel.

The first electronic transitions in NO_2 produce a complex pattern of absorption bands throughout the ultraviolet and visible regions of the spectrum; they even continue into the photographic infrared. The molecule is paramagnetic, with seventeen valence electrons. According to Walsh's predictions (see section 1.3c and Figure 1.21), it should be bent in its ground electronic state, with the odd electron occupying the $6a_1(2\pi_u)$ orbital and tending to be localized near the nitrogen atom (as indeed it is). In themselves, neither of the first two electronic transitions

$$\ldots (4b_2)^2(6a_1)^0(2b_1)^1, \tilde{A}^2B_1 \longleftarrow \ldots (4b_2)^2(6a_1)^1, \tilde{X}^2A_1$$

$(\pi^*_{N-O} \rightarrow \pi^*_{N-O}, 320\text{--}1000 \text{ nm}),$

$$\ldots (4b_2)^1(6a_1)^2, \tilde{B}^2B_2 \qquad \longleftarrow \ldots (4b_2)^2(6a_1)^1, \tilde{X}^2A_1$$

$(n_O \rightarrow \pi^*_{N-O}, 235\text{--}258 \text{ nm})$

should promote any large increase in bond length in the upper state, (cf. CO_2, N_2O etc.). In fact both transitions populate bound upper states, but each one shows a sharp predissociation limit, at 397·9 and 248 nm, respectively. The first corresponds to the dissociation

$$NO_2 \rightarrow NO(\tilde{X}^2\Pi) + O(2^3P) \quad \Delta H = +299 \text{ kJ mole}^{-1}$$

and it is possible that the second does, also (see p. 156).[90] At wavelengths between 235 and 200 nm there is a third system of diffuse bands, all of which are heavily predissociated, and photolysis in this region generates $O(2^1D)$ atoms.[96]

The predissociation limit which sets in at 397·9 nm competes against fluorescent decay of the $\tilde{A}$ state and there is an inverse correlation between the quantum efficiency of photodissociation and the fluorescence intensity (see p. 175). The quantum yield of O_2 increases from 0·0(435·8 nm) to 0·36(405 nm) to 0·96(313 nm).[97]

The decomposition at 405 nm is interesting since it proceeds at a wavelength to the red of the spectroscopic predissociation limit, and in addition the energy absorbed, 295·1 kJ mole^{-1} is slightly below the ON–O bond strength, 298·9 ± 2·1 kJ mole^{-1}. Nonetheless, atomic oxygen does appear to be formed since in the presence of $^{18}O_2$, photolysis of NO_2 at 405 nm leads to isotopic scrambling. To overcome this difficulty, Ford and Jaffé[98] suggested that the atomic oxygen might be produced through bimolecular reaction of the excited NO_2

$$NO_2(\tilde{A}^2B_1) + NO_2 \rightarrow N_2O_3 + O(2^3P)$$

In the absence of predissociation it should be possible to maintain an appreciable population of molecules in the $\tilde{A}$ state, since its fluorescent lifetime of 44 μsecs is unusually long.

As a more attractive alternative, Pitts, Sharp and Chan have persuasively argued the case for the photoselection of NO_2 molecules from upper rotational or vibrational levels of the ground electronic states[97] (cf. the photodissociation of Br_2, p. 259). Increasing the temperature from 296°K to 496°K raises the quantum yield of O_2 at 405 nm from 0·36 to 0·69, reflecting the rise in the thermal population of NO_2 molecules in excited rotational levels.

Once $O(2^3P)$ atoms have been produced they react with undecomposed NO_2 to give NO_3, NO and vibrationally excited oxygen. The transient absorption bands of NO_3 (at 662 and 624 nm), and of $(O_2)_{v'' \leqslant 12}$ have both been photographed immediately following flash photolysis of NO_2 under isothermal conditions in the near-ultraviolet.[99] Mass spectrometric analysis of the isotopic distribution among the products formed when $^{18}O(2^3P)$ atoms attack NO_2, indicates reaction through a common N-centred intermediate, i.e. a bond is formed at the site of greatest electron density.[100]

$$O(2^3P) + NO_2 \rightarrow \left[O-N \begin{array}{c} O \\ \diagdown \\ O \end{array} \right]^* \begin{array}{c} \nearrow \; NO + (O_2)_{v'' \leqslant 12} \\ \searrow \\ NO_3 \end{array}$$

The large O–O separation in the intermediate produces the high level of vibrational excitation in the O_2. Those with 11–12 vibrational quanta are endowed with virtually the entire heat of the reaction. We shall meet rather similar behaviour in the photochemistry of O_3, though the reactive atoms now occupy the 2^1D state and the geometry of the intermediate (if indeed there is one) has not been established.

As in NO_2, the first electronic transitions in O_3 extend throughout most of the visible and continue into the photographic infrared. The vapour condenses to an inky blue–black liquid at low temperatures (in the uncertain, but often brief interval between cooling and detonation). With eighteen valency electrons the outermost orbital $6a_1(2\pi_u)$, is fully occupied and the molecule has a bond angle of 117°. The dissociation

$$O_3(\tilde{X}^1A_1) \rightarrow O_2(\tilde{X}^3\Sigma_g^-) + O(2^3P)$$

into ground state products is spin-allowed. It requires only 105 kJ mole^{-1}, equivalent to absorption of a photon at 1140 nm, well into the infrared, and it is not surprising that all the absorption bands are predissociated.

Because of this their assignment to individual electronic transitions can only be speculative.

The first absorption extends from 760–438 nm and is extremely weak, with $\varepsilon_{max} = 1 \cdot 12 \, l \, mole^{-1} \, cm^{-1}$ at 602 nm. Apart from the two diffuse bands centred at 537 nm and 602 nm, it shows very little structure. A sharper and stronger system of diffuse bands lies between 353 nm and 290 nm, where it overlaps a region of strong continuous absorption extending down to 200 nm, $\varepsilon_{max} = 2,800 \, l \, mole^{-1} \, cm^{-1}$ at 255 nm. Possible assignments have been discussed by Mulliken.[101]

Light absorption invariably leads to decomposition, but the nett quantum yield $\phi(-O_3)$, depends on the absorbed energy and sometimes on the pressure, rising from $\leqslant 2$ at ~ 600 nm to 4 at 334 nm and increasing to as much as 16 at 253·7 nm.[102] If the O_3 is moist, the quantum yield at 253·7 nm may be as high as 130, while in the visible it still remains $\leqslant 2$.[103] There is no doubt that the detailed mechanism of the decomposition can be extremely complex, and is very sensitive to the experimental conditions. Depending on the absorbed wavelength, photodissociation can produce atomic and molecular oxygen both in their ground triplet states, or in any of their low-lying metastable states (i.e. ^1D or ^1S, $\tilde{a}^1\Delta_g$, $\tilde{b}^1\Sigma_g^+$ or $\tilde{A}^3\Sigma_u^+$). In addition, the primary products may be translationally excited, and flash photolysis experiments have revealed molecular oxygen carrying more than 293 kJ mole^{-1} in vibration. All are potentially reactive intermediates capable of decomposing additional ozone molecules, but their reactivity is a function of their energy contents and quantum states, and in order to unravel the sequence of elementary steps it is essential to know the relative probabilities of each one.

At $\geqslant 600$ nm, the simple sequence

$$O_3 + h\nu \rightarrow O_2(\tilde{X}^3\Sigma_g^-) + O(2^3P) \quad \Delta H = -105 \, kJ \, mole^{-1} \quad (1)$$

$$O(2^3P) + O_3 \rightarrow 2O_2 \qquad\qquad \Delta H = -397 \, kJ \, mole^{-1} \quad (2)$$

$$O(2^3P) + O_2 + M \rightarrow O_3 + M \qquad \Delta H = -105 \, kJ \, mole^{-1} \quad (3)$$

will account for $\phi(-O_3) \not> 2$. The absorbed energy is insufficient for electronic excitation of either of the primary products and since the quantum yield is not significantly reduced by addition of inert foreign gases, it is unlikely that any excited oxygen molecules are formed in step (2) either. The quantum efficiency of the primary step is almost certainly unity.

Electronic excitation becomes energetically feasible as the absorbed wavelength is reduced below 600 nm. Possible primary steps are listed below, together with the wavelengths corresponding to their threshold energies.

$$\to O_2(\tilde{a}^1\Delta_g) + O(2^3P) \quad 590\,\text{nm} \tag{4}$$

$$\to O_2(\tilde{b}^1\Sigma_g^+) + O(2^3P) \quad 460\,\text{nm} \tag{5}$$

$$\to O_2(\tilde{X}^3\Sigma_g^-) + O(2^1D) \quad 410\,\text{nm} \tag{6}$$

$$O_3 + h\nu \qquad \to O_2(\tilde{a}^1\Delta_g) + O(2^1D) \quad 310\,\text{nm} \tag{7}$$

$$\to O_2(\tilde{b}^1\Sigma_g^+) + O(2^1D) \quad 260\,\text{nm} \tag{8}$$

$$\to O_2(\tilde{X}^3\Sigma_g^-) + O(2^1S) \quad 234\,\text{nm} \tag{9}$$

There is little doubt that steps (4) and possibly (5) operate at wavelengths > 310 nm, although they are spin-forbidden. At 334 nm the quantum yield of decomposition $\phi(-O_3)$ is increased to 4, independent of the O_3 pressure.[102c] Both the $\tilde{a}^1\Delta_g$ and $\tilde{b}^1\Sigma_g^+$ electronic states of O_2 carry sufficient energy to decompose a further molecule of O_3, and the sequence

$$O_3 + h\nu \to O_2(\tilde{a}^1\Delta_g \text{ or } \tilde{b}^1\Sigma_g^+) + O(2^3P) \tag{4 or 5}$$

$$O(2^3P) + O_3 \to 2O_2 \quad \Delta H = -397\,\text{kJ mole}^{-1} \tag{2}$$

$$O_2(\tilde{a}^1\Delta_g \text{ or } \tilde{b}^1\Sigma_g^+) + O_3 \to 2O_2 + O(2^3P)$$

$$\Delta H = +(8 \text{ or } -54\,\text{kJ mole}^{-1}) \tag{10}$$

$$O(2^3P) + O_3 \to 2O_2 \quad \Delta H = -397\,\text{kJ mole}^{-1} \tag{2}$$

removes four O_3 molecules from the system. The decomposition by $O_2(\tilde{a}^1\Delta_g)$ is relatively slow, since the reaction is 8 kJ mole^{-1} endothermic. We shall see that $O(2^1D)$ atoms are not produced until the absorbed wavelengths lie < 310 nm, and there is no evidence of any contribution from step (6).

As the wavelength is reduced below 310 nm, the quantum yield of decompostion begins to rise with the O_3 pressure, although its limiting value at infinitely low pressures remains equal to 4.[102c] The increase must be associated with a chain reaction sequence and many years ago Schumacher, Heidt and coworkers suggested that the chain carriers were excited oxygen molecules.[104] The type of excitation was unspecified. In 1957, Norrish and McGrath[105] discovered that the flash photolysis of O_3 at wavelengths > 200 nm under isothermal conditions produced oxygen molecules excited into high vibrational levels of their ground electronic state. Levels up to $v'' = 23$ can be seen in absorption, though the bands originating from $v'' > 17$ are weak. They reach a maximum intensity at the peak of the photolysis flash, but Basco and Norrish[106] have presented a variety of arguments to exclude the possibility of vibrationally excited oxygen being a primary product, (cf. the ultraviolet photodissociation of the isovalent molecule NOCl, which produces NO in levels

$v'' \leqslant 11$). The following sequence was proposed,

$$O_3 + h\nu \longrightarrow O_2(\tilde{a}^1\Delta_g \text{ or } \tilde{b}^1\Sigma_g^+) + O(2^1D) \qquad (7 \text{ or } 8)$$

$$O_2(\tilde{a}^1\Delta_g \text{ or } \tilde{b}^1\Sigma_g^+) + O_3 \longrightarrow 2O_2 + O(2^3P) \qquad (10)$$

$$O(2^3P) + O_3 \longrightarrow 2O_2 \qquad (2)$$

$$\text{chain} \begin{cases} O(2^1D) + O_3 \longrightarrow O_2 + (O_2)_{v'' \leqslant 23} & \Delta H = -577 \text{ kJ mole}^{-1} \quad (11) \\ (O_2)_{v'' > 17} + O_3 \longrightarrow 2O_2 + O(2^1D) & \Delta H = +293 \text{ kJ mole}^{-1} \quad (12) \end{cases}$$

$$(O_2)_{v'' \leqslant 23} + \text{gas/wall} \longrightarrow (O_2)_{v'' \leqslant 17} \qquad (13)$$

with $O(2^1D)$ atoms producing vibrationally excited oxygen in step (11) and being regenerated in the chain propagation step (12). How plausible is it?

(i) There is no doubt that $O(2^1D)$ atoms are involved in the chain reaction. In the red, where the energy is insufficient to excite $O(2^1D)$ there is no reaction chain. The quantum yield begins to rise at the threshold wavelength for the photodissociation step (7). Ultraviolet flash photolysis of moist O_3 produces vibrationally excited OH radicals in place of oxygen; only the levels $v'' \leqslant 2$ are populated. The reaction

$$O(2^3P) + H_2O \longrightarrow \cdot OH + \cdot OH \quad \Delta H = +63 \text{ kJ mole}^{-1}$$

is endothermic, but with $O(2^1D)$ sufficient energy is released to excite ·OH into its second vibrational level. In the red, where $O(2^3P)$ atoms are produced, moisture has no effect on the decomposition of ozone.

(ii) The evidence for the primary production of O_2 in the $\tilde{a}^1\Delta_g$ or $\tilde{b}^1\Sigma_g^+$ states is strong, but remains indirect at present. $O_2(\tilde{a}^1\Delta_g)$ has been detected in emission during the ultraviolet photolysis of O_3/O_2 mixtures, but its phosphorescence is excited through energy transfer

$$O(2^1D) + O_2(\tilde{X}^3\Sigma_g^-) \longrightarrow O(2^3P) + O_2(\tilde{a}^1\Delta_g)$$

rather than photodissociation.[107] A transient system of absorption bands photographed in the vacuum ultraviolet during the flash photolysis of O_3, may be due to $O_2(\tilde{a}^1\Delta_g)$.[108]

(iii) The proposed sequence of reaction steps gives the expression

$$\phi(-O_3) = 4 + \frac{2k_{12}}{k_{13}}[O_3]$$

for the quantum yield of decomposition, i.e. it increases with the ozone concentration, and $\underset{[O_3] \to 0}{\text{Lim}} \phi(-O_3) = 4$ in agreement with observation.

(iv) O_2 molecules with > 17 vibrational quanta carry sufficient energy to supply the 293 kJ mole^{-1} required in step (12), though Fitzsimmons and Bair[109] believe their population is too small for vibrationally excited oxygen to play a major role in the decomposition. It is conceivable that the chain carrier produced in (11) could be electronically excited oxygen, perhaps in the metastable $\tilde{A}^3\Sigma_u^+$ state. There is no clear answer at present. The truth may be that there is no single mechanism that operates to the exclusion of all the others.

(iii) Simple polyatomic molecules

Although CH_4 seems a simple enough polyatomic molecule its photochemistry is far from simple. Several alternative primary steps have been identified but their relative contributions, their dependence on the absorbed wavelength and the energy distribution among the primary products all remain to be established.

The continuous absorption which begins ~ 146 nm and reaches a maximum ~ 93 nm, is associated with the electronic transition

$$\ldots (2a_1)^2(1t_2)^5(3a_1)^1, \tilde{A}^1T_2 \leftarrow \ldots (2a_1)^2(1t_2)^6, \tilde{X}^1A_1$$

(part $\sigma \rightarrow \sigma^*$, part Rydberg—see p. 137), but probably includes other transitions as well. At 123·6 nm the molecular elimination of H_2

$$CH_4 \rightarrow :CH_2^* + H_2 \quad \Delta H = +(438 \pm 25)\,\text{kJ mole}^{-1}$$

is much more efficient than the radical dissociation

$$CH_4 \rightarrow \cdot CH_3^* + H\cdot \quad \Delta H = +435\,\text{kJ mole}^{-1}$$

since equimolar mixtures of CH_4 and CD_4 produce H_2 and D_2 (with the same quantum efficiency, 0·58), but little HD.[110] Ketene (CH_2CO) is formed when CH_4 is photolysed in the presence of CO,[110a] and photolysis in a N_2 matrix at 14°K produces CH_2N_2.[111] Photons at 123·6 nm carry 966 kJ mole^{-1} and with so large an excess energy it is possible that both the $:CH_2$ and $\cdot CH_3$ radicals are produced in electronically excited states (see p. 173). The suggestion is supported by the results of a Valence Bond calculation which was used to map out the potential energy surfaces of dissociating CH_4 molecules,[112] but as yet no direct experimental evidence has been presented. If CH_2 is produced in its first excited singlet state, it might be possible to observe red emission bands associated with their fluorescent decay, $\tilde{a}^1A_1 \leftarrow \tilde{b}^1B_1$.[112] The analogous emission has been observed from excited $\cdot NH_2$ radicals produced in the vacuum ultraviolet photolysis of $NH_3\cdot$.

When CH_4 is subjected to high intensity photolysis, the picture becomes much more involved. Flash photolysis through LiF windows (transmitting wavelengths $\geqslant 110$ nm) under isothermal conditions produces $\cdot \check{C}H$, detected as a transient in ultraviolet absorption, and greatly increases the final yield of H_2.[113] HD is a major product of the flash photolysis of equimolar CH_4–CD_4 mixtures (cf. the low intensity result). At the highest flash energies, the yields of $\cdot\ddot{C}H$ and HD both fall with an increase in the pressure of added argon, but when the energy is reduced the $\cdot\ddot{C}H$ yield becomes independent of the pressure. These and other results suggest that absorption of light $\geqslant 110$ nm produces $\cdot\ddot{C}H$ radicals directly in the primary step

$$CH_4 \rightarrow \cdot\ddot{C}H + H\cdot + H_2 \quad \Delta H = +899 \text{ kJ mole}^{-1}$$

and also in some secondary process(es) involving excited precursors which are sensitive to collisional deactivation.[113] The precursors might be electronically excited $:CH_2$ or $\cdot CH_3$, for example

$$2CH_2^* \rightarrow \cdot CH_3 + \cdot\ddot{C}H$$

Reaction of $\cdot\ddot{C}H$ with undecomposed CH_4 produces 'hot' $\cdot C_2H_5$ which subsequently dissociates to give C_2H_4.

$$\cdot\ddot{C}H + CH_4 \rightarrow C_2H_5^* \qquad \Delta H = -410 \text{ kJ mole}^{-1}$$

$$\cdot C_2H_5^* \rightarrow C_2H_4 + H\cdot \quad \Delta H = +167 \text{ kJ mole}^{-1}$$

As a final complication there is also some evidence for the primary production of atomic carbon[113]

$$CH_4 \rightarrow C + 2H_2 \quad \Delta H = +803 \text{ kJ mole}^{-1}$$

The absorbed energy is more than sufficient to produce carbon in its lowest singlet state 2^1D, which would avoid violation of the spin correlation rule (the ground state is 2^3P).

The first electronic transition in NH_3, $(n_N \rightarrow \sigma_{N-H}^*)$,

$$\ldots (3a_1)^1(4a_1)^1, \tilde{A}^1A_1(\tilde{A}^1A_2'') \leftarrow \ldots (3a_1)^2, \tilde{X}^1A_1$$

populates a planar excited state (see p. 156).† The associated absorption consists of a long progression in the out-of-plane deformation frequency of the excited molecule, extending from 217–170 nm with a maximum intensity ~ 190 nm. With the equilibrium N–H bond length only increasing from 1·02 to 1·08 Å,[90] the transition cannot lead to direct dissociation, but all of its vibronic levels are heavily predissociated and no rotational structure can be resolved (see p. 156). Absorption at >170 nm produces $\cdot NH_2$ and $H\cdot$, exclusively in their ground electronic states and probably with unit quantum yield

$$NH_3 \rightarrow \cdot NH_2(\tilde{X}^2B_1) + H(1^2S) \quad \Delta H = +431 \text{ kJ mole}^{-1}$$

---

† Under the symmetry operations of the planar, D_{3h} point group, the electronic state becomes $\tilde{A}^1A_2''$.

Predissociation into excited $\cdot NH_2(\tilde{A}^2 A_1)$

$$NH_3 \rightarrow \cdot NH_2(\tilde{A}^2 A_1) + H(1^2 S) \quad \Delta H = +552 \, kJ \, mole^{-1}$$

while energetically feasible, fails to satisfy the requirements of symmetry conservation; the products correlate with the *ground* electronic state of NH_3, i.e. one of A_1 symmetry.[90]

Absorption below 170 nm populates a series of more highly excited electronic states. The transitions probably involve the promotion of an electron from the non-bonding $3a_1$ orbital, into a series of Rydberg orbitals also centred on the nitrogen atom. They alter the equilibrium geometry from pyramidal to planar, but as before, direct dissociation would not be expected. Indeed, rotational structure can be resolved in the $\tilde{B} \leftarrow \tilde{X}$ band system (169–140 nm), but it is broadened and all the excited states suffer predissociation. Transient absorption from $NH(\tilde{X}^3 \Sigma^-)$ (and $NH_2(\tilde{X}^2 B_1)$) has been photographed during the isothermal flash photolysis of NH_3 at $<155 \, nm$,[114] while continuous photolysis in the vacuum ultraviolet excites fluorescence from both $\cdot NH_2$ and $:NH$[115,116]

$$NH_2(\tilde{A}^2 A_1) \rightarrow NH_2(\tilde{X}^2 B_1) + h\nu_{fl}$$

$$NH(\tilde{c}^1 \Pi) \rightarrow NH(\tilde{a}^1 \Delta) + h\nu_{fl}$$

The following primary steps have been identified

$$NH_3 \begin{cases} \xrightarrow{\lambda \leqslant 169 \, nm} NH_2(\tilde{A}^2 A_1) + H(1^2 S) & \Delta H = +431 \, kJ \, mole^{-1}, \\ \xrightarrow{\lambda < 155 \, nm} NH(\tilde{X}^3 \Sigma^-) + H_2(\tilde{X}^1 \Sigma_g^+) & \Delta H = \sim +376 \, kJ \, mole^{-1}, \\ \xrightarrow{\lambda \leqslant 133 \, nm} NH(\tilde{c}^1 \Pi) + H_2(\tilde{X}^1 \Sigma_g^+) & \Delta H = \leqslant +899 \, kJ \, mole^{-1}\dagger \end{cases}$$

The second one is spin-forbidden; it is possible that atomic rather than molecular hydrogen is produced, though this is unlikely to become energetically feasible until wavelengths $<140 \, nm$. By adding $C_2 D_4$ to scavenge atomic hydrogen McNesby, Tanaka and Okabe[117] estimate that $\frac{1}{7}$th of the H_2 recovered after photolysis at 123·6 nm, is formed through molecular elimination in the primary process.

The $n_O \rightarrow \pi^*_{C-O}$ electronic transition in $H_2 CO$, excites long progressions of absorption bands in the near-ultraviolet, mainly associated with a change in the equilibrium geometry from planar to pyramidal and a stretching of the C–O bond‡ (see pp. 76, 138).

$$(353-230 \, nm, \varepsilon_{max} \sim 10 \, l \, mole^{-1} \, cm^{-1}) \quad \tilde{A}^1 A''(^1 A_2) \leftarrow \tilde{X}^1 A_1$$
$$(397-360 \, nm, \varepsilon_{max} \sim 10^{-3} \, l \, mole^{-1} \, cm^{-1}) \quad \tilde{a}^3 A''(^3 A_2) \leftarrow \tilde{X}^1 A_1$$

† Unfortunately, accurate values for the excitation energies of $NH(\tilde{a}^1 \Delta$ and $\tilde{c}^1 \pi)$ are unavailable at present, so the energy requirements can only be estimated.

‡ Under the reduced symmetry of the excited molecule, the electronic states become $^1 A''$ and $^3 A''$, respectively.

The bands remain sharp under medium resolution, until ~ 275 nm but predissociation actually sets in at much longer wavelengths.[90] The fluorescent lifetime of $2 \cdot 3 \times 10^{-7}$ sec is much shorter than the value calculated from the integrated absorption[118] and no emission has been detected from molecules with one or more quanta of the deformation frequency.[119] At 313 nm (and 195°C), the two predissociations

$$H_2CO \xrightarrow{\qquad} \begin{array}{l} H \cdot + H\dot{C}O \quad \Delta H = \sim +368 \text{ kJ mole}^{-1} \\ H_2 + CO \quad\quad \Delta H = \sim +4 \text{ kJ mole}^{-1} \end{array}$$

have a combined quantum efficiency of unity,[120] and the transient spectrum of $H\dot{C}O$ can be photographed in absorption in the visible, following isothermal flash photolysis of the vapour.[121]

The nett decomposition follows a radical chain mechanism at more elevated temperatures due to secondary decomposition of the $H\dot{C}O$ radical.[122]

$$\text{chain} \begin{cases} H\dot{C}O \rightarrow H \cdot + CO \quad\quad \Delta H = +75 \text{ kJ mole}^{-1} \\ H \cdot + H_2CO \rightarrow H_2 + H\dot{C}O \quad \Delta H = -67 \text{ kJ mole}^{-1} \end{cases}$$

Increasing the temperature (or reducing the absorbed wavelength) also lowers the quantum efficiency of the molecular predissociation. At 365 nm and 140°C, photolysis of an equimolar mixture of H_2CO and D_2CO produces HD as well as H_2 and D_2.[123] Their relative yields indicate contributions from both the molecular and radical predissociation steps. At 350°C, the nett quantum yield of decomposition rises to ~ 200 and the relative yield of HD now indicates the complete absence of any molecular dissociation.

Abrahamson, Littler and Vo[124] suggest that the predissociating states are of the type $^1(n_O, \sigma^*_{C-H_2})$, where an electron has been promoted into orbitals that are $C-H_2$ antibonding and either $H-H$ bonding (molecular split) or antibonding (radical split). Figure 1.23 shows that the CH_2 group orbitals labelled $7a_1$ and $3b_2$ would fulfil the requirements. Admittedly they would lie well above the $2b_1(\pi^*_{C-O})$ orbital in the planar molecule, but the pyramidal conformation of the excited state would allow overlap with the $2p_y$ and $2p_x$ orbitals centred on the oxygen atom and their energies would be lowered. The two orbitals are represented in Figure 3.9, using the localized orbital description. Abrahamson and coworkers conclude that the probable predissociation paths are

$$^1A'', \, ^1(n_O, \sigma^*_{C-H_2}) \rightarrow H_2(\tilde{X}^1\Sigma_g^+) + CO(\tilde{X}^1\Sigma^+)$$
$$\Big\uparrow$$
$$H_2CO(\tilde{X}^1A_1) \rightarrow \tilde{A}^1A'', \, ^1(n_O, \pi^*_{C-O})$$
$$\Big\downarrow$$
$$^1A', \, ^1(n_O, \sigma^*_{C-H_2}) \rightarrow H\cdot(1^2S) + HCO(\tilde{X}^2A')$$

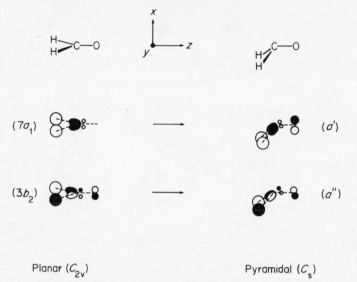

Planar (C_{2v}) Pyramidal (C_s)

Figure 3.9 Localized C–H$_2$ antibonding orbitals in H$_2$CO (schematic). (After E. W. Abrahamson, J. G. F. Littler and K.-P. Vo, ref. 124.)

Internal conversion into high vibrational levels of the ground state is likely to be much slower in view of the large energy gap (see section 2.5c). The same is true of intersystem crossing to the $\tilde{a}^3A''$ state; $^1(n_O \rightarrow \pi^*_{C-O})$ absorption excites fluorescence but not phosphorescence. This could be due to rapid radiationless decay in the triplet state, of course, but since phosphorescent emission can be excited by an electrodeless discharge, the possibility is excluded.

In ketene, the carbonyl chromophore is modified by interaction with the adjacent ethylenic double bond. The orbital that was labelled n_O in H$_2$CO becomes conjugated with p_π orbitals on the neighbouring carbon atoms. The low-lying π-orbitals are represented in Figure 3.10. Excitation of an electron into the $2b_2$–π^*_{C-O} antibonding orbital causes the molecule to bend about the carbonyl carbon atom (cf. CO$_2$ and H$_2$CO). A long progression of weak diffuse absorption bands which runs from 380 nm to 260 nm ($\varepsilon_{max} \simeq 121 \, \text{mole}^{-1} \, \text{cm}^{-1}$ at 330 nm) is associated with the transition $(\pi \rightarrow \pi^*_{C-O})^{125a}$

$$\ldots (1b_2)^2(2b_1)^1(2b_2)^1, \tilde{A}^1A_2(^1A'') \leftarrow \ldots (1b_2)^2(2b_1)^2, \tilde{X}^1A_1$$

The band spacing corresponds to excitation of a C–C–O bending mode in the upper state. A much weaker series of diffuse bands extending into the visible from 380 nm to $\geqslant 473.5$ nm has been assigned to the singlet–

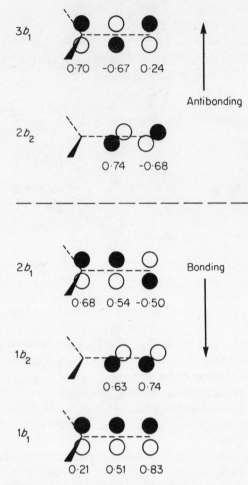

Figure 3.10 Low-lying orbitals in $H_2C{=}CO$ (orbital coefficients calculated by R. N. Dixon and G. H. Kirby, ref. 125a).

triplet transition $\tilde{a}^3A'' \leftarrow \tilde{X}^1A_1$, which also appears to rattle the C–C–O skeleton out of its linear conformation.[125a]

The complete lack of any resolvable rotational structure, or of any recorded emission spectrum, either fluorescent or phosphorescent, implies rapid predissociation. The primary products of predissociation are exclusively CO and either singlet or triplet CH_2; since the absorption remains diffuse out to wavelengths $\geqslant 473.5$ nm, $D(CH_2–CO) \leqslant 255$ kJ mole^{-1}.[125a] The quantum efficiency of the predissociation is a sensitive function of the absorbed wavelength and of the pressure and temperature. For example, at 366 nm and 25° it increases from 0·007 at 10 cm mercury pressure to 0·04 as the pressure tends to zero. At 313 nm it increases from 0·7 to unity, while at 270 nm the quantum efficiency is always unity regardless of the temperature or pressure.[125b]

The pressure dependence at the longer wavelengths reflects competition between the unimolecular decomposition of vibrationally excited ketene in the upper electronic state and the transfer of excess vibrational energy in collisions which carry the excited molecule into levels lying below the critical energy for decomposition E^* (see pp. 167 and 203).

$$(CH_2CO)^*_{E > E^*} \nearrow \quad CH_2 + CO$$
$$\xrightarrow{(+M)} \quad (CH_2CO)^*_{E < E^*} \quad \text{(a multistep process when } E \gg E^*)$$

This mechanism taken as it stands predicts a unit quantum efficiency of decomposition at infinitely low pressures, provided the absorbed photon carries the molecule into vibrational levels in the upper electronic state with $E > E^*$. Since the absorption spectrum is completely diffuse this criterion would seem to be met at all wavelengths, and the low quantum yield at the longer wavelengths presents something of a puzzle. It is possible that internal conversion is fast enough to compete against dissociation when $E \simeq E^*$. As an alternative, Bowers[126] has suggested that there is an energy deficit when ketene is dissociated at long wavelengths, a deficit which is made up by photoselection of thermally excited molecules from the ground state (cf. the photodissociation of NO_2). As the temperature is raised, both the proportion of thermally excited molecules and the quantum efficiency increase. Unfortunately, neither explanation can explain the total lack of resolvable rotational structure.

The nett quantum yield of decomposition is twice the primary quantum efficiency. Both singlet and triplet CH_2 can decompose a second molecule of ketene

$$CH_2 + CH_2CO \rightarrow C_2H_4 + CO$$

Primary production of triplet CH_2 tends to be favoured at longer wavelengths, though still within the $\tilde{A}^1A_2(^1A'') \leftarrow \tilde{X}^1A_1$ band system.[127] The relative yields of singlet and triplet CH_2 at any given wavelength have been estimated by analysing the final products of photolysis in the presence of reactive substrates such as *cis* or *trans*-butene-2, or paraffins or CO.[128] Hopefully, the spread of products depends unambiguously on the spin multiplicity of the CH_2, allowing differentiation between the reactions of each species. This is not necessarily justified. For example, the dimethyl-cyclopropanes formed by addition of CH_2 across the double bond in cis or trans-butene-2 are vibrationally excited since the addition is strongly exothermic. Unless collisionally deactivated they may isomerize; if the ratio of the cis and trans-isomers varies with the total pressure it cannot be used as a reliable measure of stereospecific addition by singlet CH_2. To make matters more suspect, the substrate may accept energy from the excited ketene altering both the absolute rate of predissociation and conceivably the relative proportions of singlet and triplet CH_2 in the primary products. Because of such difficulties, estimates of the primary quantum yields must be treated with some caution. For this reason, we have noted only that the formation of triplet CH_2 is favoured as the absorbed wavelength increases (more chance of intersystem crossing during the lifetime of the excited ketene?), and avoided quoting quantitative data.

3.3 MORE COMPLEX POLYATOMIC MOLECULES

In the vapour phase at low pressures, light is absorbed by isolated atoms or molecules; at higher pressures their energy levels can be slightly perturbed by the fields of neighbouring molecules at the instant of absorption. This is the origin of the collisional broadening of spectral lines. So far all the photochemical changes we have considered in the present chapter have been in the vapour phase but as molecules grow more complex they also become less volatile. In most cases their photochemistry has to be studied in condensed media, most commonly in dilute solution though the photochemistry of the solid state promises to be an important growth point during the next few years. In the liquid phase all the molecules are embedded in the potential field of their neighbours, which fluctuates with the random Brownian motion and continuously perturbs the wave functions and energy levels of each molecule. This may influence the course of a photochemical change by altering the relative spacings of the molecular electronic states or the relative rates of intra-molecular radiationless transitions between them; vibrational relaxation is very much faster in a liquid medium than in the vapour phase. Specific molecular interactions can lead to association and light absorption may populate electronic states of a complex rather than the individual molecules (or ions) which form it (see pp. 148 and 215). For example, electron transfer within an excited complex can lead to permanent photochemical change. If the excited molecule dissociates into reactive radical fragments, they are much more likely to react together within the solvent cage than to diffuse out of it. Anyone who has attempted to cross Piccadilly Circus during the evening rush-hour will appreciate that the rate of movement is a function of the density of the medium through which the body is diffusing. If one should encounter an acquaintance in the jostling crowd it will be a relatively slow process to escape from them afterwards, particularly if you are attracted to them. Transferring the analogy to molecular diffusion in a liquid medium, the duration of a typical encounter might be 10^{-10}–10^{-11} sec. A pair of geminate free radicals produced by photodissociation will not escape from each other if they react in that time. If they combine, the excess energy is readily transferred to the dense but mobile surrounding medium, and under such circumstances the nett quantum yield of decomposition can be orders of magnitude lower in solution than in the vapour phase.[129]

Much of the most interesting photochemistry of complex polyatomic molecules does not involve photodissociation at all, but proceeds through intramolecular rearrangements, or intermolecular reactions of the photo-excited molecules (though not necessarily in the electronic state initially populated by light absorption). Photoexcitation alters the electron distribution around the atomic framework and in a sense each electronic state is a 'new molecule'. It can have a distinctive pattern of chemical reactivity, or in the absence of bimolecular reaction it may relax into new (often more strained and less thermodynamically stable) structures. Because the terms of reference of this book (and the energy of its author) are strictly limited, only a few representative examples will be described. Many comprehensive reviews are now available for the more enthusiastic reader.[49,130]

3.3a Experimental: primary processes in condensed media

Actinometry
The quantum efficiency ϕ_λ of a primary photochemical process is equal to the quantum yield of the product it forms, measured at the absorbed wavelength λ,

$$\phi_\lambda = \frac{\text{Rate of formation of primary product (molecules } l^{-1}\text{ sec}^{-1})}{\text{Rate of absorption of light (quanta } l^{-1}\text{ sec}^{-1}\text{ at wavelength } \lambda)}$$

Unless the primary product is stable, its rate of formation is not easy to measure, but any detailed speculation about the mechanism of the primary process is of little value unless the primary quantum efficiencies are known. When the primary product is unstable (for example, a free radical), it may be possible to derive its quantum yield by measuring those of the final products, provided the sequence of secondary reactions which produce them is well established from kinetic studies. Sometimes the secondary reactions can be prevented, by carrying out the photolysis in the presence of a chemically reactive substrate, which quantitatively traps or scavenges the primary products.

Fortunately, measuring the absorbed light intensity has become a much more routine procedure, at least in the quartz ultraviolet and much of the visible, since Parker and Hatchard's development of a simple, accurate and sensitive chemical actinometer, potassium ferrioxalate.[131] This utilizes a photochemical reaction of known quantum efficiency as a 'quantum counter'. The ferrioxalate is irradiated in aqueous solution at room temperature: absorption of wavelengths < 550 nm promotes electron transfer from an oxalate ligand to the central ferric ion, reducing it to ferrous

$$Fe^{3+}(C_2O_4)^{2-} \xrightarrow{hv} Fe^{2+} + (C_2O_4)^{\cdot-}$$

$$(C_2O_4)^{\cdot-} + Fe^{3+}(C_2O_4)^{2-} \rightarrow 2CO_2 + Fe^{2+} + (C_2O_4)^{2-}$$

After irradiation the yield of ferrous ions can be found at very low concentrations by measuring the optical density of the intensely red complex formed on addition of 1,10-phenanthroline. Parker and Hatchard measured the *quantum* yield of Fe^{2+} by using a thermopile, previously calibrated against the output from a standard lamp, to measure the incident radiant energy.

Ferrioxalate actinometry has become a standard photochemical technique, since it has many desirable qualities. The ferrioxalate ion absorbs strongly over a wide range of wavelengths while its photoproducts remain transparent; their quantum yield remains constant over most of the actinometer's working range (see Figure 3.11). The technique is sensitive yet simple to use.

Several other alternative types of actinometer are described in the text books written by Parker,[132] and by Calvert and Pitts.[49]

Laser flash photolysis
The flash photolysis technique can be used to study photochemical processes in liquid, or even in solid solution, as well as in the gas phase (though with conventional flash lamps the technique is still restricted to monitoring changes which occur in the microsecond region). Admittedly it is a simple matter to reduce the half-life of the emission from a capacitor discharge to a few nanoseconds (10^{-9} sec), i.e. a lifetime comparable to those of many excited or tautomeric intermediates first produced in photochemical reactions. All that is necessary is a sufficient reduction in the capacitance and inductance of the discharge circuit. Unfortunately, the intensity of

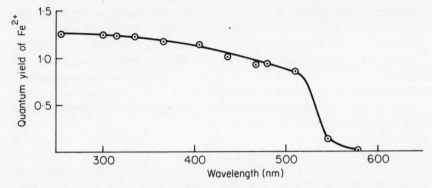

Figure 3.11   Quantum yield of the ferrioxalate actinometer. (C. F. Hatchard and C. A. Parker, ref. 131.)

the minuscule pulse of light that emerges is far too small to produce transient photoproducts in any concentration that would be detectable in absorption (though it is possible to monitor fluorescent emission with a sensitive photomultiplier detector). The flash of light emitted by a 'Q-switched' laser also decays within a few nanoseconds (see p. 80), but is many orders of magnitude brighter and there is no difficulty in producing a 0·1 J pulse of light in the near-ultraviolet. Over a period of 10 nsec, this corresponds to a mean power of 10 Mw of light! Porter, Windsor and others incorporated the laser into flash photolysis systems;[55] Porter and Topp's method is particularly simple and elegant, and can be used both to photograph transient absorption spectra or record their arrival and decay photoelectrically.

The light pulse is produced at 347·1 nm from a frequency doubled, Q-switched ruby laser. Part of the pulse is directed into the absorption cell, while a partially reflecting mirror diverts the remainder to another plane mirror some metres distant (see Figure 3.12 (plates XII–XIV of plate section)). On its return the light is focussed into a fluorescent solution which provides an intense luminescence against which any transient absorption spectra lying in the visible or very near-ultraviolet can be photographed. The fluorescent background decays within ~ 2 nsec. The delay between the photolysis pulse and the background pulse is generated by the time taken for the light to travel to the far mirror and back again. At a velocity of 3×10^{8} msec^{-1}, each metre travelled introduces a delay of 3·33 nsec. In comparison with time of this order the laser pumping flash, which is produced by a capacitor discharge and lasts for 1 msec, is virtually a steady light source. It is also very much brighter than a continuous light source and provides an excellent background for photoelectric monitoring of the transient changes in absorption. Porter and Topp[55] showed that the rate of decay of the short-lived absorption produced in solutions of 3,4-benzopyrene, for example, matched the rate of development of the triplet–triplet absorption (Figure 3.12), and also the fluorescent decay rate, confirming its assignment to absorption from the first excited singlet state. The technique is certain to produce much more detailed information about the rates of primary photochemical processes than has been possible up to the present.

Photolysis at low temperatures

Lowering the temperature of a solution offers the possibility of trapping unstable photoproducts, both free radical, molecular and ionic (see pp. 253). Ultraviolet, visible, infrared and magnetic resonance spectroscopy can be used to identify them and follow their kinetic behaviour. By way of illustration, the ultraviolet absorption bands of the benzylic radical produced by photoexcitation of an alkyl substituted benzene frozen into a hydrocarbon glass at 77°K are shown in Figure 3.13, and Figure 3.14 shows infrared absorption which has been assigned to the ketene (I), an intermediate in the photoisomerization of the substituted dienone (II). At 77°K the ketene has a half-life of ∼ 1 hr, but at room temperature its thermal cyclization either to the final product (III) or back to the starting material is so rapid that the intermediate is undetectable.[133]

In the first example the high viscosity of the matrix prevents bimolecular association of the benzylic radicals, but in the second the intermediate is stabilized because of the activation energy required for the unimolecular cyclization.

The tacit assumption that lowering the temperature does not affect the primary photochemical process must be treated cautiously; in many systems it is patently untrue. For example, biphotonic photochemical processes are common in glassy solutions, where the lack of diffusional quenching extends the lifetime of molecules excited into the lowest triplet state. Sufficient steady state concentrations can accumulate to allow secondary excitation into higher triplet states, and the concentration of perhaps another 250–290 kJ mole^{-1} into the system. Benzylic radicals are produced through this type of mechanism; the reaction is thought to

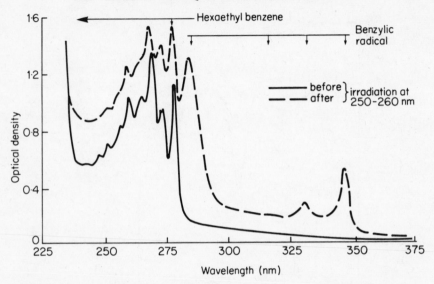

Figure 3.13 Photolysis of hexaethyl benzene in a paraffin hydrocarbon glass at 77°K.

involve energy transfer from the doubly excited aromatic hydrocarbon to the solvent which dissociates into radicals. These subsequently 'bite back' at the unexcited state abstracting a hydrogen atom from the aromatic side chain.[134] In many aromatic amines, biphotonic excitation leads to photoionization[135] and amine radical cations can be detected in absorption, frozen in a glassy matrix. Their absorption increases with the square of the light intensity. Photoexcitation of benzene in fluid solution in the near-ultraviolet leads to isomerization, but when frozen in glassy solution in a saturated hydrocarbon or alcohol the excited benzene reacts with a neighbouring molecule of the solvent to produce a substituted hexatriene,[136] a reaction which plays no part in its photochemistry in fluid solutions. The rigidity rather than the temperature of the solvent is critical and although the reaction probably involves the lowest triplet state, the intensity dependence shows it to be monophotonic.

Weak intermolecular interactions can lead to association or aggregation at low temperatures so that the photosensitive species may become specifically oriented molecular complexes, or aggregates of like molecules. Entirely novel photoprocesses can result.[137] For example, polyhalomethanes dissociate in the gas phase, following excitation of the $^1(n \rightarrow \sigma^*)$ electronic transition. In a glassy hydrocarbon matrix at 77°K the same transition results in the rapid development of intense colour centres.[138]

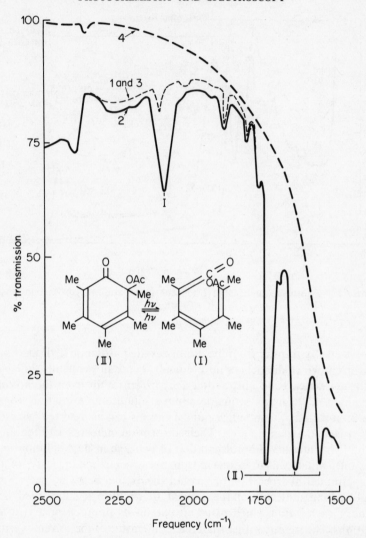

Figure 3.14 Molecular rearrangement of a 2,4-cyclohexadienone (II) to a ketene (I) in paraffin at 77°K, (1) and (2) before and after $1\frac{1}{2}$ h irradiation at 300 nm, 77°K; (3) after warming solution; (4) background transmission (through sapphire windows).

They fade when the glass softens or on exposure to visible light and the colours are thought to be due to absorption by charged species, produced through photoelectron transfer between neighbouring aggregated molecules. Claridge and Willard[139] have been able to record the e.s.r. spectra of alkyl radicals frozen in glassy matrices, produced as a result of photoelectron transfer within excited charge-transfer complexes of halomethanes and aromatic amines.

$$\geqslant N \ldots RX \xrightarrow{h\nu} \geqslant N^+ \ldots RX^- \longrightarrow \geqslant N^{+\cdot} + R\cdot + X^-$$

Study of the photochemistry of complex molecules at low temperatures is still in its infancy; its development should provide exciting entertainment for the enthusiast.

Isotopic labelling

The photochemistry of polyatomic molecules frequently involves some kind of unimolecular rearrangement, often stereospecific because possible reaction paths can be restricted by the requirements of orbital symmetry correlation (see p. 305). The reaction mechanism can often be clarified by 'stamping' some of the atoms with an isotopic label, and determining where they are placed in the photoproduct(s) isolated at the end of the experiment. Determination of the structure of the final products usually depends on the results of n.m.r., infrared and/or mass spectrometric analysis. Photochemical rearrangements of the benzene ring provide a good illustration of the technique.

Irradiation of mesitylene in the near ultraviolet promotes a $1 \rightarrow 2$ shift of the methyl substituents. By labelling the substituted carbon atom with the ^{14}C isotope, Kaplan and Wiltzbach[140] revealed that the apparent $1 \rightarrow 2$ migration was really the result of a rearrangement of the ring carbon atoms.

Subsequently they irradiated benzene-1,3,5-d_3 to establish that benzene itself undergoes the same type of rearrangement.[141]

Both involve valence isomerization in the excited aromatic ring, to produce isomers such as benzvalene (subsequently isolated as a primary photoproduct) and 'Dewar' benzene, and the isotopic scrambling follows their re-isomerization back to the aromatic structure. For example,

Dynamic nuclear spin polarization

This sounds forbidding, but promises to be something with which photochemists will have to be familiar. Spin angular momentum in an atomic nucleus generates a magnetic moment which can couple to the fields of other magnetic nuclei or to an externally applied field, to produce a set of different spin states with slightly different energies. Under conditions of thermal equilibrium their populations follow a Boltzmann distribution and if the external field is sufficiently intense, magnetic dipole transitions between the levels can be detected in absorption at radio frequencies. In 1967 it was discovered that the products of some fast free radical reactions (for example, those produced from the thermal decomposition of peroxides) gave intense proton n.m.r. signals in emission, reflecting a population inversion among the nuclear spin states. The overpopulation is caused by magnetic coupling between the nuclear spins and those of the free electrons in the radical precursors. As the electron spin states relax into their Boltzmann distribution the coupling disturbs the nuclear spin states and, because these relax relatively slowly (typical relaxation times ~ 2 sec), temporary population inversions can result. Alternatively, if the coupling causes an increased population in a lower energy spin state some of the n.m.r. signals will appear with an enhanced absorption intensity. Because the effect is a dynamic one, depending on a relaxation process it was termed *chemically induced dynamic nuclear spin polarization*.

The effect has induced an overpopulation of excited mental states also as it has been realized that the n.m.r. spectrum of the product reflects mechanistic details of the route by which it was formed. When a photoproduct shows nuclear spin polarization, there is unambiguous evidence of unpaired electron spin(s) in its immediate precursor(s). To illustrate the power of the technique, Closs and coworkers[142] recorded the n.m.r. spectra of the products of hydrogen atom abstraction from toluene by triplet diphenyl methylene (produced by direct photolysis of diphenyl diazomethane) and triplet benzophenone. For example

$$\phi_2CO(T_1) + H_3C\phi \xrightarrow{h\nu} \phi_2\dot{C}\text{-OH} + \dot{C}H_2\phi \longrightarrow$$

$$\begin{array}{cc} OH & OH \\ | & | \\ \phi_2C & \!\!\!-\!\!\!-C\!-\phi_2 \\ \nearrow & \\ \rightarrow \phi_2C(OH)CH_2\phi \\ \searrow & \\ \phi CH_2CH_2\phi \end{array}$$

In both cases, nuclear spin-polarized spectra were detected only in the methylene group protons in the unsymmetrical radical association product, i.e. the one formed by rapid combination of the newly formed radicals within the solvent cage. The transition from the radical pair, initially in a triplet state, to the singlet molecular product produces an over-population in one of the nuclear spin states. Closs and Trifunac[142b] established that over-populations also result during the combination of radical pairs initially formed in a singlet state (for example, through hydrogen atom abstraction by an excited singlet molecule), but the nuclear spin polarization is now different. This should allow the technique to be used to establish the electronic spin state of the precursor to the radical pair.

3.3b Results and discussions

(i) Cis–trans *isomerization in substituted olefins*
The long progressions of diffuse bands that are superimposed on the broad ultraviolet absorption continua of ethylene reflect the large geometrical changes that follow electronic excitation. Their diffuseness makes interpretation of the spectra difficult but there is general agreement that all the lower-lying transitions are associated with excitation of electrons out of the C–C bonding π-orbital, i.e. the one which keeps the molecule planar (see the excellent discussion of Merer and Mulliken[143]). When the electron is promoted into the antibonding π-orbital, the molecule can relieve interelectronic repulsion by twisting into a 90° conformation about the C–C axis, while the C–C distance increases to a length near that of a single σ-bond. The average vibrational spacings in the absorption

Very weak diffuse bands, 350–270 nm $\quad \tilde{a}^3B_{1u}$

$\tilde{X}^1A_{1g}$

Intense absorption, reaching a max. at $\quad \tilde{A}^1B_{1u}$
162 nm, diffuse vibrational structure
from 215–175 nm

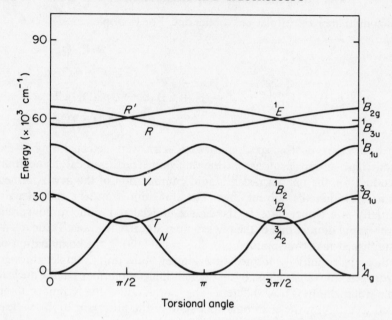

Figure 3.15 Potential energies of the lower-lying electronic states of ethylene as a function of the torsional angle about the C–C axis. (Taken from A. J. Merer and R. S. Mulliken, *Chem. Revs.*, **69**, 639, (1969).)

associated with the two $\pi \rightarrow \pi^*$ transitions (see pp. 49–51) lie in the range 850–1,000 cm^{-1}. Although appropriate to a σ_{C-C} stretching frequency, the structure is complicated by strong coupling between the stretching and torsional modes and in C_2D_4, where the bands are much sharper, the spacings are quite irregular. It is also possible that the CH_2 groups adopt a pyrimidal conformation when the molecule twists (cf. N_2H_4). Since the pair of $2p_\pi$ orbitals become degenerate in the 90° conformation, the '90° ground state' is a triplet, and the energy of the $^3B_{1u}$ level falls below that of the 'ground' singlet state ^1A_g. This is shown in Figure 3.15, which also includes the angular dependence of the energies of some neighbouring electronic levels. R and R' represent Rydberg states, in which an electron has been excited into orbitals of large volume mainly composed of $3s$ a.o.'s centred on the carbon atoms.

Because of the large change in equilibrium geometry, the 'vertical' $^1(\pi \rightarrow \pi^*)$ transition lies at short wavelengths, and produces a highly energized molecule. The origin of the associated absorption band probably lies at an energy ~ 447 kJ mole^{-1} (because of the vanishingly small Franck–Condon factor it cannot be located precisely), while the vertical transition requires a further 293 kJ mole^{-1}. In the vapour phase the majority of excited molecules dissociate producing both atomic and molecular hydrogen as primary photoproducts,[144] although some *cis–trans* isomerization has been observed after vacuum ultraviolet irradiation of CHD=CHD.† Isomerization would be expected to have a much greater chance of competing against dissociation in the less energetic $^3(\pi,\pi^*)$ state. One way of controlling the excess vibrational energy concentrated into the triplet olefin is to populate it by energy transfer from a series of excited triplet photosensitizers. We have already noted that substitution of $Cd(5^3P_1)$ for $Hg(6^3P_1)$ virtually eliminates sensitized photodissociation in CHD=CHD (section 3.1b). Alternatively, the photoisomerization can be effected in solution, where vibrational energy transfer is rapid and the solvent is a very efficient heat sink. Direct excitation into the triplet state becomes a possibility when the transition is intensified by a high pressure of dissolved oxygen. This technique has been used[145] to study *cis–trans* isomerization in 1,2-dichloroethylene, and is discussed below. For illustration, we also discuss the photoisomerization of azoalkanes and stilbene.

Liquid *cis* and *trans*-CHCl=CHCl are virtually transparent at wavelengths > 250 nm, but when saturated with O_2 at pressures ⩽ 130 atmospheres, both develop an intense absorption in the range 300–400 nm. Grabowski and Bylina[145] discovered that irradiation at 313 nm and 366 nm interconverts the two isomers with a quantum efficiency,

$$\phi_{cis \to trans} = 0.45 \pm 0.06 \quad \text{and} \quad \phi_{trans \to cis} = 0.61 \pm 0.07$$

Within experimental error the combined quantum efficiency is unity. It was concluded that both isomers are excited into a common triplet state which rapidly relaxes into a twisted equilibrium conformation and then decays through radiationless intersystem crossing into either of the two possible isomers. The S_0–T_1 energy gap will be slightly smaller for the less thermodynamically stable *cis*-isomer, which may account for the slight preference for decay into the *cis*-conformation. It is quite common for photochemical transformations to lead to the least thermodynamically stable products, and we have seen that radiationless transitions are favoured by a small energy gap and the least change in equilibrium geometry (section 2.5c). *cis*-CHCl=CHCl is thought to be slightly nonplanar in its equilibrium conformation.

Azo-alkanes, RN=NR, are iso-electronic with alkyl substituted butenes, RCH=CHR. Both exist in *cis* or *trans* conformations but the ethylenic

---

† The large change in geometry also prevents luminescent decay, because of the very poor Franck–Condon overlap between the planar and twisted conformations.

σ_{C-H} orbitals are replaced by two non-bonding m.o.'s centred on the nitrogen atoms. The two orbitals are probably almost degenerate since they are unlikely to overlap appreciably. A weak $^1(n \rightarrow \pi^*)$ absorption band ($\varepsilon \simeq 10 \,l \,mole^{-1} \,cm^{-1}$), centred around 340–350 nm, now complements the intense $^1(\pi \rightarrow \pi^*)$ absorption band lying at wavelengths < 210 nm. Both bands are completely continuous and devoid of structure, and at low pressures in the gas phase the azo-alkanes photodissociate with near unit quantum efficiency.

$$RN{=}NR + hv \rightarrow 2R\cdot + N_2$$

If the azo-alkane is diluted with CO_2 or some other chemically inert foreign gas, the quantum efficiency of dissociation falls and a proportion of the excited molecules suffer *cis–trans* isomerization instead. From the pressure dependence, Steel and Milne[146] have established that azoisopropane excited at 366 nm, has a dissociative lifetime $> 5 \times 10^{-10}$ sec in the excited state, remarkably long in view of the structureless absorption band. In solution the quantum efficiency of dissociation falls to zero, while $\phi_{cis \rightarrow trans} \simeq \phi_{trans \rightarrow cis} \simeq 0.5$. No fluorescent emission can be detected either in the gas phase or in solution, indicating rapid intersystem crossing from the $^1(n,\pi^*)$ state initially populated, presumably into a neighbouring $^3(\pi,\pi^*)$ level (see discussion on p. 189). By analogy with ethylene, this level should be common to both isomers, with a '90° equilibrium conformation', and its subsequent decay should lead to the ground states of either *cis* and *trans*-azo-alkane with near equal probability (cf. 1,2-dichloroethylene). In contrast, the $^3(n,\pi^*)$ levels are likely to remain planar, or nearly so, since they still have two electrons occupying the N–N π-bonding m.o.; distinct *cis* and *trans*-$^3(n,\pi^*)$ states, which still retain a potential barrier to rotation about the N–N axis, are to be expected.

In support of this model, Steel and Milne[146] discovered that although triplet benzophenone and benzaldehyde are efficiently quenched by azo-alkanes, triplet energy transfer does not lead to isomerization. On the other hand, singlet energy transfer from excited naphthalene does lead to isomerization, and the sensitized molecules behave in much the same way as those excited by direct absorption. It was concluded that triplet energy transfer populates the two $^3(n,\pi^*)$ levels, both of which lie below a common $^3(\pi,\pi^*)$ state, but that neither $^3(n,\pi^*)$ level is accessible through intersystem crossing from the $^1(n,\pi^*)$ states. Figure 3.16 summarizes the suggested photochemical scheme; like many another it is plausible and accommodates the experimental observations, but like all postulates in science it should be treated with constructive scepticism, as a hypothesis to be measured against future observation.

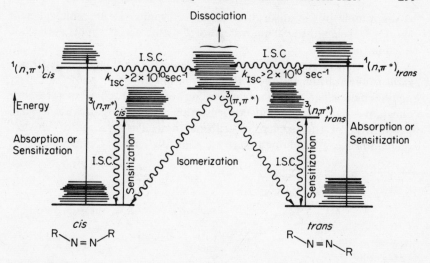

Figure 3.16 Photochemistry of azo-alkanes: schematic state diagram. (After C. Steel and G. S. Milne, ref. 146.)

In stilbene ϕCH=CHϕ, the central ethylenic π-system can become conjugated with the aromatic π-orbitals, and it is not immediately obvious that the central bond remains 'ethylenic'. Rest assured that it does, at least in the ground electronic state, because the activation energy required for thermal *cis–trans* isomerization is ~ 167 kJ mole^{-1}. The highest occupied and lowest vacant π-m.o.'s in stilbene are represented in Figure 3.17. The occupied m.o. is bonding between the central carbon atom but in the excited orbital they are separated by a nodal plane, and just as in ethylene the first $\pi \rightarrow \pi^*$ transition would be expected to cause rotation about the central C–C bond. In both isomers the first $^1(\pi \rightarrow \pi^*)$ absorption bands (measured in solution) are very broad continua, extending from 240–350 nm, and shifted to the red of both the ethylenic and aromatic chromophores by the π-electron conjugation. A progression of diffuse vibrational bands is superimposed on the continuum in *trans*-stilbene but no structure

Highest occupied π-orbital   Lowest vacant π-orbital

Figure 3.17 Molecular orbitals in *trans*-stilbene.

is discernible in the absorption of the *cis*-isomer. Its maximum lies at shorter wavelengths (280 nm instead of 294 nm), and it almost about one third as intense.†

Promotion of either isomer into the lowest $^1(\pi,\pi^*)$ state leads to *cis–trans* isomerization, and prolonged irradiation leads to a photostationary equilibrium where the relative concentrations of each isomer reach a constant value.[148] The nett concentration slowly falls because a small proportion of the excited *cis*-stilbene molecules cyclize to dihydro-phenanthrene.[149] Although there are some dissenters,[150] the majority of

research publications favour a triplet mechanism for the *cis–trans* interchange. A chart which summarizes many of the observations is shown in Figure 3.18; some of the results on which it is based are as follows:

(i) *cis–trans* isomerization can be promoted by energy transfer from triplet photosensitizers.[151] The relative photostationary equilibrium concentrations depend on the relative rates of energy transfer to each isomer, but if both direct and sensitized isomerizations proceed through a common intermediate, the two different equilibrium values should be directly related. This has been experimentally verified.[152]

(ii) Although neither isomer is phosphorescent, their lowest $^3(\pi,\pi^*)$ states can be located from the $S_0 \rightarrow T_1$ absorption bands induced by oxygen, or heavy atom perturbation.[153] Since it is not possible to identify the spectral origin with any certainty when there is a large change in the molecular geometry, the energies of the potential minima in the triplet states are uncertain.

(iii) Photosensitized isomerization in the sense *cis* $\rightarrow$ *trans* continues to be effective even when the energy of the triplet sensitizer falls below the minimum 238 kJ mole^{-1} needed to excite the $S_0 \rightarrow T_1$ absorption band of *cis*-stilbene. Hammond[152] has suggested that 'non-vertical' energy transfer, produces a non-planar triplet molecule, (see p. 213).

(iv) It is likely that the lowest triplet level reaches a potential minimum when the molecule is twisted through approximately 90° about the central

---

† The difference is caused by steric crowding in *cis*-stilbene, which twists the aromatic rings out of the molecular plane. Rotation about the α_{C-C} bonds stabilizes the lower orbital and shifts the excited orbital to higher energy.[147]

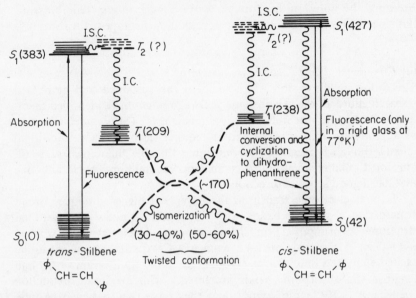

Figure 3.18 Photo-isomerization of stilbene: schematic state diagram. (Energies are given in brackets, in kJ mole^{-1}; they refer to the near planar *cis* or *trans*-conformations. The states T_2 are conjectural). (After D. Gegion, K. A. Muszkat and E. Fischer, ref. 148.)

bond, but it is also possible that at least one other potential energy minimum exists, probably in a transoid conformation of the triplet molecule. In the *trans*-isomer the $S_0 \rightarrow T_1$ transition shows a progression of diffuse vibrational maxima; the absorption band of *cis*-stilbene is structureless. A transient absorption spectrum photographed during flash photolysis of frozen solutions of *trans*-stilbene has been attributed to the transoid triplet, stabilized in the rigid medium at low temperature.[154] The observations are also supported by MO calculations of the changes in orbital and electronic state energies caused by rotation about the central C–C bond.[155]

(v) There are no experimental data to locate the excited triplet levels T_2, but MO calculations suggest they lie close to S_1. A small activation energy, ~ 8–12 kJ mole^{-1}, for the unsensitized photoisomerization of *trans*-stilbene may be associated with the rate of intersystem crossing from S_1 into a slightly elevated triplet level T_2.[148] This hypothesis is supported by the photochemical behaviour of the isomer in rigid glassy solution at 93°K. In brief, its photochemical behaviour virtually disappears! The quantum efficiency of isomerization falls to 0·006 while the fluorescence efficiency increases to 0·75, compared with 0·08 in a liquid solution at 30°C.

Freezing the solution has obviously prevented intersystem crossing, but since T_1 lies well below S_1, it suggests that the impeded transition is $S_1 \rightsquigarrow T_2$.

(ii) Conjugated dienes: valence isomerization

The following discussion owes much to the sustained and combined research efforts of those who some years ago might have identified themselves as organic, theoretical or photochemists. The detailed charting of a photochemical transformation requires each of their traditional talents, though it is as well to remember that the theoretical cart still runs most reliably when pulled by an experimental horse, preferably one that has a good sense of direction!

The first electronic transition in conjugated dienes produces broad structureless absorption bands, with very shallow 'ripples' of vibrational structure superimposed on the continuum. Their average spacing $\sim 1440 \text{ cm}^{-1}$ is appropriate for a symmetric $C{=}C$ stretching mode in the excited state, and implies a marked change in the equilibrium C–C bond lengths. The absorption bands are intense with maximum extinction coefficients $\sim 10^4 \text{ l mole}^{-1} \text{ cm}^{-1}$ and they have been assigned to fully allowed ${}^1(\pi \rightarrow \pi^*)$ transitions (see below). The simplest compound, 1,3-butadiene has its absorption maximum at 210 nm but increasing substitution shifts the maximum to longer wavelengths. A set of empirical 'rules' for estimating the magnitude of the shifts was developed by Woodward, Fieser and Fieser and others[156] as an aid to identifying the molecular structures of steroidal dienes and trienes from their ultra-violet absorption spectra. There is a particularly large shift when the 1,3-diene is held in a cis-conformation by incorporation in a ring system. For example, 1,3-cyclohexadiene has a maximum absorption at 257 nm.

The nodal surfaces of the four intravalency shell π-m.o.'s in butadiene are shown in Figure 3.19. Their electronic configuration in the ground electronic state is $(\pi_1)^2(\pi_2)^2$ and there can be little π-bonding between the central carbon atoms, although unexcited 1,3-butadiene evidently favours the trans-conformation (described as s-trans since it involves rotation about an essential single bond). In the first excited state, where the electronic configuration is $(\pi_1)^2(\pi_2)^1(\pi_3)^1$ the potential barrier to rotation about the central bond should rise, while the barriers to rotation about the terminal C–C bonds should fall. In view of these prognostications it is not surprising to find that irradiation of a solution of trans,trans-2,4-hexadiene (I) in its first ${}^1(\pi \rightarrow \pi^*)$ absorption band leads to a photostationary state in which there are equal concentrations of the cis,cis (III), cis,trans (II) and trans, trans isomers.[157]

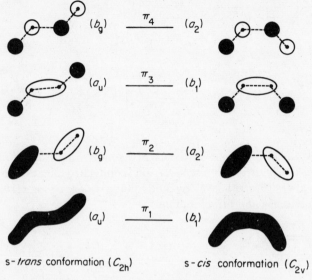

Figure 3.19 Nodal surfaces of the π-m.o's in butadiene.

The isomerization very probably takes place within the excited singlet state since conjugated dienes do not phosphoresce and there is no evidence of any measurable rate of intersystem crossing into the lowest $^3(\pi,\pi^*)$ level. The level is readily populated through triplet energy transfer, however, since the 'vertical' $S_0 \rightarrow T_1$ transition† absorbs an energy of only 210–250 kJ mole^{-1} (cf. the 500–590 kJ mole^{-1} needed to excite the $S_0 \rightarrow S_1$ transition), and photosensitization experiments in solution have shown that efficient cis–trans isomerization also occurs in the lowest triplet level.[151] When the diene is sufficiently concentrated some of the triplet dienes react with unexcited molecules to form cyclic dimers.[151] For example, 1,3-butadiene gives two isomeric, divinyl cyclobutanes and a vinyl cyclohexene following the initial formation of dimeric, biradical intermediates.

† Seen in absorption under oxygen perturbation.

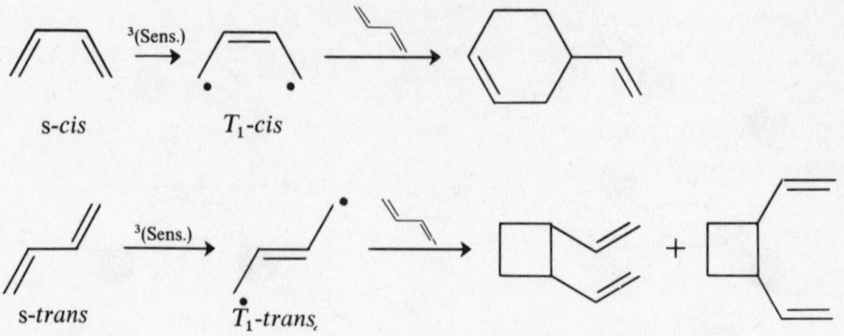

s-cis T_1-cis

s-trans T_1-trans

The behaviour of molecules in the excited singlet state is quite different. As well as the *cis–trans* interconversion, conjugated dienes can escape from the lowest $^1(\pi,\pi^*)$ level by undergoing valence isomerization to yield cyclobutenes and bicyclobutanes.[158]

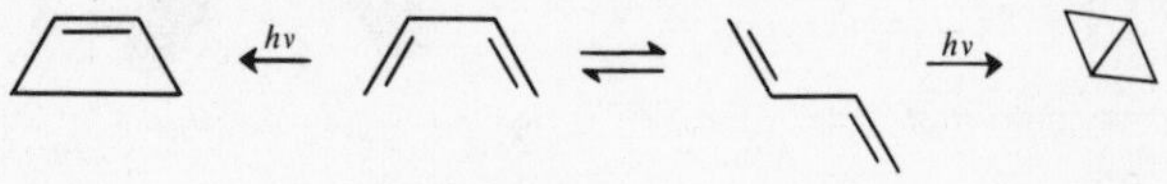

They can also fragment into radical and molecular products but only in the vapour phase. In solution, excess vibrational energy is so rapidly conducted away that the fragmentation path is closed. The valence isomerization into a cyclobutene is the simplest case of a general class of intramolecular reactions termed *electrocyclic* by Woodward and Hoffmann. They all involve the formation of a single bond between the terminal atoms of a linear system containing $k\pi$-electrons or the converse ring opening. In a 1,3-diene, $k = 4$; in the cyclization of *cis*-stilbene to dihydrophenanthrene, or of 1,3,5-hexatriene to 1,3-cyclohexadiene $k = 6$ and so on. The remarkable feature of the cyclization is their strict stereospecificity. For example, the photocyclization of *trans,trans*-2,4-diene leads exclusively to the *cis*-3,4-dimethylcyclobutene: the *trans*-isomer is not formed.[157] On the other hand thermal ring opening of *cis*-3,4-dimethylcyclobutene leads exclusively to the *cis,trans* isomer of hexa-2,4-diene.[159]

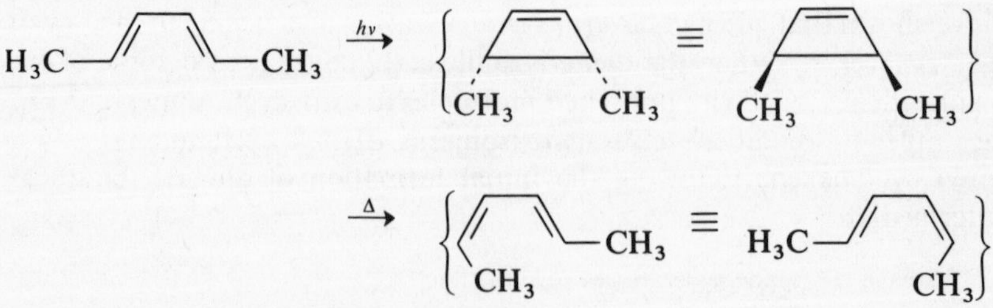

As the photocyclization proceeds the two methyl groups rotate out of the molecular plane in a *disrotatory* fashion, i.e. they move in opposite senses, while the thermal ring opening involves rotation in the same sense, i.e. their motion is *conrotatory*. In 1965, Woodward and Hoffmann,[160] and Longuet-Higgins and Abrahamson[161] developed theoretical treatments which could rationalize the strictly observed stereochemical control. They were based on the requirement that orbital symmetry be conserved during a concerted reaction, i.e. one which proceeds 'smoothly' in a single step; the m.o.'s of the product have to correlate with those of the reactant. Within

disrotatory motion   conrotatory motion

the space of three years or so, their ideas were expanded, tested, re-formulated and codified into sets of predictive rules appropriate to particular types of reaction (for example, electrocyclic) occurring through concerted mechanisms. They are summarized in an excellent review by Woodward and Hoffmann[162] published in 1969, to which the reader is referred. It lists references to no less than eleven other reviews and demonstrates the tremendous impact which the notion of orbital corre-lation has had, and continues to have, in mechanistic organic chemistry. To illustrate the method, consider the conversion of butadiene to cyclo-butadiene.

The four intravalency shell π-m.o.'s in butadiene were represented in Figure 3.19. After cyclization they must transform into the four m.o.'s represented in Figure 3.20. Although both molecules belong to the C_{2v} point group their symmetry must be lowered during the process of cycliza-tion or ring opening. Conrotatory motion leaves only the two-fold axis, disrotatory motion leaves the vertical plane, and the resulting orbital correlation diagram is as shown in Figure 3.20. In a concerted process the orbital symmetries must not change. The orbitals are said to transform adiabatically from those of one molecule into those of the other. We have already met a similar type of diagram in section 1.3c where Walsh's diagram (Figure 1.20) was used to discuss the changes in the m.o.'s of a linear triatomic molecule as the bond angle altered from $180°$ to $90°$. On the basis of Figure 3.20, Longuet-Higgins and Abrahamson made the following predictions,

(i) Thermal cyclization should occur in the *conrotatory* sense, since the ground state electronic configuration of butadiene $(\pi_1)^2(\pi_2)^2$ correlates

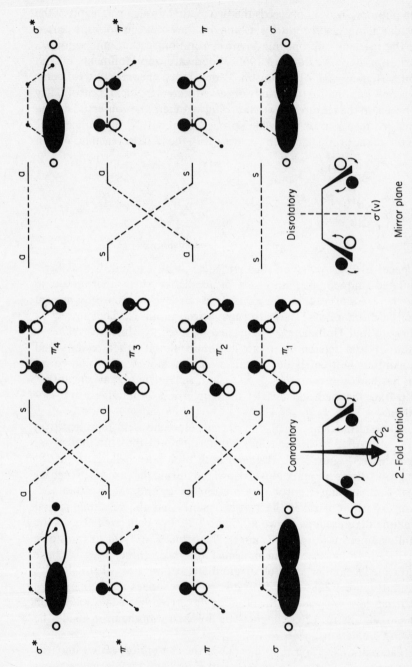

Figure 3.20 Schematic diagram showing orbital correlation in the cyclization of butadiene. 's' and 'a' indicate symmetric or anti-symmetric behaviour under rotation about the two-fold axis (conrotatory motion), or mirror plane (disrotatory motion).

with the ground electronic state of cyclobutene $(\sigma)^2(\pi)^2$. Disrotatory motion correlates with the excited configuration $(\sigma)^2(\pi)^1(\pi^*)^1$; it would only be possible to reach the ground state through a symmetry-forbidden radiationless transition along the reaction path, and even if this did occur, the cyclization would probably be impeded by a large potential barrier. The arguments apply with equal force, to the reverse reaction and we have seen that the thermal ring opening of 3,4-dimethyl butene does proceed in a conrotatory fashion.

(ii) Photocyclization from the lowest excited state should be allowed under *disrotatory* motion (again as observed) since this permits correlation of the excited configuration $(\pi_1)^2(\pi_2)^1(\pi_3)^1$ with the singly excited electronic configuration of the cyclobutene. Conrotatory motion leads to the doubly excited configuration $(\sigma)^2(\pi)^1(\sigma^*)^1$.

(iii) The correlation diagram implies that the crucial factor controlling the stereochemistry of the products is the symmetry of the highest occupied orbital; for one type of motion its energy rises, for the other it falls (but see below). Similar types of argument can be developed for the cross-cyclization of 1,3-dienes to bicyclobutanes, assuming a concerted reaction mechanism. The assumption can be checked by examining the stereochemical structure of the products.

The model is elegant and its predictions are accurate, but in the absence of quantitative data it must beg a number of questions about the detailed reaction mechanism. Does the cyclization occur in the first excited state or do the potential energy surfaces of the first excited singlet state of butadiene and the ground singlet state of cyclobutene intersect at some intermediate nuclear configuration(s)? Is there an intersystem crossing, perhaps into a lower lying triplet state somewhere along the reaction path? How do the orbital and nett potential energies of the molecule change as the nuclear configuration of butadiene is distorted towards that of cyclobutene? One might conclude from Figure 3.20 (fallaciously) that the excited singlet level falls, but the conclusion is quite unjustified since the figure is purely a correlation diagram. It has no coordinates and there is no geometric or energetic scale. More data are needed to plot the likely reaction path. They have been provided by Van-der-Lugt and Oosterhoff.[163]

Using the valence bond treatment they have calculated the potential energies of the ground and low-lying electronic states of s-*cis* butadiene and cyclobutene as a function of the interbond angle for disrotatory and and conrotatory motion of the terminal groups. They neglect the triplet states, since Srinivasan[158] has found that neither O_2 nor NO are able to quench the unsensitized photochemical reactions of excited butadiene, and there is no evidence that intersystem crossing is able to compete against alternative radiationless decay processes. The results of the calculations are illuminating. As the bond angle falls from 120° towards 90°, the ground state shows a large potential barrier toward disrotatory motion but only a small barrier for motion in the opposite sense. Thermal cyclization should be conrotatory, in agreement with the prediction based on orbital symmetry. The first excited singlet state is found to have the electron configuration $(\pi_1)^2(\pi_2)^1(\pi_3)^1$ as expected, but as the ring closes its energy rises continuously toward the more elevated level in cyclobutene. Its potential curve cannot intersect that of the ground state, after all. Instead, 'the driving force' toward the photoexcited disrotatory ring closure is provided by a deep minimum in the potential energy of the second excited singlet state. Its potential curve rapidly drops below that of the first, almost intersects the potential maximum in the ground state and then climbs back to its rightful place as the second excited state of cyclobutene. There is a simple reason for the near intersection. The second excited singlet has in large measure the electronic configuration $(\pi_1)^2(\pi_3)^2$; both orbitals are doubly occupied and the state is totally symmetric $^1A_1$. So is the ground state, since it has the closed shell configuration $(\pi_1)^2(\pi_2)^2$. The barrier in the potential curve of the ground state is associated with the 'attempted' intersection of two states of the same symmetry (see p. 131).

Van-der-Lugt and Oosterhoff[163] conclude that photocyclization must involve a radiationless transition into the second excited singlet state, and follows the path

$$B(S_0) \xrightarrow{hv} B(S_1) \rightsquigarrow B(S_2)_{vib} \rightsquigarrow B(S_2) \underset{\searrow}{\overset{\nearrow}{}} \begin{array}{l} B(S_0)_{vib} \rightsquigarrow B(S_0) \\ c\text{-}B(S_0)_{vib} \rightsquigarrow c\text{-}B(S_0) \end{array}$$

B and c-B represent the diene and the cyclobutene. The possibility of direct excitation from $S_0 \rightarrow S_2$ is excluded since the transition is forbidden. In general, if a thermal conrotatory (or disrotatory) process has a large potential barrier, it must be due to an avoided intersection with an excited state of the same symmetry and the excited state must have a potential minimum (see Figure 2.24). The photoinitiated process becomes favoured when the thermal process is not. This must be true of all electrocyclic reactions, and since the ground state is totally symmetric, the photochemical reaction pathway has to involve a doubly excited electronic state, i.e. one in which all the orbitals are doubly occupied.† The final result is the same as predicted on the basis of orbital symmetry correlation

---

†In a closed shell molecule, the first excited state can not have the same symmetry as the ground state.

but for different reasons. It is not the symmetry of the highest occupied orbital in the excited diene which is responsible for the stereochemical control of the photochemical reaction, but a kind of 'feedback' from the orbital symmetry control of the thermal reaction.

(iii) *Carbonyl compounds*

When the terminal methylene group in an olefin is replaced by an oxygen atom, a localized non-bonding orbital is introduced at a level lying between the highest occupied and lowest vacant π-m.o.'s (see Figure 3.21). The lowest excited singlet state is reached by $n_O \rightarrow \pi^*$ excitation, which transfers an electron from an orbital localized around the oxygen atom to one delocalized in a π-system. The $^3(n,\pi^*)$ state of simple aliphatic and aromatic ketones resembles an alkoxy or aryloxy radical, and if the excited molecule does not dissociate it can undergo typical free radical reactions at the electron deficient oxygen atom (for example, inter- or intramolecular hydrogen atom transfer—see p. 214). The $^1(n_O \rightarrow \pi^*_{C-O})$ transition in formaldehyde ($\lambda_{max} \sim 280$ nm) absorbs an energy some 210 kJ mole^{-1} less than the $^1(\pi \rightarrow \pi^*)$ transition in ethylene, but conjugation of the two π-systems in an $\alpha\beta$-unsaturated aldehyde or ketone, splits the π-m.o.'s. Conjugation lowers the energy of both the $^1(n_O \rightarrow \pi^*_3)$ and $^1(\pi_2 \rightarrow \pi^*_3)$ transitions, but the change is more marked in the latter since conjugation cannot affect the non-bonding orbital lying in an orthogonal plane. Acrolein ($CH_2{=}CHCHO$), like butadiene, adopts an s-*trans* conformation in its ground electronic state, and the $^1(n_O \rightarrow \pi^*_3)$ transition transfers double bond character from the terminal bonds to the central bond of the $-C-C-C-O$ chain. When the chain is extended the weak $^1(n_O \rightarrow \pi^*)$ absorption 'disappears' under the much more intense $^1(\pi \rightarrow \pi^*)$ band.[164] In $\beta\gamma$-unsaturated ketones, the ethylenic and carbonyl chromophores are separated by a methylene group, but there is frequently a considerable intensification of the $^1(n_O \rightarrow \pi^*)$ absorption. The intensity is 'borrowed' from an intramolecular charge transfer transition, in which electronic charge migrates in the excited state, from the $>C{=}C<$ bonding π-orbital to the antibonding $>C{=}O$ π-orbital.[164] Intramolecular charge transfer states have also been identified in aromatic ketones which have an electron donating substituent in one of the rings (see p. 189), but in these the electronic charge is redistributed solely within the π-system. The state designated 'C–T' by Porter and Suppan in a molecule such as *p*-amino benzophenone, probably involves electron transfer from the lone pair orbital centred on the nitrogen atom into the aromatic π-system, i.e. it is populated through a $^1(l \rightarrow a_\pi)$ transition (see pp. 143–144). In strongly acidic media the nitrogen atom is protonated and the $^1(l \rightarrow a_\pi)$ absorption band disappears.

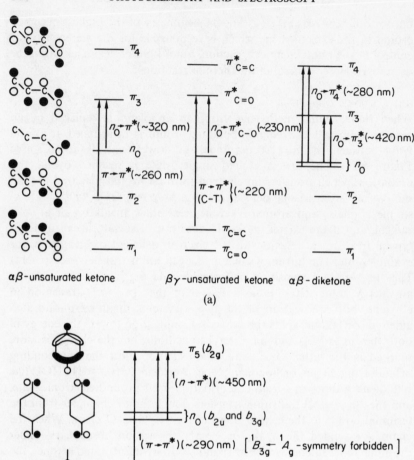

Figure 3.21 Electronic orbitals and low-lying transitions in carbonyl compounds
(a) unsaturated, and $\alpha\beta$-dicarbonyl compounds. (b) p-Benzoquinone.

When there are two adjacent carbonyl groups (for example, in glyoxal or biacetyl), their π-orbitals conjugate to give orbitals of the same form as those in an $\alpha\beta$-unsaturated aldehyde or ketone, but the lowest $^1(n \rightarrow \pi^*)$ transition, $^1(n_O \rightarrow \pi_3^*)$, shifts into the visible: biacetyl is a yellow liquid (with a rather nauseating smell of concentrated margarine which seems to permeate everything). Glyoxal is known to retain a planar *trans*-conformation following $^1(n_O \rightarrow \pi_3^*)$ excitation,[90] and there can be little interaction between the two non-bonding orbitals. This seems to be generally true. A second, equally weak absorption centred around 280 nm is due to the $^1(n_O \rightarrow \pi_4^*)$ transition. *p*-Benzoquinone also has weak absorption maxima at 450 nm and 290 nm; the first is associated with the lowest $^1(n_O \rightarrow \pi_5^*)$ transition, but the other has been assigned to the $^1(\pi_4 \rightarrow \pi_5^*)$ transition, forbidden in the centrosymmetric molecule by the g ↮ g selection rule (see Figure 3.21(b)).

The great majority of carbonyl compounds leave the lowest excited singlet state by intersystem crossing into a neighbouring triplet level. Many are phosphorescent in a rigid medium but they rarely fluoresce, and even when they do the fluorescence efficiency is usually small. Any photochemical change that follows excitation is very likely to have been initiated in the lowest triplet state. The generalization can break down occasionally, for example when light absorption carries the molecule into heavily predissociated vibronic levels of the $^1(n,\pi^*)$ state. When acetaldehyde absorbs light at 254 nm, it is excited into vibronic levels lying 130 kJ $mole^{-1}$ above the potential minimum of its $^1(n,\pi^*)$ state. In the vapour phase most of the excited molecules predissociate into molecular products.[165]

$$CH_3CHO \xrightarrow{254\ nm} CH_4 + CO; \quad \Delta H = -21\ kJ\ mole^{-1}$$

(Note that the primary products carry away ~ 493 kJ $mole^{-1}$!) Molecules excited at 334 nm carry 17 kJ $mole^{-1}$ in vibration. They still predissociate, but only after intersystem crossing into the $^3(n,\pi^*)$ state, and the primary products are free radical instead of molecular.[165]

$$CH_3CHO \rightarrow \cdot CH_3 + H\dot{C}O; \quad \Delta H = +309\ kJ\ mole^{-1}$$

The primary quantum yields have been estimated by adding radical scavengers such as NO or I_2. Unfortunately, the scavengers can also be effective in quenching the excited molecules, particularly those in the triplet state, so that the estimates have to be taken with caution.

It is almost always the case that the lowest excited singlet state has $n-\pi^*$ character. The first $^1(\pi,\pi^*)$ state usually lies some way above it, but the order can be reversed when the carbonyl group is bonded to a large aromatic ring system, or when the molecule is dissolved in highly polar or hydrogen-bonding solvents (see p. 189). Because of the greater singlet–triplet separation in the $\pi-\pi^*$ states, the lowest $^3(n,\pi^*)$ and $^3(\pi,\pi^*)$ states usually lie much closer together and their order is much more easily reversed by perturbation from the solvent or substituent groups. Bimolecular reactions of excited carbonyl compounds such as hydrogen atom abstraction or oxetane ring formation, are typical of their $^3(n,\pi^*)$ electronic states.

$$>C{=}O[^3(n,\pi^*)] + HR \rightarrow >\dot{C} - OH + R{\cdot};$$

If the character of the lowest triplet level is predominantly $^3(\pi,\pi^*)$ the carbonyl group loses its 'biradical' character and the reactions are very inefficient.

Chemical reactions involving *excited* triplet states have been proposed by a number of researchers. Although internal conversion to the lowest triplet level is likely to be very rapid, a very efficient chemical process might intercept some of the excited molecules before they relax. The lowest excited singlet and triplet states in 9-anthraldehyde both have (π,π^*) character. Irradiation at >410 nm produces a dimer, bridged at the 9,10 positions (cf. anthracene p. 214), a reaction characteristic of the $^1(\pi,\pi^*)$ state, but excitation in the presence of 2-methyl butene-2 at <410 nm produces an oxetane ring adduct, a reaction typical of $^3(n,\pi^*)$ states.[166] Dimerization does not compete.

The results suggest that intersystem crossing into the excited $^3(n,\pi^*)$ level may not occur until the absorbed light has a wavelength <410 nm. The cycloaddition of cyclopentenone to olefins which produces a cyclobutane derivative rather than an oxetane ring system may also involve the reaction of the ketone in an excited triplet state (T_2).[167] For example

The possible chemical consequences of electronic excitation in aldehydes and ketones are many and varied; they are also sensitive to environmental changes which can disturb the delicate balance between competing primary steps. In the vapour phase the dominant primary chemical process is always a strong function of the vibrational energy content of the excited molecule, and its quantum efficiency depends on the absorbed wavelength, the pressure and the temperature. Unimolecular processes may have to compete against electronic energy transfer or bimolecular chemical reaction of the excited molecule. In condensed media, dissociation into free radicals is inhibited by the cage effect, by the rapid transfer of vibrational energy and possibly by changes in the rates of radiationless transitions that lead to predissociation or to internal conversion. Both the electronic character of the lowest excited state and the nature of the photochemical change can be altered by the solvent polarity. The photochemistry of carbonyl compounds often seems a speculative science when mechanistic details come to be discussed. Ideally mechanistic schemes should be based on kinetic and quantum yield measurements and experimental variables should be altered one at a time while the others are held constant; however, this can take many months and when photochemists discover something new many of them find it difficult to control their impatience to communicate the discovery as soon as possible. As a result some suggested 'mechanisms' are based on very inadequate evidence.

The main photochemical reactions of carbonyl compounds include,

(i) Predissociation at an α_{C-C} bond, (originally classified by Norrish[168] as the Type I process),

(ii) Transfer of a γ–hydrogen atom to the carbonyl oxygen, followed by cleavage of the β_{C-C} bond (Norrish Type II process) or ring closure to form a cyclobutanol,

(iii) α-cleavage together with an intramolecular hydrogen atom transfer (Norrish Type III process),

(iv) Photoenolization, for example

(v) Photoreduction by intermolecular hydrogen atom (or sometimes electron) transfer, for example

(vi) Addition across a π-bond (both inter- and intramolecular), for example

(vii) Molecular rearrangements and isomerizations, for example*

Many examples of each type of reaction are known: our discussion of specific carbonyl compounds will include a very small selection of them.

Alkyl aldehydes and ketones

α-Cleavage of simple aldehydes and ketones in the vapour phase has been identified in both the $^1(n,\pi^*)$ and $^3(n,\pi^*)$ states but the relative quantum efficiencies of dissociation, intersystem crossing, internal conversion and so on, are all strongly dependent on the vibrational energy of the excited molecule. They vary with the absorbed wavelength, pressure and temperature. For example, when acetone is irradiated at 313 nm, in the presence of triplet energy acceptors such as biacetyl (which also accumulates as a reaction product), or cis-butene-2, its phosphorescence is completely quenched and its photodecomposition almost stops.[170] A weak fluorescence still glimmers through. In the absence of quenching the majority of excited molecules reach the triplet state and either dissociate or return to the ground state; relatively few can dissociate in the $^1(n,\pi^*)$

---

* The mechanistic schemes are due to Zimmerman.[169] The valence bond structures are representations of the possible electronic distributions at intermediate stages along the reaction path. They are an attempt to represent graphically 'electronic wave functions' of the vibrating molecule and must not be taken too literally, since they are based on good chemical intuition rather than incontrovertible scientific fact.

state before intersystem crossing takes place. The competing steps are summarized in the chart on page 317.

At 254 nm, the fluorescence and phosphorescence efficiencies fall to zero, the $^1(n,\pi^*)$ absorption profile is completely continuous and the quantum efficiency of dissociation is probably near unity.[171] Dissociation is too fast at the shorter wavelength to permit competition from intersystem crossing or fluorescence and all the molecules dissociate from upper vibrational levels of the $^1(n,\pi^*)$ state. Reducing the wavelength (or increasing the temperature) also reduces the yield of biacetyl because of the secondary decomposition of vibrationally excited $CH_3\dot{C}O$ radicals.

$$CH_3COCH_3 \xrightarrow{h\nu} \cdot CH_3 + (CH_3\dot{C}O)_{vib} \rightarrow 2 \cdot CH_3 + CO$$

When acetone is irradiated in the vapour phase at 313 nm in the presence of HBr, the quantum yields of CO and C_2H_6 are reduced almost to zero but isopropyl alcohol, acetate and bromide are found among the final products. Instead of dissociating, the triplet acetone is chemically quenched by the HBr and a hydrogen atom is transferred to it,[172]

$$\begin{array}{c} CH_3 \\ \diagdown \\ CO[^3(n,\pi^*)] + HBr \rightarrow \\ \diagup \\ CH_3 \end{array} \qquad \begin{array}{c} CH_3 \\ \diagdown \\ \dot{C}-OH + Br\cdot \\ \diagup \\ CH_3 \end{array}$$

Hydrogen atom transfer is much more favoured in the liquid phase, where the rate of dissociation is greatly reduced. The e.s.r. spectra of $(CH_3)_2\dot{C}OH$ and $\cdot CH_2COCH_3$ radicals have been recorded during the ultraviolet irradiation of liquid acetone itself, indicating hydrogen atom transfer between excited and unexcited molecules.[173] Irradiation in isopropanol yields pinacol as a photochemical product.

Carbonyl compounds endowed with a γ–hydrogen atom readily undergo the molecular, 'Norrish Type II', elimination reaction, both in the liquid and vapour phases. The reaction probably involves the intermediate formation of a six-membered ring system, followed by dissociation into the enol form of a lower ketone, and an olefin. McMillan, Calvert and Pitts[174] observed transient infrared absorption from the enol form of acetone during photolysis of 2-pentanone†

$$\begin{array}{ccccc} CH_3 & O & & CH_3 & OH \\ \diagdown & \diagup\!\!\diagup & & \diagdown & \diagup \\ & C & H & & C \\ & | & | & \xrightarrow{h\nu} & \| \qquad + CH_2{=}CH_2 \\ H_2C & CH_2 & & CH_2 \\ & \diagdown\ \diagup \\ & CH_2 \end{array}$$

† Using a very long absorbing path in the vapour.

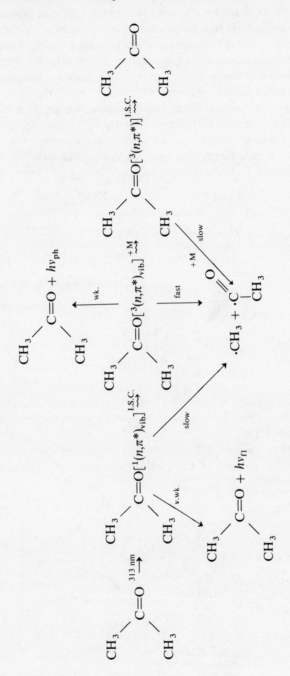

Yang[175] finds that the efficiency of transfer of the γ-hydrogen atom increases as the C–H bond changes from primary to secondary to tertiary. At the same time, the efficiency of intersystem crossing from the $^1(n,\pi^*)$ state is reduced (measured by quenching the triplet state with biacetyl or *cis*-piperylene). If the hydrogen atom is transferred sufficiently quickly, it can take place before intersystem crossing, while the excited molecule still occupies the singlet state, but otherwise, the reaction occurs in the $^3(n,\pi^*)$ state. The rate of hydrogen transfer increases as the strength of the C–H bond is reduced. γ-hydrogen atom transfer may also lead to cyclobutanol ring formation. For example, irradiation of the optically active aldehyde (I) leads to the products[176]

The optical activity is preserved in the cyclobutanol, indicating very fast ring closure, or possibly a concerted process which does not pass through a 'biradical' stage.

In methyl cyclopropyl ketone, the major primary photochemical process is an opening of the cyclopropyl ring. α-Cleavage is relatively inefficient.[49]

A plausible mechanism might include the following sequence (cf. the isomerization of the bicyclic ketone described on p. 315).

Alicyclic ketones

In the vapour phase $^1(n \rightarrow \pi^*)$ excitation into high vibrational levels leads to efficient decarbonylation; the reaction is especially fast in small ring systems and is not inhibited by radical scavengers or triplet quenchers such as O_2 or NO, for example

$$CH_2=CH_2 + CH_2=C=O \qquad \text{(minor process)}$$

The cyclobutane dione (I) fragments exclusively from the $^1(n,\pi^*)$ state, even in solution.[177] The quantum efficiency is ~ 1, and is not reduced by triplet quenchers.

(I)

An alternative reaction that leads to the formation of an ω-alkenal in five and six membered ring systems. For example

is relatively inefficient in the vapour phase because the competing fragmentation is too rapid, but in solution the situation is reversed. Molecular
rearrangement becomes the dominant photochemical process.[178] The
reaction is quenched by triplet quenchers such as piperylene, can be
photosensitized by triplet energy donors such as benzene, and must follow
intersystem crossing into the $^3(n,\pi^*)$ state. The most attractive mechanism
involves the rapid disproportionation of an intermediate biradical formed
through α-cleavage, producing either the alkenal or a ketene.

In the particular case of cylcohexanone, the ketene has not been detected
in its pristine state, but only after conversion to a carboxylic acid through
nucleophilic attack in an aqueous solvent.

A similar sequence accounts for the photoproducts obtained in solution
from the cyclic keto-ether, γ-butyrolactone[179]

The loss of CO_2 is unaffected by triplet quenchers and must occur in the
$^1(n,\pi^*)$ state, but again hydrogen atom transfer probably takes place after
intersystem crossing.

Aryl ketones
When the carbonyl group is bonded directly to an aromatic ring system
the α-cleavage reaction becomes very inefficient (if they are isolated from
each other α-cleavage is still efficient and in dibenzyl acetone, alternative

primary photochemical processes cannot compete). The main photo-chemical reactions involve hydrogen atom transfer, both inter- and intramolecular, or cycloaddition to unsaturated bonds. The intermolecular reduction of the carbonyl group has already been discussed.

Intramolecular hydrogen atom transfer in a long chain alkyl aryl ketone such as butyrophenone leads to the Norrish Type II process,[180]

For R = H or CH$_3$, the elimination reaction has a quantum efficiency of 0·4, but an amino substituent reduces it to zero. It also changes the character of the lowest triplet state from $^3(n,\pi^*)$ to $^3(\pi,\pi^*)$, since the phosphorescent lifetime increases by several orders of magnitude, con-firming the assignment of reactivity to the $^3(n,\pi^*)$ state.[180] Other sub-stituents can not only suppress the elimination, but replace it with entirely new photochemical processes. p-Bromobutyrophenone dissociates at the C–Br bond with a quantum efficiency of 0·25, and an o-hydroxy substituent introduces photoenolization as the primary photochemical reaction.[180] Photoenolization often leads to reversible photochromism. For example, o-methylbenzophenone develops an intense yellow colour on exposure to near-ultraviolet light, though the colour is probably due to a cyclized secondary product rather than the enol itself.[181]

Note that *cis–trans* isomerization accompanies the enolization. The hydrogen transfer takes place in the $^3(n,\pi^*)$ state of the ketone, and if there is no simultaneous intersystem crossing the enol will first appear in a triplet state, where the equilibrium geometry may well be twisted about the exocyclic double bond.[181]

The photoaddition of aryl (and alkyl) ketones across a $>C=C<$ bond is a general process, known to organic photochemists as the Paterno–Büchi reaction in honour of those who discovered it.[182] It probably involves the 'biradical' mechanism

though detailed 'mechanistic' arguments based solely on the relative yields of the alternative products, rather than kinetic measurements should be treated with scepticism. Cycloaddition may have to compete against triplet energy transfer, or even hydrogen atom transfer if the molecule under attack has a relatively weak C–H bond (for example, in an allylic group). Benzophenone ($T_1 = 293$ kJ mole^{-1}) readily adds to norbornene

but acetophenone ($T_1 = 309$ kJ mole^{-1}) only sensitizes its dimerization, implying that the lowest triplet level of norbornene lies at an energy somewhere between 293 and 309 kJ mole^{-1}.[183]

Unsaturated aldehydes and ketones

The promotion of an electron into the first antibonding π-orbital of an $\alpha\beta$-unsaturated aldehyde or ketone (π_3^* in Figure 3.21) lowers the potential barrier to rotation about the C=C bond. When either *cis* or *trans* isomers can exist they are readily interconverted in the photoexcited state.

When the molecule adopts a *cis*-configuration a γ-hydrogen atom can be transferred to the oxygen atom. For example, transient infrared absorption bands of the enol form of crotonaldehyde, and also of ethyl ketene have been recorded during the photolysis of *trans*- crotonaldehyde in the vapour phase.[184] Their production can be understood in terms of the following scheme

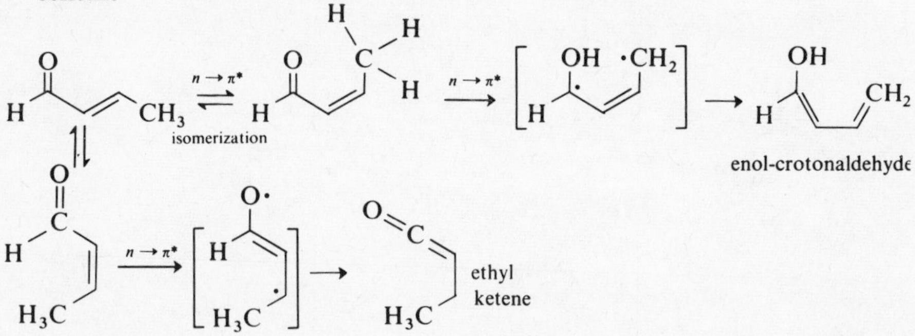

enol-crotonaldehyde

ethyl ketene

In some ketones the enol form reverts (in the dark) to the βγ-unsaturated isomer,

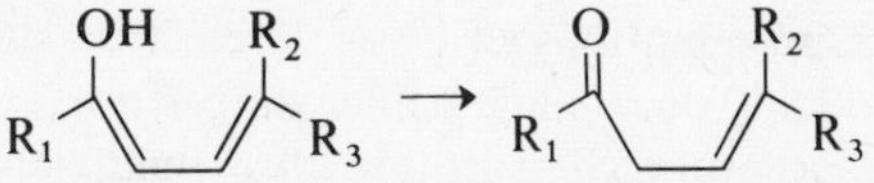

Cyclic enones undergo α-cleavage only when the rings are small. Rings of five or more carbon atoms usually suffer molecular rearrangement, or cycloaddition across the C=C bond to produce dimers or unsymmetrical adducts. Intramolecular cycloadditions have also been reported. They probably involve the formation of biradical intermediates following reaction of the triplet ketone, but the character of the reactive triplet state has not often been positively identified. The dimerization of cyclo-hexanone is thought to involve its $^3(\pi,\pi^*)$ state,[185]

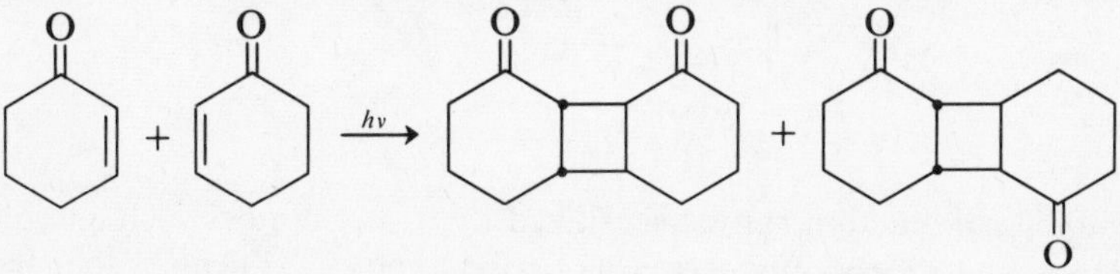

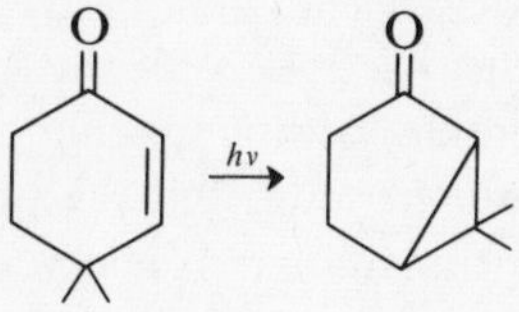

Alkyl substitution in the 4-position permits the molecular rearrangement

to take place,[186] even in the vapour phase. Some think the electron distribution takes on a 'biradical' character following excitation; others favour a 'zwitterionic' type of distribution, since the rearrangements are often reminiscent of reactions occurring in the ground state via a polar type of mechanism.

It is best to keep an open mind at the present time, and perhaps await the discovery of a better method of describing the reaction pathway than a series of valence representations of possible electron distributions. The rearrangement probably occurs in a triplet state, since it can be quenched by triplet energy acceptors. Substitution of a methylene group for the carbonyl oxygen atom produces an isovalent diene, which does not undergo the rearrangement. Since its lowest triplet state has to have (π,π^*) character, it might be the $^3(n,\pi^*)$ state that is reactive in the ketone.[187]

The same type of rearrangement also occurs in cross-conjugated, cyclohexa-2,5-dienones,[169] but with a much higher quantum efficiency (typically ~ 0.8 rather than 0.008), for example

(I)

The identical reaction can be sensitized by triplet donors such as acetophenone and it probably occurs in a triplet state.[188] On the other hand triplet quenchers do not reduce the quantum yield and rearrangement must be a very rapid process. The dienone (I) phosphoresces in a rigid glass at low temperature. Its spectral profile shows the same type of banded structure as in benzophenone, with a spacing characteristic of the C–O stretching mode in the excited state, and the phosphorescence has been assigned to a $^3(\pi^* \rightarrow n)$ transition.[188] The rearrangement presumably begins in the $^3(n,\pi^*)$ state, though there is likely to be a $^3(\pi,\pi^*)$ state not too far above it. Whether these intersect and the extent to which their characters may mix as the molecular symmetry is distorted is not known.

Cyclohexa-2,4-dienones readily undergo α-cleavage to produce open chain ketenes which may subsequently cyclize to bicyclohexenones[189] (see example on p. 290).

Quinones

Photoexcited quinone molecules behave chemically as biradical species.[190] They readily abstract a hydrogen atom from an alcoholic solvent, and form cycloadducts at both the C=O and C=C bonds. Apart from the one case of duroquinone (the fully methylated derivative of p-benzoquinone), the hydrogen abstraction follows intersystem crossing into the $^3(n,\pi_5^*)$ state. Duroquinone reacts while still in its singlet state. It was one of the first organic photochemical reactions to which Porter applied the flash spectroscopic technique, and in the case of duroquinone Bridge and Porter[191] were able to assign transient absorption bands photographed in acid or alkaline solutions to the semiquinone radical (QH·) and its ion (Q·$^-$), respectively.

Cycloaddition of alkenes and dienes can involve direct reaction of the excited quinone[192] but when its triplet energy is of the same order or slightly greater than the lowest triplet (π,π^*) state of the hydrocarbon, it is possible for addition to proceed through triplet energy transfer followed by a back reaction of the triplet acceptor.[193] p-Benzoquinone forms adducts at both the C=O and C=C bonds, to produce spirooxetanes and cyclobutane ring systems, and when irradiated on its own p-benzoquinone forms a cage dimer in low yield.

REFERENCES

1. G. C. Eltenton, *J. Chem. Phys.*, **15,** 455 (1947)
2. F. P. Lossing and A. Tickner, *J. Chem. Phys.*, **20,** 907 (1952)
3. F. P. Lossing, D. G. H. Marsden and J. B. Farmer, *Can. J. Chem.*, **34,** 701 (1956)
4. H. E. Gunning, *Can. J. Chem.*, **36,** 39 (1958).
5. C. C. McDonald, D. J. R. McDowell and H. E. Gunning, *Can. J. Chem.*, **37,** 1432 (1959)
6. D. G. Horne, R. Gosavi and O. P. Strausz, *J. Chem. Phys.*, **48,** 4578 (1968)
7. A. B. Callear and R. M. Hedges, *Nature*, **218,** 163 (1968)
8. (a) E. Gaviola and R. W. Wood, *Phil. Mag.*, **6,** 1191 (1928)
   (b) L. O. Olsen, *J. Chem. Phys.*, **6,** 307 (1938)
   (c) C. R. Masson and E. W. R. Steacie, *J. Chem. Phys.*, **18,** 210 (1950)
9. D. H. Volman, *Advances in Photochemistry*, Vol. I, Interscience, New York, 1963, p. 52
10. H. E. Gunning and O. P. Strausz, *Advances in Photochemistry*, Vol. I, Interscience, New York, 1963, p. 209
11. O. P. Strausz and H. E. Gunning, *Can. J. Chem.*, **39,** 2459 (1961)
12. H. E. Gunning and R. Pertel, *Can. J. Chem.*, **37,** 35 (1959)
13. R. J. Cvetanovic, *J. Chem. Phys.*, **23,** 1203 (1955)
14. M. Z. Hoffman and R. B. Bernstein, *J. Chem. Phys.*, **33,** 526 (1960)
15. R. J. Cvetanovic, *Advances in Photochemistry*, Vol. I, Interscience, New York, 1963, p. 115
16. J. E. Cline and G. S. Forbes, *J. Amer. Chem. Soc.*, **61,** 716 (1939)
17. O. P. Strausz and H. E. Gunning, *Can. J. Chem.*, **39,** 2244 (1961)
18. E. Jakubowski, P. Kebarle, O. P. Strausz and H. E. Gunning, *Can. J. Chem.*, **45,** 2287 (1967)
19. S. Penzes, O. P. Strausz and H. E. Gunning, *J. Chem. Phys.*, **45,** 2322 (1966)
20. K. Yang, *J. Amer. Chem. Soc.*, **89,** 5344 (1967)
21. C. C. McDonald and H. E. Gunning, *J. Chem. Phys.*, **23,** 532 (1955)
22. M. Z. Hoffman, M. Goldwasser and P. L. Damour, *J. Chem. Phys.*, **47,** 2195 (1967)
23. D. J. LeRoy and E. W. R. Steacie, *J. Chem. Phys.*, **10,** 676 (1942)
24. A. B. Callear and R. J. Cvetanovic, *J. Chem. Phys.*, **23,** 1182 (1956)
25. J. R. Majer, B. Mile and J. C. Robb, *Trans. Faraday Soc.*, **57,** 1342 (1961)
26. R. J. Cvetanovic, *Progress in Reaction Kinetics* (Ed. G. Porter), Vol. 2, Pergamon Press, Oxford, 1964, p. 39
27. H. E. Hunziker, *J. Chem. Phys.*, **50,** 1268 (1969)
28. B. deB. Darwent, *J. Chem. Phys.*, **19,** 258 (1951)
29. A. B. Callear and R. J. Cvetanovic, *J. Chem. Phys.*, **24,** 873 (1956)
30. P. Kebarle and M. Avrahami, *J. Chem. Phys.*, **38,** 700 (1963)
31. R. J. Cvetanovic and L. C. Doyle, *J. Chem. Phys.*, **37,** 543 (1962)
32. D. W. Setser, B. S. Rabinovitch and E. G. Spittler, *J. Chem. Phys.*, **35,** 1840 (1961)
33. R. A. Holroyd and G. W. Klein, *J. Phys. Chem.*, **69,** 2129 (1965)
34. J. Collin and F. P. Lossing, *Can. J. Chem.*, **35,** 778 (1957)
35. S. Boué and R. Srinivasan, 5th International Photochemistry Symposium, IBM Research Centre, New York, (1969)
36. R. Srinivasan and K. H. Colclough, *J. Amer. Chem. Soc.*, **89,** 4932 (1967)

37. C. S. Burton and H. E. Hunziker, 5th International Photochemistry Symposium IBM Research Centre, New York (1969)
38. (a) P. Kebarle and F. P. Lossing, *Can. J. Chem.*, **37**, 389 (1959)
    (b) A. G. Harrison and F. P. Lossing, *Can. J. Chem.*, **37**, 1478 (1959)
    (c) A. G. Harrison and F. P. Lossing, *Can. J. Chem.*, **37**, 1696 (1959)
    (d) A. G. Harrison and F. P. Lossing, *Can. J. Chem.*, **38**, 544 (1960)
39. R. G. W. Norrish and R. P. Wayne, *Proc. Roy. Soc.*, *A*, **284**, 1 (1965)
40. (a) H. E. Avery and R. J. Cvetanovic, *J. Chem. Phys.*, **48**, 380 (1968)
    (b) F. J. Duncan and R. J. Cvetanovic, *J. Amer. Chem. Soc.*, **84**, 3593 (1962)
    (c) H. M. Frey and R. Walsh, *Chem. Comm.*, 158 (1969)
41. N. L. Rutland and R. Pertel, *J. Amer. Chem. Soc.*, **87**, 4213 (1965)
42. R. Pertel, 5th International Photochemistry Symposium, IBM Research Centre, New York (1969)
43. L. W. Sieck, *J. Chem. Phys.*, **50**, 1748 (1969)
44. W. Groth, W. Pessara and H. J. Rommel, *Z. Phys. Chem. (Frankfurt)*, **32**, 192 (1962)
45. W. Groth, *Z. Phys. Chem. (Frankfurt)*, **1**, 300 (1954)
46. W. Groth and O. Oldenburg, *J. Chem. Phys.*, **23**, 729 (1955)
47. G. H. Miller and J. R. Dacey, *J. Phys. Chem.*, **69**, 1434 (1965)
48. (a) I. Tanaka and E. W. R. Steacie, *Can. J. Chem.*, **35**, 821 (1957)
    (b) I. Tanaka and J. R. McNesby, *J. Chem. Phys.*, **36**, 3170 (1962)
49. J. G. Calvert and J. N. Pitts, '*Photochemistry*', Wiley, New York (1966)
50. (a) R. G. W. Norrish and G. Porter, *Nature*, **164**, 658 (1949)
    (b) G. Porter, *Proc. Roy. Soc.*, *A*, **200**, 284 (1950)
    (c) G. Porter, '*Photochemistry and Reaction Kinetics*' (Ed. P. G. Ashmore, F. S. Dainton and T. M. Sugden), Cambridge University Press, 1967, Ch. 5
51. Described by G. Wettermark, *Arkiv. f. Kemi.*, **18**, 1 (1961)
52. G. Herzberg, *Proc. Roy. Soc.*, *A*, **262**, 291 (1961)
53. (a) J. Koszewski and Z. R. Grabowski, *Bull. L'Acad. Pol. Sci., Ser. sci. chim.*, **11**, 1165 (1963)
    (b) K. C. Herr and G. C. Pimentel, *Appl. Opt.*, **4**, 25 (1965); G. A. Carlson and G. C. Pimentel, *J. Chem. Phys.*, **44**, 4053 (1966)
54. (a) K. H. Welge, F. Wanner, F. Stuhl and A. Heindricks, *Rev. Sci. Instr.*, **39**, 126 (1968)
    (b) R. J. Donovan, *Trans. Faraday Soc.*, **65**, 1419 (1969)
    (c) W. Braun, A. M. Bass and A. E. Ledford, *Appl. Optics*, **6**, 47 (1967)
55. (a) J. R. Novak and M. Windsor, *J. Chem. Phys.*, **47**, 3075 (1967)
    (b) G. Porter and M. R. Topp, *Nature*, **220**, 1228 (1968); *Proc. Roy. Soc. A*, **315**, 163 (1970)
    (c) R. Bonneau, J. Faure, J. Joussot-Dubien, L. Lindqvist and C. Barthele, *Compt. Rend. B*, **267**, 412 (1968)
56. D. A. Ramsay, *J. Chem. Phys.*, **21**, 165 (1953)
57. H. S. Johnston, *Proc. Nat. Acad. Sci. U.S.*, **57**, 1146 (1967)
58. G. N. Lewis, D. Lipkin and T. T. Magel, *J. Amer. Chem. Soc.*, **63**, 3005 (1941)
    G. N. Lewis and D. Lipkin, *J. Amer. Chem. Soc.*, **64**, 2801 (1942)
59. (a) G. C. Pimentel, '*Formation and Trapping of Free Radicals*' (Ed. A. M. Bass and H. P. Broida), Academic Press, New York, 1960
    (b) I. Norman and G. Porter, *Proc. Roy. Soc. A*, **230**, 399 (1955)
60. D. E. Milligan and M. E. Jacox, *J. Chem. Phys.*, **47**, 5146 (1967)
61. G. E. Ewing, W. E. Thompson and G. C. Pimentel, *J. Chem. Phys.*, **32**, 927 (1960)

62. R. S. Mulliken, *Phys. Revs.*, **51,** 310 (1937)
63. R. J. Donovan and D. Husain, *Trans. Faraday Soc.*, **62,** 1050, 2643 (1966)
64. P. Cadman, I. W. M. Smith and J. C. Polanyi, *J. Chim. Phys.*, 111 (1967).
65. A. Kupperman and J. M. White, *J. Chem. Phys.*, **44,** 4352 (1966)
66. R. A. Ogg and R. R. Williams, *J. Chem. Phys.*, **13,** 586 (1945); **15,** 691 (1947)
67. W. H. Hamill, H. A. Schwarz and R. R. Williams, *J. Amer. Chem. Soc.*, **74,** 6007 (1952)
68. J. L. Holmes and P. Rodgers, *Trans. Faraday Soc.*, **64,** 2348 (1968)
69. K. R. Wilson, *Symposium on Excited State Chemistry, Pacific Conference on Chemistry and Spectroscopy*, Anaheim, Calif., U.S.A., October, 1969
70. G. B. Kistiakowsky and J. C. Sternberg, *J. Chem. Phys.*, **21,** 2218 (1953)
71. W. B. Tiffany, *J. Chem. Phys.*, **48,** 3019 (1968)
72. E. Rabinowitch and W. C. Wood, *Trans. Faraday Soc.*, **32,** 907 (1936); *J. Chem. Phys.*, **4,** 497 (1936)
73. G. Porter, *Discussions Faraday Soc.*, **33,** 198 (1962)
74. L. Y. Nelson and G. C. Pimentel, *J. Chem. Phys.*, **47,** 3671 (1967)
75. S. J. Rand and R. L. Strong, *J. Amer. Chem. Soc.*, **82,** 5 (1960)
76. (a) T. A. Gover and G. Porter, *Proc. Roy. Soc. A*, **262,** 476 (1961)
    (b) G. Porter, Z. G. Szabo and M. G. Townsend, *Proc. Roy. Soc. A*, **270,** 493 (1962)
77. E. W. Abrahamson, D. Husain and J. R. Weisenfeld, *Trans. Faraday Soc.*, **64,** 833 (1968)
78. A. G. Briggs and R. G. W. Norrish, *Proc. Roy. Soc. A*, **278,** 27 (1964)
79. G. Porter and F. Wright, *Discussions Faraday Soc.*, **14,** 23 (1953)
80. E. D. Morris, jun., and H. S. Johnston, *J. Amer. Chem. Soc.*, **90,** 1918 (1968)
81. M. A. A. Clyne and J. A. Coxon, *Trans. Faraday Soc.*, **62,** 1175 (1966)
82. G. Burns and R. G. W. Norrish, *Proc. Roy. Soc. A*, **271,** 289 (1963)
83. S. N. Filseth and J. E. Willard, *J. Amer. Chem. Soc.*, **84,** 3806 (1962)
84. A. B. Callear and J. F. Wilson, *Trans. Faraday Soc.*, **63,** 1358, 1983 (1967)
85. C. J. Hochanadel, J. A. Ghormley and J. W. Boyle, *J. Chem. Phys.*, **48,** 2416 (1968)
86. S. V. Filseth and K. H. Welge, *J. Chem. Phys.*, **51,** 839 (1969)
87. J. N. Murrell and J. M. Taylor, *Mol. Phys.*, **16,** 609 (1969)
88. (a) G. Black and G. Porter, *Proc. Roy. Soc., A,* **266,** 185 (1962)
    (b) F. Stuhl and K. H. Welge, *J. Chem. Phys.*, **46,** 2440 (1967); **47,** 332 (1967)
89. (a) J. R. McNesby, I. Tanaka and H. Okabe, *J. Chem. Phys.*, **36,** 605 (1962)
    (b) L. J. Stief, *J. Chem. Phys.*, **44,** 277 (1966)
90. G. Herzberg, '*Electronic Spectra and Electronic Structure of Polyatomic Molecules*', Van Nostrand, Princeton, 1966
91. P. Fowles, M. deSorgo, A. J. Yarwood, O. P. Strausz and H. E. Gunning, *J. Amer. Chem. Soc.*, **89,** 1352 (1967)
92. R. J. Donovan, *Trans. Faraday Soc.*, **65,** 1419 (1969)
93. R. J. Donovan and D. Husain, *Chem. Revs.*, **70,** 489 (1970)
94. see R. A. Young, G. Black and T. G. Slanger, *J. Chem. Phys.*, **50,** 309 (1969) for leading references.
95. (a) N. G. Moll, D. R. Clutter and W. E. Thompson, *J. Chem. Phys.*, **45,** 4469 (1966)
    (b) E. Weissberger, W. H. Breckenridge and H. Taube, *J. Chem. Phys.*, **47,** 1764 (1967)
96. K. F. Preston and R. J. Cvetanovic, *J. Chem. Phys.*, **45,** 2888 (1966)

97. J. N. Pitts, J. H. Sharp and S. I. Chan, *J. Chem. Phys.*, **39**, 238 (1963); **40**, 3655 (1964)

98. H. W. Ford and S. Jaffé, *J. Chem. Phys.*, **38**, 2935 (1963)

99. (a) F. J. Lipscomb, R. G. W. Norrish and B. A. Thrush, *Proc. Roy. Soc. A*, **233**, 455 (1956)

   (b) D. Husain and R. G. W. Norrish, *Proc. Roy. Soc., A*, **273**, 165 (1963)

100. M. A. A. Clyne and B. A. Thrush, *Trans. Faraday Soc.*, **58**, 511 (1962)

101. R. S. Mulliken, *Can. J. Chem.*, **36**, 10 (1958)

102. (a) E. Castellano and H. J. Schumacher, *J. Chem. Phys.*, **36**, 2238 (1962); *Z. Phys. Chem.*, (*Frankfurt*), **34**, 198 (1962)

   (b) R. G. W. Norrish and R. P. Wayne, *Proc. Roy. Soc., A*, **288**, 200, 361 (1965)

   (c) I. T. N. Jones and R. P. Wayne, *J. Chem., Phys.*, **51**, 3617 (1969)

103. G. S. Forbes and L. J. Heidt, *J. Amer. Chem. Soc.*, **56**, 1671 (1934)

104. (a) L. J. Heidt, *J. Amer. Chem. Soc.*, **57**, 1710 (1935)

   (b) H. J. Schumacher and U. Beretta, *J. Amer. Chem. Soc.*, **52**, 2377 (1930); U. Beretta and H. J. Schumacher, *Z. Phys. Chem.*, *B*, **17**, 405, 417 (1932)

105. R. G. W. Norrish and W. D. McGrath, *Proc. Roy. Soc.*, *A*, **242**, 265 (1957); **254**, 317 (1960)

106. N. Basco and R. G. W. Norrish, *Discussions Faraday Soc.*, **33**, 99 (1962)

107. T. P. Izod and R. P. Wayne, *Proc. Roy. Soc.*, *A*, **308**, 81 (1968)

108. R. E. Huffman, J. C. Larrabee and V. C. Baisley, *J. Chem. Phys.*, **50**, 4594 (1969)

109. R. V. Fitzsimmons and E. J. Bair, *J. Chem. Phys.*, **40**, 451 (1964)

110. (a) B. H. Mahan and R. Mandel, *J. Chem. Phys.*, **37**, 207 (1962)

   (b) A. H. Lauffer and J. R. McNesby, *J. Chem. Phys.*, **49**, 2272 (1968)

111. D. E. Milligan and M. E. Jacox, *J. Chem. Phys.*, **47**, 5146 (1967)

112. S. Karplus and R. Bersohn, *J. Chem. Phys.*, **51**, 2040 (1969)

113. W. Braun, J. R. McNesby and A. M. Bass, *J. Chem. Phys.*, **46**, 2071 (1967)

114. K. Bayes, K. H. Becker and K. H. Welge, *Z. Naturforsch.*, **17a**, 676 (1962)

115. A. N. Terenin, *Usp. Fiz. Nauk.*, **36**, 292 (1948)

116. H. Okabe and M. Lenzi, *J. Chem. Phys.*, **47**, 5241 (1967)

117. J. R. McNesby, Y. Tanaka and H. Okabe, *J. Chem. Phys.*, **36**, 605 (1962)

118. M. Jeunehomme and A. B. F. Duncan, *J. Chem. Phys.*, **41**, 1692 (1964)

119. J. C. Brand and R. I. Reed, *J. Chem. Soc.*, 2386 (1957)

120. J. G. Calvert and E. W. R. Steacie, *J. Chem. Phys.*, **19**, 176 (1952)

121. D. A. Ramsay, *J. Chem. Phys.*, **21**, 960 (1953)

122. E. I. Akeroyd and R. G. W. Norrish, *J. Chem. Soc.*, 890 (1936)

123. R. Klein and L. J. Schoen, *J. Chem. Phys.*, **24**, 1094 (1956); **29**, 953 (1958)

124. E. W. Abrahamson, J. G. F. Littler and K.-P. Vo, *J. Chem. Phys.*, **44**, 4082, (1966)

125. (a) R. N. Dixon and G. H. Kirby, *Trans. Faraday Soc.*, **62**, 1406 (1966)

   (b) W. A. Noyes, Jr., and I. Unger, *Pure and Applied Chem.*, **9**, 461 (1964)

126. P. G. Bowers, *J. Chem. Soc., A*, 466 (1967)

127. Shih-Yeng Ho and W. A. Noyes, Jr., *J. Amer. Chem. Soc.*, **89**, 5091 (1967)

128. See the discussion by J. A. Kerr, *Ann. Reps.*, **65**, 189 (1968)

129. (a) J. Franck and E. Rabinowitch, *Trans. Faraday Soc.*, **30**, 120 (1934)

   (b) R. M. Noyes, *Progress in Reaction Kinetics*, (Ed. G. Porter), Vol. 1, Pergamon Press, Oxford, 1961, p. 129

130. (a) For example 'Molecular Photochemistry', N. J. Turro, W. A. Benjamin, Inc., New York, 1965
    (b) Advances in Photochemistry (Ed. W. A. Noyes, G. S. Hammond and J. N. Pitts), Vols. 1–7, Wiley, New York
    (c) Organic Photochemistry (Ed. O. L. Chapman), Marcel Dekker Inc., New York, 1967
    (d) Reactivity of the Photoexcited Organic Molecule, Interscience, London, 1967
    (e) Annual Reports of the Chemical Society, London
131. (a) C. G. Hatchard and C. A. Parker, Proc. Roy. Soc., A, 235, 518 (1956)
    (b) C. A. Parker and C. G. Hatchard, J. Phys. Chem., 63, 22 (1959)
132. C. A. Parker, Photoluminescence of Solutions, Elsevier, Amsterdam, (1968)
133. A. J. Waring, personal communication.
134. B. Brocklehurst, W. Gibbons, F. T. Lang, G. Porter and M. Savadatti, Trans. Faraday Soc., 62, 1793 (1966)
135. (a) Kh. S. Bagdarsar'yan, and V. A. Kondratiev, Kinetika i Kataliz, 6, 777 (1965)
    (b) K. Cadogan and A. C. Albrecht, J. Chem. Phys., 43, 2550 (1965)
136. E. Migirdicyan, J. Chim. Phys., 63, 520 (1966)
137. N. C. Perrins and J. P. Simons, Trans. Faraday Soc., 65, 390 (1969)
138. G. P. Brown and J. P. Simons, Trans. Faraday Soc., 65, 3245 (1969)
139. R. F. C. Claridge and J. E. Willard, J. Amer. Chem. Soc., 87, 4992 (1965)
140. L. Kaplan, K. E. Wilzbach, W. G. Brown and S. S. Yang, J. Amer. Chem. Soc., 87, 675 (1965)
141. (a) K. E. Wilzbach, C. Harkness and L. Kaplan, J. Amer. Chem. Soc., 90, 1116 (1968)
    (b) L. Kaplan and K. E. Wilzbach, J. Amer. Chem. Soc., 90, 3291 (1968)
142. (a) G. L. Closs and M. E. Closs, J. Amer. Chem. Soc., 91, 4549, 4550 (1969)
    (b) G. L. Closs and A. D. Trifunac, J. Amer. Chem. Soc., 91, 4554 (1969)
143. A. J. Merer and R. S. Mulliken, Chem. Revs., 69, 639 (1969)
144. (a) H. Okabe and J. R. McNesby, J. Chem. Phys., 36, 301 (1962)
    (b) M. C. Sauer and L. M. Dorfman, J. Chem. Phys., 35, 497 (1961)
145. Z. R. Grabowski and A. Bylina, Trans. Faraday Soc., 60, 1131 (1964)
146. C. Steel and G. S. Milne, Chemical Society Anniversary Meeting, Nottingham, April 1969
147. R. I. T. Cromartie and J. N. Murrell, J. Chem. Soc., 2063 (1961)
148. See D. Gegiou, K. A. Muszkat and E. Fischer, J. Amer. Chem. Sov., 90, 3907 (1968), and refs. cited therein
149. S. Malkin and E. Fischer, J. Phys. Chem., 66, 2482 (1962)
150. Notably J. D. Saltiel and E. D. Megarity, J. Amer. Chem. Soc., 91, 1265 (1969)
151. P. J. Wagner and G. S. Hammond, Advances in Photochemistry, Vol. 5, Wiley, New York, 1968, p. 21
152. G. S. Hammond, Reactivity of the Photoexcited Organic Molecule, Interscience, London, 1967
153. (a) D. F. Evans, J. Chem. Soc., 1351 (1957)
    (b) R. H. Dyck and D. S. McClure, J. Chem. Phys., 36, 2326 (1962)
154. (a) G. Heinrich, H. Blume and D. Schulte-Frohlinde, Tetrahedron Letters, 4693 (1967)
    (b) W. G. Herkstroeter and D. S. McClure, J. Amer. Chem. Soc., 90, 4522 (1968)

155. P. Borrell and H. H. Greenwood, *Proc. Roy. Soc., A,* **298,** 453 (1967)
156. See discussion by H. H. Jaffé and M. Orchin, *Theory and Applications of Ultra-Violet Spectroscopy,* Wiley, New York, 1962
157. R. Srinivasan, *J. Amer. Chem. Soc.,* **90,** 4498 (1968)
158. R. Srinivasan, *Advances in Photochemistry,* Vol. 4, Wiley, New York, 1966, p. 113
159. H. M. Frey and R. Walsh, *Chem. Revs.,* **69,** 103 (1969)
160. R. B. Woodward and R. Hoffman, *J. Amer. Chem. Soc.,* **87,** 395, 2046, 2511, 4388, 4389 (1965)
161. H. C. Longuet-Higgins and E. W. Abrahamson, *J. Amer. Chem. Soc.,* **87,** 2045 (1965)
162. R. B. Woodward and R. Hoffman, *Angewandte Chem. (Intl. Ed.),* **8,** 781 (1969)
163. W. T. A. Van-der-Lugt and L. J. Oosterhoff, *J. Amer. Chem. Soc.,* **91,** 6042 (1969)
164. J. N. Murrell, *The Theory of the Electronic Spectra of Organic Molecules,* Methuen, London, 1963
165. C. S. Parmenter and W. A. Noyes, Jr., *J. Amer. Chem. Soc.,* **85,** 416 (1963)
166. N. C. Yang, *Pure and Applied Chemistry,* **9,** 591 (1964)
167. P. DeMayo, J-P. Pete and M. Tchir, *J. Amer. Chem. Soc.,* **89,** 5712 (1967)
168. R. G. W. Norrish, *Trans Faraday Soc.,* **33,** 1521 (1937)
169. (a) H. E. Zimmerman and J. S. Swenton, *J. Amer. Chem. Soc.,* **89,** 906 (1967)
    (b) H. E. Zimmerman, K. G. Hancock and G. C. Licke, *J. Amer. Chem. Soc.,* **90,** 4892 (1968)
170. A. S. Davies and R. B. Cundall, *Progress in Reaction Kinetics,* (Ed. G. Porter), Vol. 4, Pergamon Press, Oxford, 1967, p. 149
171. C. S. Parmenter, *J. Chem. Phys.,* **41,** 658 (1964)
172. C. W. Larson and H. E. O'Neal, *J. Phys. Chem.,* **70,** 2475 (1966)
173. H. Zeldes and R. Livingston, *J. Chem. Phys.,* **45,** 1946 (1966)
174. G. R. McMillan, J. G. Calvert and J. N. Pitts, Jr., *J. Amer. Chem. Soc.,* **86,** 3602 (1964)
175. N. C. Yang, S. P. Elliott and B. Klein, *J. Amer. Chem. Soc.,* **91,** 7551 (1969)
176. K. H. Schulte-Elte and G. Ohloff, *Tetrahedron Letters,* 1143, (1964)
177. P. Wagner, C. A. Stout, S. Searless, Jr., and G. S. Hammond, *J. Amer. Chem. Soc.,* **88,** 1242 (1966)
178. R. Srinivasan, *Advances in Photochemistry,* Vol. 1, Interscience, New York, 1963, p. 83
179. R. Simonaitis and J. N. Pitts Jr., *J. Amer. Chem. Soc.,* **90,** 1389 (1968)
180. E. J. Baum, J. K. S. Wan and J. N. Pitts, Jr., *J. Amer. Chem. Soc.,* **88,** 2652 (1966)
181. (a) K. R. Huffman, M. Loy and E. F. Ullman, *J. Amer. Chem. Soc.,* **87,** 5417 (1965)
    (b) E. F. Ullman and K. R. Huffman, *Tetrahedron Letters,* 1863 (1965)
182. (a) E. Paterno and G. Chieffi, *Gazzetta,* **39,** 341 (1909)
    (b) G. Büchi, C. G. Inman and E. S. Lipinsky, *J. Amer. Chem. Soc.,* **76,** 4327 (1954)
183. D. R. Arnold, R. L. Hinman and A. H. Glick, *Tetrahedron Letters,* 1424 (1964)
184. J. W. Coomber, J. N. Pitts, Jr., and G. S. H. Schrock, *Chem. Comm.,* 190 (1968)
185. E. Y. Lam, D. H. Valentine and G. S. Hammond, *J. Amer. Chem. Soc.,* **89,** 3482 (1967)

186. O. L. Chapman, T. A. Rettig, A. A. Griswald, A. I. Dutton and P. Fitton, *Rec. Chem. Progr.*, **28**, 167 (1967)

187. W. G. Dauben and W. A. Spitzer, *J. Amer. Chem. Soc.*, **90**, 802 (1968)

188. (a) H. E. Zimmerman and J. S. Swenton, *J. Amer. Chem. Soc.*, **86**, 1436 (1964)
    (b) H. E. Zimmerman and D. I. Schuster, *J. Amer. Chem. Soc.*, **84**, 4527 (1962)

189. (a) D. H. R. Barton, *J. Chem. Soc.*, 1, (1960)
    (b) J. Griffiths and H. Hart, *J. Amer. Chem. Soc.*, **90**, 3297 (1968)

190. J. M. Bruce, *Quarterly Revs.*, **21**, 405 (1967)

191. N. K. Bridge and G. Porter, *Proc. Roy. Soc.*, A, **244**, 259 (1958)

192. J. A. Barltrop and B. Hesp, *J. Chem. Soc.*, 5782 (1965)

193. J. Saltiel, R. M. Coates and W. G. Dauben, *J. Amer. Chem. Soc.*, **88**, 2745 (1966)

PROBLEMS

1. Chlorine dioxide (ClO_2) exhibits a discrete, banded absorption spectrum in the vapour phase, lying in the visible and near ultraviolet regions. At $\lambda < 375$ nm, however, the rotational structure becomes diffuse. Following flash photolysis of the vapour under isothermal conditions at 20°C, the banded spectrum is almost completely removed, being replaced by the transient spectra of $ClO\cdot$ and vibrationally excited oxygen (in its ground electronic state). Levels up to $v'' = 8$ ($\equiv 142$ kJ mole^{-1}) are populated. What conclusions can be drawn from these observations?

$$(1 \text{ cm}^{-1} \equiv 11.95 \text{ J mole}^{-1}, \qquad D_{O-O} = 493 \text{ kJ mole}^{-1},$$

$$D_{O-ClO} = 251 \text{ kJ mole}^{-1}, \qquad D_{Cl-O} = 263 \text{ kJ mole}^{-1}).$$

2. t-Butyl nitrite absorbs in the near-ultraviolet. The vapour has been irradiated at different wavelengths in the near-ultraviolet and acetone has been detected among the photoproducts. Its quantum yield increases steadily as the absorbed wavelength is reduced, but is almost zero at the longest wavelengths. Comments?

3. If you were required to investigate the near-ultraviolet photolysis of acetone vapour and had access to any experimental equipment you might want, how would you investigate,
    (a) the nature of the primary process,
    (b) the identity of the primary products, and
    (c) their subsequent reactions?

4. Anthracene and maleic anhydride form a weak charge transfer complex in dioxane solution. Irradiation at 365 nm, where both anthracene and the complex absorb, results in the formation of a Diels–Alder adduct. How might you attempt to investigate the mechanism of the photoaddition?

5. Infrared laser emission has been detected from vibrationally excited HCl and HF, during isothermal flash photolysis of $CH_2{=}CHF$ or $CH_2{=}CHCl$ in the vapour phase ($\lambda > 165$ nm). What inferences might be drawn regarding the photodissociation of the vinyl halides?

6. The photodissociations

$$N_3CN(\tilde{X}^1\Sigma^+) \xrightarrow{h\nu} N_2(\tilde{X}^1\Sigma_g^+) + NCN(^1\Delta_g)$$

$$OC{=}C{=}CO(\tilde{X}^1\Sigma_g^+) \xrightarrow{h\nu} CO(\tilde{X}^1\Sigma^+) + C_2O(^1\Delta_g)$$

both satisfy the spin conservation rule. On the basis of the Walsh correlation diagram (Figure 1.20), would you expect the radicals NCN and C_2O to have singlet or triplet ground states?

APPENDIX I

A reinvestigation of the quenching of excited mercury atoms in the presence of N_2 has revealed unexpected subtleties in the interpretation of the original experiments.† When the mercury vapour is excited by a flash of monochromatic mercury resonance radiation at 253·7 nm rather than the original polychromatic flash, the decay of the transient absorption due to $Hg(6^3P_0)$ is accelerated by one or two orders of magnitude. The slow decay recorded in the earlier experiments which was attributed to inefficient quenching of $Hg(6^3P_0)$ by N_2, was actually due to repopulation of the $Hg(6^3P_0)$, probably by energy transfer from long-lived, metastable triplet N_2 molecules. Although the first interpretation was perfectly plausible on the basis of the data then available, and was widely quoted, the later experiments show that it was incorrect. There is no disgrace— this is the scientific method in practice! The triplet N_2 could be excited by a polychromatic flash in a two-photon sequence such as

$$Hg(6^1S_0) + h\nu \xrightarrow{\lambda = 253\cdot7\text{ nm}} Hg(6^3P_1)$$

$$Hg(6^3P_1) + M \xrightarrow{\hspace{2cm}} Hg(6^3P_0) + M$$

$$Hg(6^3P_0) + h\nu \xrightarrow{\lambda \neq 253\cdot7\text{ nm}} Hg^{**} \text{ (doubly excited triplet)}$$

$$Hg^{**} + N_2(\tilde{X}^1\Sigma_g^+) \xrightarrow{\hspace{2cm}} Hg(6^1S_0) + N_2(\tilde{A}^3\Sigma_u^+)$$

Neither $Hg(6^3P_1)$ nor $Hg(6^3P_0)$ atoms have sufficient energy to excite N_2 into its lowest triplet state and when the flash source only emits light at 253·7 nm there is no way of producing doubly excited atoms.

† A. B. Callear and J. C. McGurk, *Chem. Phys. Letters*, **6**, 417 (1970).

INDEX

The letter F *indicates a figure illustration*